RESEARCH METHODOLOGY IN THE MEDICAL AND BIOLOGICAL SCIENCES

RESEARCH METHODOLOGY IN THE MEDICAL AND BIOLOGICAL SCIENCES

Edited by

Petter Laake

Haakon Breien Benestad

and

Bjørn Reino Olsen

AMSTERDAM • BOSTON • HEIDELBERG • LONDON • NEW YORK • OXFORD
PARIS • SAN DIEGO • SAN FRANCISCO • SINGAPORE • SYDNEY • TOKYO
Academic Press is an imprint of Elsevier

Academic Press is an imprint of Elsevier
84 Theobald's Road, London WC1X 8RR, UK
30 Corporate Drive, Suite 400, Burlington, MA 01803, USA
525 B Street, Suite 1900, San Diego, CA 92101-4495, USA

First published as *Forskningsmetode i medisin og biofag* by Gyldendal Norsk Forlag 2004

English translation by M. Michael Brady copyright © Gyldendal Norsk Forlag 2007
Published by Elsevier Ltd

British Library Cataloguing in Publication Data
A catalogue record for this book is available from the British Library

Library of Congress Cataloguing in Publication Data
A catalogue record for this book is available from the Library of Congress

ISBN: 978-0-12-373874-5

For information on all Academic Press publications
visit our web site at http://books.elsevier.com

Working together to grow
libraries in developing countries

www.elsevier.com | www.bookaid.org | www.sabre.org

ELSEVIER   BOOK AID
            International   Sabre Foundation

Transferred to Digital Printing in 2011

CONTENTS

CONTENTS

CONTENTS

PREFACE

Change is the watchword of our times. The pace of graduate studies has accelerated, as less time is available for pursuing master's and doctoral degrees. At the same time, schooling is as demanding as ever, in the form of formal coursework in scientific theory and practice. We hope that this textbook will make academic life easier for graduate students.

The history of this book began in 2000 in a Norwegian national programme for medical faculty studies leading to PhD degrees, which in part called for a compulsory course for all students enrolled in doctoral programmes at medical faculties. The course covers a broad range of topics and now has been held several times. Its curriculum comprises the basis of this book, which was first published in Norwegian by Gyldendal Akademisk. This edition in English is a revised and expanded version of the Norwegian edition.

There are several reasons for the broad range of topics included in the courses and in this book. Doctoral studies should qualify the candidate for research as well as for other professional pursuits requiring scientific insight, not least in the basics of the research process, in both depth and breadth. Doctoral students at medical faculties have a variety of backgrounds that may include undergraduate studies in medicine, the natural sciences, the social sciences or nursing. Many of these students may have gained only narrow scientific capabilities, so that even after finishing their graduate studies, they will barely be able to work with or be advised by colleagues other than those within their own narrow fields. Consequently, it is essential that studies leading to a doctoral degree include schooling in general scientific research, regardless of the specific discipline pursued.

Entire books have been written on each of the topics treated in the chapters of this book. Some of these books are referenced and listed under Further Reading at the end of the chapters. But in our experience, few doctoral candidates read professional books, mostly because the time allocated for studies is so short that extracurricular reading is limited to journals at best. Hence, we believe that there is a need for a book such as this one, comprehensive yet succinct.

This book is intended for students with various professional backgrounds. Multidisciplinary communication and research cooperation are increasingly important, so scientists in any one scientific community should be familiar with and respect the traditions of other scientific communities. Simply put, this book outlines the curriculum for doctoral candidates in medicine, but is also intended for students in the biological and biomedical fields. We hope that it can give the reader a broad scientific perspective. We believe that everyone in the medical and biomedical sciences should know a little about all the topics of this book, and that all should have deeper knowledge of topics relative to their disciplines.

We have tried to make the material easily accessible for a broad group of readers, in part through including many examples. For instance, Chapter 11 on Statistical Issues includes examples of statistical data analyses. We have made the data we have used accessible online at http://books.elsevier.com/9780123738745. Along with these data files, we have included some statistical tables. We chose the SPSS statistical program package for presenting examples. Other application packages could have served equally well.

ACKNOWLEDGEMENTS

We are grateful to Kristian Laake for compiling and editing the data files, tables and software used in Chapter 11. We extend our thanks to Bruce Reed for his critical reading of the manuscript of Chapter 15 and for offering expert advice to grant applications.

We gratefully acknowledge the translation and editing by M. Michael Brady. The translation has been published with the financial support of NORLA (Norwegian Literature Abroad, Fiction & Non-fiction). Finally, we extend our deepest thanks for the constructive cooperation of the Gyldendal Akademisk editors, including Editorial Director Thore Lie, Editor-in-Chief Torhild Bjerkreim and Foreign Rights Manager Oliver Møystad, whose efforts resulted in our book becoming an international text.

Petter Laake, Haakon Breien Benestad and Bjørn Reino Olsen

CONTRIBUTORS

Professor **Haakon Breien Benestad** MD (b. 1940) has been a staff member of the Institute of Basic Medical Sciences, Department of Physiology, University of Oslo, since 1968; until 1988 he also held a part-time post as a general practitioner. He has lectured students of medicine, odontology, nutrition and physiotherapy, and has been an adviser for master and graduate students at the Faculty of Medicine, the Faculty of Mathematics and Natural Sciences at the University and at the Norwegian University of Sport and Physical Education. For many years, he has been the faculty consultant in university teaching. Jointly with Jens-Gustav Iversen, he has held various basic courses for undergraduate and graduate students, compiled compendia for these courses and written a textbook of anatomy, physiology and immunology for social and health studies in upper secondary schools.

Professor **Heidi Kiil Blomhoff** PhD (b. 1958) has been a staff member of the Institute of Basic Medical Sciences, Department of Biochemistry, University of Oslo, since 1997. Before then, for 13 years she was on the staff of the Department of Immunology at the Norwegian Radium Hospital. She has advised many master students and graduate students at the Faculties of Medicine and of Mathematics and Natural Sciences of the University of Oslo. She is an experienced lecturer in immunology and cell biology at both faculties, and she has also taught graduate and postgraduate courses for bioengineers and medical doctors. In recent years, she has been involved in the teaching of lecture techniques in the graduate studies programme at the Faculty of Medicine, University of Oslo.

Director of Public Health **Peter Bradley** MD, MA, MPH (b. 1965) is a specialist in public health (England and Norway) and general practice (England) and is now director of public health in Suffolk West Primary Care Trust. He has previously worked in the Norwegian Medicines Agency, Norwegian Institute of Public Health and Norwegian Directorate for Health and Social Welfare (Norwegian National Council for Healthcare Priorities) and as a public health doctor and general practitioner in England. He has 10 years' experience teaching and supporting evidence-based practice at all levels and has completed a PhD on the implementation of evidence-based practice. He has presided over a number of international courses in evidence-based medicine in cooperation with the CASP international network. From 1999 to 2004 he contributed to teaching on the graduate studies programme at the medical faculty at the University of Oslo.

Chief Librarian **Ellen Christophersen** (b. 1961) is a staff member of the University of Oslo Library for Medicine and Healthcare, Educational Section. She has had 15 years of versatile practice in specialized libraries, mostly medical libraries. For the past five years, she has worked specifically with courses and teaching schemes for staff and students at the National Hospital and at the Faculty of Medicine, University of Oslo. Since 2002, she has taken part in courses on medical research methods as a lecturer and adviser in systematic literature search and reference processing.

Professor **Erik Dissen** MD (b. 1963) joined the staff of the Institute of Basic Medical Sciences, Department of Anatomy, University of Oslo, in 2000. Since 1987, he has worked with molecular biology methods, with immunology as a principal interest. He has been an adviser for master and graduate students and researchers at the Faculty of Medicine and the Faculty of Mathematics and Natural Sciences, and has been lecturing in cell biology, immunology and histology.

Senior Researcher **Thore Egeland** PhD (b. 1960) is a senior scientist at the Department for Medical Genetics of the Ullevål University Hospital. He previously led the biostatistics group at the National Hospital and has been a Professor at the Institute of Basic Medical Sciences, Department of Biostatistics, University of Oslo. Moreover, he has been on the staff of Veritas Research AS, the Norwegian

Computing Centre and the Norwegian Cancer Hospital. He has lectured on statistics for a broad range of students at all levels, including master and graduate courses in statistics, as well as applied courses for lawyers, engineers and medical doctors. He is currently concerned principally with genetic linkage and association studies aimed to localize disease-predisposing variants.

Professor **Sigbjørn Fossum** MD (b. 1945) has since 1973 been on the staff of the Institute of Basic Medical Sciences, Department of Anatomy, University of Oslo. He started an activity now known as the Immunobiological Laboratory, in which the principal research entails molecular biological studies of how leucocytes recognize target cells. He has lectured students of medicine, odontology, nutrition and physiotherapy, and has been an adviser for master and graduate students at the Faculty of Medicine, the Faculty of Mathematics and Natural Sciences at the University. Jointly with Per Brodal and Hans A. Dahl, he has written a textbook in anatomy and physiology for healthcare personnel. He has written a compendium in cell biology, and he is a frequent lecturer in ongoing education and graduate studies for medical doctors, engineers and other healthcare personnel.

Professor **Harald Grimen** PhD (b. 1955) is on the staff of the Centre for the Study of Professions (CSP) at Oslo University College and on the staff of the Section for Medical Anthropology, University of Oslo. He was previously a professor of philosophy at the University of Tromsø and an Associate Professor at the Centre for the Study of the Sciences and the Humanities, University of Bergen. For several years, he has lectured in the philosophy of science and in methods for various disciplines.

Chief Librarian **Anne-Marie Baune Haraldstad** (b. 1952) is a staff member of the University of Oslo Library for Medicine and Healthcare, as the leader of the library's educational section. She has more than 20 years of versatile experience in medical libraries, and for the past 12 years has been responsible for courses and teaching schemes for staff and students at the National Hospital and at the Faculty of Medicine, University of Oslo. Since 2002, she has taken part in courses on medical research methods as a lecturer and adviser in systematic literature search and reference processing.

Post.doc. **Bjørn Hofmann** PhD (b. 1964), is a postdoctoral fellow and Professor of the Section for Medical Ethics, University of Oslo, and is a Senior Adviser at the National Knowledge Centre for Healthcare and a Professor at the College at Gjøvik. He has conducted research in the philosophies of technology and science, ethics and research methods, and has lectured and advised in several healthcare disciplines at all levels, from undergraduate to postgraduate programmes. He has been responsible for the graduate course in the philosophy of science and research ethics for healthcare professionals at the Section for Medical Ethics.

Professor **Søren Holm** MD (b. 1963) is a Professor at the Section for Medical Ethics, Faculty of Medicine, University of Oslo, and a Professorial Fellow in Bioethics at the Cardiff Centre for Ethics, Law and Society. His education is in medicine and medical ethics, and he has lectured in medical ethics, philosophy of science and research methods at universities in Norway, Denmark and the United Kingdom.

Professor of Medical Anthropology **Benedicte Ingstad** (b. 1943) has been on the staff of the Institute for General Practice and Community Medicine, University of Oslo, since 1976. She established and leads the Section for Medical Anthropology. She has taught students of medicine, odontology and nutrition, as well as master's programme students in international public health. Her specialist fields include qualitative methods in medical research, in which she has arranged several graduate courses. She has conducted research work in southern Africa, Greenland and Norway.

Professor **Jens-Gustav Iversen** MD (1935–2006) was a staff member of the Institute of Basic Medical Sciences, Department of Physiology, University of Oslo, from 1965 until his death in 2006, save for the 1971–1975 academic years, when he was on the Faculty of the then newly established University of Tromsø. He lectured students of medicine, odontology and nutrition, and was an adviser for master and graduate students at the Faculty of Medicine, the Faculty of Mathematics and Natural Sciences at the University. Jointly with Haakon B. Benestad, he held various basic courses for undergraduate and graduate students, compiled compendia for these courses and wrote a textbook of anatomy, physiology and immunology for social and health studies in upper secondary schools.

Professor **Petter Laake** PhD (b. 1947) is a staff member of the Institute of Basic Medical Sciences, Department of Biostatistics, University of Oslo. He has previously been on the staff of Statistics Norway and the Department of Economics, University of Oslo, lecturing on statistical research. He has 20 years of experience in lecturing, communicating and advising in statistics, at all levels and for various target groups. In cooperation with the Norwegian Medical Association, he initiated the development and operation of a programme of teaching medical statistics via the Internet for medical doctors and bioengineers. Since 2001, he has worked with the graduate studies programme at the Faculty of Medicine, University of Oslo. He now is the coordinator of the PhD programme at the Faculty of Medicine and is in charge of the mandatory basic course in research methods.

Professor **Bjørn Reino Olsen** MD (b. 1940) is Hersey Professor of Cell Biology at Harvard Medical School and Professor of Developmental Biology, Harvard School of Dental Medicine. Since 2005, he has also served as Dean of Research at Harvard School of Dental Medicine. Before assuming the post at Harvard in 1985, for 10 years he was Professor of Biochemistry at UMDNJ-Rutgers (now known as Robert Wood Johnson) Medical School. He is a member of the Norwegian Academy of Sciences and Letters and has honorary doctoral degrees from the University of Oslo and University of Medicine and Dentistry of New Jersey, USA. He has lectured worldwide and mentored and trained a large number of students and postdoctoral fellows. He is a member of and has held leadership positions in several professional organizations, has served on several Editorial Boards, and is currently Editor-in-Chief of Matrix Biology and BioMed Central's Journal of Negative Results in Biomedicine.

Professor **Ole Petter Ottersen** MD (b. 1955) has been on the staff of the Institute of Basic Medical Sciences, Department of Anatomy, University of Oslo, since 1978. He has broad research experience, has taught students of medicine and physiotherapy, and has advised numerous master and graduate students. He has held many key positions in medical research, including being Dean for Research at the Faculty of Medicine, University of Oslo, and for many years the board chairman for the National Programme for Research in Functional Genomics

(FUGE). He is now Director of the Centre for Molecular Biology and Neuroscience (DMBN), one of the 13 Centres of Excellence (CoE) in Norway. The Centre has the status of a Research School at the University of Oslo and, in particular, is responsible for graduate and postgraduate teaching.

Senior adviser **Liv Merete Reinar** (b. 1958) is on the staff of the Norwegian Knowledge Centre for the Health Services. She is a midwife, and in 2003 earned a Master of Science in Primary Care from University College, London. She is experienced in lecturing and in communicating research. This work includes involvement with research comparisons in the form of knowledge summaries and systematic reviews. The target groups have been staff members in health and social services or administration. She has also taught at various levels at universities and colleges.

Professor **Eva Skovlund** PhD (b. 1959) is a member of the staff of the Norwegian Medicines Agency (NoMA) and of the School of Pharmacy, Department of Biosciences, University of Oslo. She holds a Bachelor of Pharmacy and a Doctorate of Statistics. She has previously been a member of staff of the Department of Mathematics, and the Institute of Basic Medical Sciences, Department of Biostatistics, University of Oslo, and has been a research adviser for the Norwegian Cancer Association. She has more than 15 years of experience in lecturing and advising in applied statistics, mainly in clinical cancer research. Moreover, she has advised in and taught statistics for pharmacists, dentists and medical doctors at all levels from student lecturing to graduate studies programmes.

Professor **Dag S. Thelle** MD (b. 1942) is currently on the staff of the Institute of Basic Medical Sciences, Department of Biostatistics, University of Oslo. Until 2007 he was professor of cardiovascular epidemiology and preventive cardiology at Sahlgrenska Academy, Institute of Medicine, Göteborg University. He has previously been on the faculties of the University of Tromsø, the Nordic School of Public Health, Göteborg, and the Centre for Epidemiological Research, University of Oslo. He has long experience in teaching and advising in many sectors and for many target groups. He has also been on the faculty of the World Heart Federation's Teaching Seminar in Epidemiology and Prevention 1986–2000, and at the Erasmus Summer Programme in Rotterdam since 2004. His research interests include factors and conditions that influence the risk of cardiovascular disease and other chronic diseases. He is now associated to the PhD programme at the Faculty of Medicine, University of Oslo, and teaches the mandatory basic course in research methods.

Professor **Morten H. Vatn** MD (b. 1942) is senior registrar in gastroenterology at the Medical Department, National Hospital, and a Professor at the University of Oslo, associated with the Epi-Gen Institute at the Akershus University Hospital. He has extensive experience, including having led basic courses in clinical research from 1988 to 2001. His activities also include research administration, advising and active participation in projects in clinical and epidemiological studies of gastrointestinal cancer development, chronic inflammation of the intestines and clinical nutrition, in part in EU projects. He has led several international clinical trials.

LIST OF ABBREVIATIONS AND SYMBOLS

The abbreviations and symbols most used in the chapters of this book are listed below. Commonplace abbreviations, such as of SI units and the names of countries and major international organizations, are not listed.

Abbreviation/symbol	Definition
a, b, c, d, x, y, z	Observed values, integer or real numbers
A and $\bar{A}$	Event and its complement in a trial
ANOVA	Analysis of variance
AR	Attributable risk
ARR	Absolute risk reduction
bp	Base pair
c	Number of columns in a table
CHD	Coronary heart disease
c_{max}	Maximum plasma concentration of a drug
CV	Coefficient of variation
d	Half-length of confidence interval
D	Difference between observations
df	Degree of freedom
DNA	Deoxyribonucleic acid
e	Base of natural logarithms
E	Expected
EBHC	Evidence-based healthcare

EBM	Evidence-based medicine
ED	Effective dose
F	Fisher distributed test statistic
H_0	Null hypothesis
H_A	Alternative hypothesis
ICC	Intraclass correlation coefficient
ICD	International Classification of Diseases
IF	Impact factor
IR	Incidence rate
IRB	Institutional review board
IRES	Internal ribosomal entry site
IRR	Incidence rate ratio
ln	Natural logarithm
MCS	Multiple cloning site
N	Number, as of observations
N	Normal distribution
NNT	Number needed to treat
O	Observed
OR	Odds ratio
p	Unknown probability
$P(A)$	Probability of an event A
PAR	Population attributable risk
PCR	Polymerase chain reaction
PEFR	Peak expiratory flow rate
pH	Acidity or alkalinity of solution, scale 1–14
PI	Principal investigator
r	Number of rows in a table
R	Risk
R&D	Research and development
RCT	Randomized controlled trial
RD	Risk difference
RE	Restriction endonuclease
REB	Research ethics board
REC	Research ethics committee

RIA	Radio immunoassay
RNA	Ribonucleic acid
RR	Relative risk
s	Empirical standard deviation
s^2	Empirical variance
SD	Empirical standard deviation
SE	Standard error
SEM	Standard error of the mean
SPSS	Statistical Package for the Social Sciences
SRG	Scientific review group
t	Time
T	Test statistic
t_{obs}	Observed test statistic
W	Rank sum
X, Y, Z	Variables
$\bar{X}$	Mean of observations
Z	Test statistic, often given as a ratio of effect to standard error
z_{obs}	Observed test statistic
α	Regression coefficient (the constant)
β	Regression coefficient
Δ	Difference
μ	Location of normal distribution
σ	Standard deviation
σ^2	Variance
χ^2	Chi-squared distributed variable
χ^2_{obs}	Observed value of a chi-squared variable

PHILOSOPHY OF SCIENCE

Bjørn Hofmann, Søren Holm and Jens-Gustav Iversen

Sciences provide different approaches to the study of man: man can be scrutinized in terms of molecules, tissues and organs, as a living creature, and as a social and a spiritual person. Correspondingly, philosophy of science investigates the philosophical assumptions, foundations, and implications of the sciences. It is an enormous field, covering sciences such as mathematics, computer sciences and logic (the formal sciences), social sciences, the natural sciences, and also methodologies of some of the humanities, such as history. Against the backdrop of the sweep of the field, this chapter comprises a brief overview of the philosophical aspects salient to research in the medical and biological sciences. Consequently, discussion is limited to the natural sciences (Section 1.1) and the social sciences (Section 1.2). The formal sciences, such as logic and mathematics, are not discussed.

1.1 PHILOSOPHY OF THE NATURAL SCIENCES

What do we mean when we say that 'smoking is the cause of lung cancer'? What counts as a scientific explanation? What is science about, e.g. what *is* a cell? How do we obtain scientific evidence? How can we reduce uncertainty? What are the limits of science? These are only a few of the issues discussed in philosophy of the natural sciences, which are to be discussed in this chapter.

Traditional philosophy of science

The traditional philosophy of science has aimed to put forth logical analyses of the premises of science, and in particular the logical analysis of the syntax of basic scientific concepts. In the following sections, the principal traditional issues concerning reason, method, evidence and the object of science (the world) are discussed.

The glue of the world: causation

A pivotal task of the biomedical sciences is to find the causes of phenomena, such as disease. However, what is the implication of saying that something is the cause of a disease? According to Robert Koch (1843–1910), who was awarded the Nobel Prize in physiology or medicine for finding the tuberculosis bacillus in 1905, a parasite can be seen as the cause of a disease if it can be shown that the presence of the parasite is not a random accident. Such random accidents may be excluded by satisfying the (Henle–) Koch postulates:

- The organism must be found in all animals suffering from the disease, but not in healthy animals.
- The organism must be isolated from a diseased animal and grown in pure culture.
- The cultured organism should cause disease when introduced into a healthy animal.
- The organism must be reisolated from the experimentally infected animal.

As became clear to Koch, these criteria are elusive. If such postulates are considered to be general criteria for something to be a cause in the biomedical sciences, causation is unlikely.

Acknowledging that overly stringent criteria for causation minimize the chance of identifying causes of disease, the British medical statistician Austin Bradford Hill (1897–1991) outlined tenable minimal conditions germane to establishing a causal relationship between two entities. Nine criteria were presented as a way to determine the causal link between a specific factor (such as cigarette smoking) and a disease (such as emphysema or lung cancer):

- *Strength of association*: the stronger the association, the less likely the relationship is due to chance or a confounding variable.
- *Consistency of the observed association*: has the association been observed by different people, in different places, circumstances and times (similar to the replication of laboratory experiments)?
- *Specificity*: if an association is limited to specific people, sites and types of disease, and if there is no association between the exposure and other modes of dying, then the relationship supports causation.
- *Temporality*: the exposure of interest must precede the outcome by a period of time consistent with any proposed biological mechanism.
- *Biological gradient*: there is a gradient of risk associated with the degree of exposure (dose–response relationship).
- *Biological plausibility*: there is a known or postulated mechanism by which the exposure might reasonably alter the risk of developing the disease.
- *Coherence*: the observed data should not conflict with known facts about the natural history and biology of the disease.
- *Experiment*: the strongest support for causation may be obtained through controlled experiments (clinical trials, intervention studies, animal experiments).
- *Analogy*: in some cases, it is fair to judge cause–effect relationships by analogy: 'With the effects of thalidomide and rubella before us, it is fair to accept slighter but similar evidence with another drug or another viral disease in pregnancy'.

Hence, with regard to causality, these criteria are less pretentious than the Koch postulates. Nonetheless, there are many cases where we might refer to 'the cause of the disease', but where the criteria do not apply. However, the Bradford–Hill criteria admit that causation in the biomedical sciences is far from deterministic (as in the Koch postulates), and that it is an amalgam of more general criteria.

However, if the causes of phenomena studied in the biomedical sciences are not deterministic, what then are they? That is, what is the true nature of the causation with which we deal? In the deterministic version of causation, we know both *the necessary and the sufficient conditions* for an event. The Koch postulates require that there are no cases of disease without the parasites, and there are no parasites without the manifestation of disease. As Koch realized when he discovered the asymptomatic carriers of cholera, the requirement of both necessary and sufficient conditions for causation is overly rigorous.

Whenever there is a *necessary* but not *sufficient condition* for an event, we do not say that it causes the event. For example, having an arm is a necessary condition for having an inflammation of it, but having an arm is not said to be the cause of the inflammation. In this case, multiple factors are prerequisites for something happening, but no one of them alone is sufficient. There may be many necessary conditions for an event that are not considered causes of it. Nevertheless, necessary conditions are germane to causation, as without them, the event will not occur. Hence, necessary conditions are relevant through their absence: we can eliminate tuberculosis by eliminating one of its necessary conditions: *Mycobacterium tuberculosis*. It is also important to notice that necessity can mean two different things. Necessity can mean irreplaceability, that is nothing else than A could have resulted in B,[1] e.g. the modification of the Huntington gene is the only thing that results in Huntington's disease, but necessity can also mean non-redundancy, that is when many things can result in B, but one of these is A in combination with R and S (see Figure 1.1). In this case A is non-redundant. A virus infection is a non-redundant condition for having a cold, as there are many other conditions resulting in a cold, but when these are absent, and you do not have a virus infection, you will not have a cold. Under those circumstances, the virus infection is a necessary condiition for having a cold.

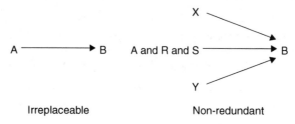

Irreplaceable   Non-redundant

Figure 1.1 Two meanings of necessity

The situation differs when there is a *sufficient* but not *necessary condition*. For example, when a person develops cancer after being exposed to ionizing radiation known to be of the sort and strength that results in cancer, we tend to say that the

1. A is an irreplaceable condition for B, if and only if nothing other than A could have resulted in B.

radiation caused the cancer. The cancer could, of course, have been caused by other factors (sufficient conditions) in the absence of the radiation. Nevertheless, if a sufficient condition is present, we know that the event (the effect) will occur. Hence, *sufficient conditions* for an event are said to be its causes. In contradistinction to necessary conditions, they work through their presence.

Another situation is when there are two factors that individually are insufficient conditions for a certain event, but which together make an event occur. For example, alone being stung by a bee or being hypersensitive to bee venom does not cause an anaphylactic reaction. However, in certain circumstances, acting jointly, both may be *sufficient* and *necessary* for an anaphylactic reaction, so both are said to cause the event. In short, each of the factors is an *insufficient* but *necessary* part of a *sufficient* and *necessary* condition for the event. Although we seldom find single factors that are both sufficient and necessary for events in the biomedical sciences, we more often find cases where multiple factors together are sufficient and necessary.

Consider the scenario when a person drinks (a lethal dose of) poison, no antidote is taken, the stomach is not pumped and the person dies. What is the cause of death? Does the person die because poison is ingested, because no antidote is taken or because the stomach is not pumped? Ingesting poison alone is not sufficient, as many people drink poison without ensuing death (because their stomachs are pumped). However, drinking the poison is part of a concert of conditions that are jointly sufficient to cause death. Moreover, given this set of conditions and not another set sufficient for death, drinking the poison was non-redundant: deaths do not occur in such circumstances when poison is not drunk.

Accordingly, drinking poison is an *insufficient* and *non-redundant* part of an *unnecessary* but *sufficient* condition for death. This is called an INUS condition (Mackie 1974). It can be argued that many relationships in biomedical sciences, regarded as causal, satisfy INUS conditions. Hence, causation is given by the conditions of an event. If the conditions are both sufficient and necessary, if they are jointly sufficient and necessary, or if they are INUS conditions, then one could argue that they are causes. However, what about smoking and lung cancer: is smoking an INUS condition for lung cancer? INUS accommodates the fact that not all smokers develop lung cancer, and not all people with lung cancer have been smokers. However, it requires a concert of conditions for which lung cancer follows when smokers, but not when non-smokers, are subjected to them.

Common to approaches defining causation in terms of sufficient (and necessary) conditions is that they hinge on scientific determinism, that is, that complex phenomena can be reduced to simple, deterministic mechanisms, and therefore in principle be predicted. In the case of smoking being an INUS condition for lung cancer, all the conditions are not known. Hence, we will have to assume the existence of hidden conditions in order to retain determinism. The belief in unidentified conditions, as well as the difficulty in explaining dose–response relationship, has challenged sufficient component conceptions of causation (sufficient condition, *insufficient* but *necessary* part of a *sufficient* and *necessary* condition, and INUS).

Rather than satisfying an INUS condition, we observe that smokers develop lung cancer at higher rates than do non-smokers. This leads us to believe that the increased probability of lung cancer among smokers is the causal link. This represents a *probabilistic approach to causation*. The central idea in probabilistic theories of causation is that causes raise the probability of their effects.

Despite their plausibility, probabilistic approaches to causation are challenged with regard to how much a probability must be raised in order to become a cause. We say that aspirin 'causes' Reye's syndrome in children and that certain tampons 'cause' toxic shock syndrome, though the probabilities are low. Accordingly, it becomes difficult to differentiate between causation and non-causal associations.

Moreover, some scientists are uncomfortable with the propensity of probabilistic approaches to abandon determinism. Events are not determined as having occurred, although there may have been (probabilistic) causes for them. This may frustrate the aim of pursuing causality: circumventing certain events (disease) and promoting others (symptom relief, health). In other words, if an event is not determined to have occurred, then nothing can be part of a sufficient condition for it. Hence, some would prefer to say that smoking is an INUS condition for lung cancer, although we do not (yet) know the concert of conditions sufficient for its occurrence.

Another approach highlights that the presence or absence of a cause 'makes a difference'. This is expressed by counterfactuals: a counterfactual draws on the contrast between one outcome (the effect), given certain conditions (the cause) and another outcome, given alternative conditions. C causes E if the same condition except C would result in a condition different from E, when all other conditions are equal (*ceteris paribus*). For example, 'if I had taken two aspirins instead of just a glass of water an hour ago, my headache would now be gone.' A counterfactual conception of causation is considered to be more precise with respect to what distinguishes causation from mere association than the probabilistic approach, while it avoids referring to hidden deterministic conditions. Counterfactuals can also be probabilistic: 'if I had taken two aspirins instead of just a glass of water an hour ago, I would be much less likely to have a headache now.' However, in practice it is not easy to satisfy the *ceteris paribus* condition. The same individual cannot be observed in exactly the same situation as both a smoker and a non-smoker.

It is important to notice that the different conceptions of causation are not mutually exclusive. For example, a probabilistic approach does not exclude sufficient conditions altogether; a sufficient cause is one that raises the probability of its effect occurring to one. A counterfactual where a factor makes all the difference[2] is equal to a *necessary condition*. (See Box 1.1 and Table 1.1.)

Box 1.1 Some criteria for causation
- Sufficient conditions for an event
- *Insufficient* but *necessary* part of a *sufficient* and *necessary* condition for an event
- *Insufficient* and *non-redundant* part of an *unnecessary* but *sufficient* condition for an event (INUS)
- Raised probability for an event (non-deterministic)
- Counterfactual: the condition (cause) makes a difference with respect to the effect

---

Table 1.1 Conceptions of causation with regards to determinism

DETERMINISTIC CONCEPTION OF CAUSATION	NON-DETERMINISTIC CONCEPTION OF CAUSATION
Sufficient condition	Probabilistic
Insufficient but necessary part of a sufficient and necessary condition	Counterfactuals 2
INUS	
Counterfactuals 1	

---

2. If C does not occur, E does not occur (*sine qua non*).

7

Might the complications of causation be avoided by referring to explanations? Might there be an imperative to find explanations of phenomena significant in the biomedical sciences?

Scientific explanation

From the time of Aristotle, philosophers have realized that a distinction could be made between two kinds of scientific knowledge; roughly, knowledge *that* and knowledge *why*. It is one matter to know *that* myocardial infarction is associated with certain kinds of pain (angina pectoris); it is a different matter to know *why* this is so. Knowledge of the former type is descriptive; knowledge of the latter type is explanatory, and it is explanatory knowledge that provides scientific understanding of the world (Salmon 1990).

How, then, do we explain the phenomena studied in the biomedical sciences? For example, how do we explain the change in haematopoietic cell growth in a medium when its temperature changes? What criteria do we have for something to be acceptable as a scientific explanation? The standard answer to questions such as these is that we explain something by showing how we could expect it to happen according to the laws of nature (*nomic expectability*) (Hempel 1965). The haematopoietic cell growth is explained by the laws that govern haematopoietic cell growth and the initial conditions, including the type of medium, the humidity and the pressure. Accordingly, a singular event is explained if (a description of) the event follows from law-like statements and a set of initial conditions.

When a phenomenon is explained by deducing it from laws or law-like statements, the sequence of deductive steps is said to follow a deductive–nomological model (DNM) that turns an explanation into an argument where law-like statements and initial conditions are the premises of a deductive argument.

Deductive nomological model of explanation:

Premise 1: Initial conditions	Type of medium, humidity, light, temperature
Premise 2: Universal law(s)	Laws of haematopoietic cell growth
Conclusion: Event or fact to be explained	Greater growth due to temperature increase

In other words, we explain a phenomenon by subsuming it in a law. For this reason DNM is often referred to as 'the covering law model of explanation'. One reason for the prominent position of DNM is its close relation to prediction. A deductive–nomological explanation of an event amounts to a prediction of its occurrence.

However, DNM incurs challenges. One is that DNM allows for symmetry. For instance, certain conditions of a growth medium for cells (temperature, humidity, light, etc.) can be explained by the growth rate of haematopoietic cells (in this medium), given the same laws. We like to think that there is an asymmetry between cause and effect (that is, what is considered to be a cause leads to an effect, and not the other way round).

Moreover, if the biomedical sciences can provide explanations only when phenomena subsume under deterministic laws of nature, then there are innumerable phenomena that cannot be explained. For instance, we tend to say that lung cancer can be explained by smoking, despite there being no strict law stating which smokers will develop lung cancer. The answer to this objection is straightforward and entails replacing deterministic laws with probabilistic statements. This engenders the deductive–statistical model (DSM) of explanation, which has the form:

Deductive statistical model of explanation:

Premise 1: Initial conditions	Having sinusitis
Premise 2: Statistical laws	Taking antibiotics probably leads to recovery
Conclusion: Event or fact to be explained	People taking antibiotics will recover

DSM is a version of DNM that supports explanations of statistical regularities by deduction from more general statistical laws (instead of deterministic laws). However, DSM cannot explain singular events, such as Mr Hanson recovering from a sinusitis after taking antibiotics. DSM can only explain why people taking antibiotics will recover (in general). In order to explain singular events in terms of statistical laws, one may refer to the inductive–statistical model (ISM) of explanation. Hence, ISM can explain likely events inductively from statistical models.

Inductive statistical model of explanation:

Premise 1: Initial conditions	Mr Hanson has sinusitis and takes antibiotics
Premise 2: Probability (r) of event, given 1	The probability of recovery in such cases $= r \approx 1$
Induction: Event or fact to be explained	Mr Hanson will recover

Table 1.2 summarizes the traditional models of explanation, DNM, DSM and ISM. Common to all the models is that explanations are arguments (deductive or inductive) and that they are based on initial conditions and on law-like statements, either deterministic or statistical (nomic expectancy). The standard form of each such argument is:

Premise 1: Initial conditions
Premise 2: Law-like statements

Implication: Event or fact to be explained

Most explanations in the biomedical sciences appear to fall under these models.

Table 1.2 Models of explanations according to Salmon (1990)

LAWS	SINGULAR EVENTS	GENERAL REGULARITIES
Universal laws	DNM	DNM
Statistical laws	ISM	DSM

However, these models of explanation incur many challenges. One is that arguments with true premises are not necessarily explanatory. For instance, if Hanson takes birth-control pills and Hanson is a man (initial conditions), and if no man who takes birth-control pills becomes pregnant (law), it leads deductively to the conclusion that Hanson will not become pregnant. According to DNM, taking birth-control pills then explains why Hanson cannot become pregnant, but it is intuitively wrong, because the premises are explanatorily irrelevant.

As already indicated, DNM permits symmetry. For example, DNM enables us to use plane geometry and the elevation of the sun to find the height of a flagpole from the length of its shadow as well as predict the length of the shadow from the height of the flagpole. However, as the length of the shadow clearly does not explain the height of the flagpole, DNM does not present a set of sufficient conditions for scientific explanation.

These challenges with regard to *relevance* and *symmetry* have made some philosophers of science argue that explanations should be based on causation: to explain is to attribute a cause. According to such a causal model of explanation (CM), one must follow specific procedures for arriving at an explanation of a particular phenomenon or event:

1. Compile a list of statistically relevant factors.
2. Analyse the list by a variety of methods.
3. Create causal models of the statistical relationships.
4. Test the models empirically to determine which is best supported by the evidence.

However, these procedures revert to some of the challenges of causation. Moreover, although it is intuitively correct that to explain a phenomenon is to find its cause, it is not necessarily so. Indeed, David Hume (1711–1776) argued that causation entails regular association between cause and effect. Hume's conception of causation as regularities adds nothing to an explanation of why one event precedes another. Accordingly, Bertrand Russel (1872–1970) claimed that causation 'is a relic from a bygone age, surviving, like the monarchy, only because it is erroneously supposed to do no harm' (Russell 1959, p. 180). Defining explanation in terms of causation would enhance our ability to predict, but not to understand the phenomena (Psillos 2002). Accordingly, explanation entails more than referring to a cause; it invokes understanding, and thus, one could argue, it must include the laws of nature.

Hence, DNM, DSM and ISM, the principal models relevant to the biomedical sciences, are but three of the many models for scientific explanation.

Modes of inference

The biomedical sciences tend to employ three modes of inference first set forth in 1903 by Charles Sanders Pierce (1839–1914): deduction, induction and abduction.

- *Deduction* entails inference from general statements (axioms, rules) to particular statements (conclusions) via logic. If all people with type 1 (insulin-dependent) diabetes are known to have deficiencies in pancreatic insulin production (rule), and Mr D has type 1 diabetes (case), then Mr D has deficiencies in pancreatic insulin production (conclusion).

- *Induction* is inference (to a general rule) from particular instances (cases). If all people observed with deficiencies in pancreatic insulin production have the symptoms of type 1 diabetes, and the people are all from the general population (that is, not selected subjects with other deficiencies causing the symptoms), we conclude that all people with deficiencies in pancreatic insulin production have the symptoms of type 1 diabetes.
- *Abduction* infers the best explanation. When we make a certain observation (case) we find a hypothesis (rule) that makes it possible to deduce (the conclusion). If Mr D has deficiencies in pancreatic insulin production, and all people with type 1 diabetes have deficiencies in pancreatic insulin production, then Mr D has type 1 diabetes.

The crucial aspect of deduction is whether the axioms hold, while both inductive and abductive inference are knowledge enhancing (ampliative inference). In induction we infer from some cases (conclusion) to the general rule, and in abduction there could of course be other rules that could explain what we observe; that is, other explanations may be even better. Table 1.3 illustrates the differences between these three modes of inference.

Table 1.3 Modes of inference

Deduction		
All Fs are Gs (rule)	All balls in this urn are red	
All Ss are Fs (case)	All balls in this particular sample are taken from this urn	
All Ss are Gs (conclusion)	All balls in this particular sample are red	
Induction		
All Ss are Gs (conclusion)	All balls in this particular sample are red	
All Ss are Fs (case)	All balls in this particular sample are taken from this urn	
All Fs are Gs (rule)	All balls in this urn are red	
Abduction		
All Fs are Gs (rule)	All balls in this urn are red	
All Ss are Gs (conclusion)	All balls in this particular sample are red	
All Ss are Fs (case)	All balls in this particular sample are taken from this urn	

What science is about

The biomedical sciences are about this world and its biomedical phenomena. However, what *is* this world? Many scientists find this question odd, even irrelevant. We deal with viruses, cells, substances and the effects of interventions, and it is clear to most of us that cells exist and that they more or less correspond to our theories. However, in history there are innumerable examples of situations where convictions of the reality of the entities of our theories, such as phlogiston (proposed by the German physician and alchemist Johann Joachim Becher, 1635–1682), ether and 'cadaver poison' (Ignaz Semmelweiss, 1818–1865), have been replaced by new entities and new convictions. How can we be sure that the world is as our scientists portray it, and how can we explain that our theories change?

Scientific realists hold that successful scientific research characteristically enhances knowledge of the phenomena of the world, and that this knowledge is largely independent of theory. Furthermore, realists hold that such knowledge is possible even in cases in which the relevant phenomena are not observable. According to scientific realism, you have good reason to believe what is written in a good contemporary medical textbook because the authors had solid scientific evidence for the (approximate) truth of the claims put forth about the existence and properties of viruses and cells and the effects of interventions. Moreover, you have good reason to think that such phenomena have the properties attributed to them in the textbook, independent of theoretical concepts in medicine.

Consequently, scientific realism can be viewed as the sciences' own philosophy of science. On the other hand, *scientific antirealism* holds that the knowledge of the world is not independent of the mode of investigation. A scientific antirealist might say that photons do not exist. Theories about them are tools for thinking. They explain observed phenomena, such as the light beam of a surgical laser. Of course, the energy emitted from a laser exists, as well as the coagulation, but the photons are held not to exist. The point is that there is no way we can know whether the world is independent of our investigations and theories.

One may distinguish several levels of scientific realism. A weak notion of scientific realism holds that there exists a real world independent of scientific scrutiny, without advancing any claim about what it is like. A stronger notion of realism argues that not only does the world exist independently of human (scientific) enquiry, but the world has a structure which is independent of this enquiry.

An even stronger notion of scientific realism holds that certain things, including entities in scientific theories such as photons and DNA, exist independently of humans and their enquiry of the world. Accordingly, the scientific realist claims that when phenomena, such as entities, states and processes, are correctly described by theories, they actually exist.

Realism is common sense (and certainly 'common science'), as we do not doubt that the phenomena we study exist independently of our investigations and theories. However, how can this intuition be justified? This is where the philosophical challenges start. Three arguments justifying scientific realism may be advanced: transcendental, high-level empirical and interventionist.

- *The transcendental argument* asks what the world must be like to make science possible. Its first premise is that science exists. Its second premise is that there must be a structured world independent of our knowledge of science. There is no way that science could exist, considering its complexity and extent, if the things science describes did not exist (Bhaskar 1997). Hence, the argument reasons from what we believe exists to the preconditions for its existence. Even when science is seen as a social activity, how could this activity exist without the precondition that the world actually exists? That is, science is intelligible as an activity only if we assume realism. However, one premise of the argument is that science expands our knowledge of the world and corrects errors. But how do we know this? Furthermore, how can we reason from what we believe to exist to the conditions of its existence? The answer is that we do so through thought experiments. We could not think of the effects of certain microbiological events without the existence of DNA. From this we argue that the existence of DNA is a *necessary condition* for the microbiological events. But what guarantees that the reason that we cannot think of the microbiological events without the existence of DNA is not due to the limits of scientific imagination?
- *The high-level empirical argument* contends that scientific theories are (approximately) true because they best elucidate the success of science. The best way to explain progress and success in science is to observe that (1) the terminology of mature sciences typically refers to real things in the world, and (2) the laws of mature sciences typically are approximately true (Putnam 1981). However, this is an abductive argument, where we argue from the conclusion (science has

success) and the rule (if science is about real things, then it has success) to the case (science is about real things). Abductive arguments are knowledge expanding, and there may be other explanations that are better, but that are yet not available to us.

- *The interventionist argument* holds that we can have well-grounded beliefs about what *is* on behalf of what we can *do* (Hacking 1983). We can use intervention to test whether the entities of our scientific theories exist. If a theoretically induced intervention does not work, it does not exist, but 'if you can spray them, then they are real'. Hence, you can test whether something is real. One problem with the interventionist argument is that it is not robust with respect to explanation. If you test whether ghosts are real by spraying 'them' with red paint, you may conclude that ghosts are not real. However, how do you know that this is the right method to show that ghosts are real? Could it not be that red paint does not adhere to ghosts, whereas yellow paint does?

Scientific realism, which most scientists find common sense, is exasperatingly difficult to justify. One could, of course, dismiss the whole question by arguing that observable results are what matters, and whether entities of our theories, be it photons or arthritis, are real does not matter. However, at certain points a scientist may reflect upon the nature of being (ontology) of the entities studied.

Scientific rationality

Rationalism is the position that reason takes precedence to other ways of acquiring knowledge. Traditionally, rationalism is contrasted with empiricism, claiming that true knowledge of the world can be obtained through sensory experience. In antiquity 'rationalism' and 'empiricism' referred to two schools of medicine, the former relying primarily on theoretical knowledge of the concealed workings of the human body, the latter relying on direct clinical experience.

One might argue that the demarcation between rationalism and empiricism remains relevant in clinical practice but not in science. There are many examples of cases in which treatments established on rationalistic ground, such as ligation of arteria mammaria interna as a treatment for angina pectoris, have been revealed by empirical studies to be without effect (beyond placebo). Correspondingly, established treatments induced from experience have been revealed to be without effect or even to be detrimental. However, modern biomedical scientists tend to rely on rationality as well as on experience in their work. Hypotheses may be generated on rational grounds (the substance S should have the effect E because it has the characteristics X, Y and Z), and theories are tested empirically, such as in animal models or in randomized clinical trials.

Nevertheless, the enduring rationalism–empiricism debate still seems relevant in the biomedical sciences because there are limits to scientific methodology. There may be ethical reasons, such as reluctance to use placebo surgery, which limits empirical research, or there may be lack of knowledge with respect to mechanisms, limiting a rationalistic approach, such as when we wish to test a substance that appears to have promise in eliciting a desired effect, but for which we lack the knowledge of why it should work.

Theory testing

The author of one of the most prominent Hippocratic writings, *The Art* (of medicine), identified three challenges to medical treatment and research: (1) the obtained effects may be due to luck or accident (and not intervention); (2) the obtained effect occurs even if there is no intervention; and (3) the effect may not be obtained despite intervention. In the terminology of causation, we are faced with the challenges that the intervention is not a necessary condition (2) and not a sufficient condition (3) for the effect, and that there may be a probabilistic relationship between intervention and effect or there may be other (unknown) causes of the effect (1). Today, almost three millennia later, we still struggle with the same kind of question: how can we be certain that our theories and hypothesis of the world are true, given the large variety of possible errors?

The standard answer to the question is to put the hypothesis to an empirical test according to the hypothetical–deductive method. The hypothetical–deductive method is the scientific method of testing hypotheses by making predictions of particular observable events, then observing whether the events turn out as predicted. If so, the hypothesis is verified (confirmed), and if not, the hypothesis is refuted (disconfirmed, or falsified). The steps of the hypothetical–deductive model are:

1. State a clear and experimentally testable hypothesis.
2. Deduce the empirical consequences of this hypothesis.

3. Perform empirical experiments (in order to compare their results with the deduced empirical consequences).
4. If the results concur with the deduced consequences, one can conclude that the hypothesis is confirmed, otherwise it is refuted.

According to the traditional interpretation of this model, hypotheses can be confirmed and scientific knowledge is accumulated through the verification of ever more hypotheses (verificationism) (Table 1.4).

Table 1.4 Simplified comparison between the structure of verification and falsification

	VERIFICATION	FALSIFICATION
1. Hypotheses	A is better than B	B is better than A
2. Deduced empirical consequences	If A is better than B, we must observe that A gives better results than B in the empirical setting	If B is better than A, we must observe that B gives better results than A in the empirical setting
3. Experiments and observations	We observe that instances where A is used obtain better results than B	We observe instances where A is used obtain better results than B
4. Conclusion	The experiment confirms the hypothesis	The experiment refutes the hypothesis, and lends support to the alternative hypothesis (A is better than B)
Logical structure	If p, then q q p (Confirming the antecedent)	If p, then q *not q* not p (Modus tollens)

However, as Karl Popper (1902–1994) showed, this approach cannot avoid the challenges mentioned above. First, the verification of a hypothesis presupposes induction, which is not warranted. Secondly, the logical form of the model is not sound.

Moreover, Popper was critical of the early twentieth century lack of standard criteria for establishing scientific truth, and of the corresponding trend to use (scientific) authority to decide what was true, which made it difficult to differentiate science from other social activities. Popper's radical turn was to avoid stating explicit (authoritative) criteria for truth and to provide stringent procedures for testing hypotheses. Furthermore, he broke with the ideal of final determination of the truth, and provided a scientific knowledge base of non-truths (falsified hypotheses). Scientific knowledge progressed through enlarging the graveyard of falsified hypotheses. The method of refutation rather than that of verification makes all truth provisional, conjectural and hypothetical. According to Popper, experiments cannot determine theory, only delimit it. Theories cannot be inferred from observations. Experiments only show which theories are false, not which theories are true. (See Box 1.2.)

Box 1.2 Popper on 'The success of refutation'
'Refutations have often been regarded as establishing the failure of a scientist, or at least of his theory. It should be stressed that this is an inductivist error. Every refutation should be regarded as a great success; not merely a success of the scientist who refuted the theory, but also of the scientist who created the refuted theory and who thus in the first instance suggested, if only indirectly, the refuting experiment.

Even if a new theory (such as the theory of Bohr, Kramers, and Slater) should meet an early death, it should not be forgotten; rather its beauty should be remembered, and history should record our gratitude to it – for bequeathing to us new and perhaps still unexplained experimental facts and, with them, new problems; and for the services it has thus rendered to the progress of science during its successful but short life.'

(Popper 1963)

Hypotheticodeductive method

In empirical fields, the hypotheticodeductive approach (see Figure 1.2) is used almost daily, often without a thought. The control experiment is a typical example. Can a possible effect or an absent effect have a trivial explanation? Might changes over time or in titrations of solvents produce effects, or might the cells have failed to respond at all? Control experiments are included to rule out such trivial explanations.

In clinical research involving trials of new drugs, patients' symptoms may be strongly influenced by the treatment situation, and a placebo may cause an effect. So, a placebo group is included to rule out (falsify) this hypothesis. Correspondingly, tests are conducted double blind, to falsify the hypothesis that the observed effect of a treatment is due to the expectations of the experimenter.

Correspondingly, statistical tests are performed to falsify the hypothesis that a result is obtained owing to biased selection (as of patients). They include assessment of whether recorded differences between groups are random. This is done by setting up the contention of a null hypothesis H_0 that there is no difference between the groups and thereafter assessing the probability for its being true. If that probability is very small, the null hypothesis is rejected, which strengthens the principal hypothesis that there is a real, not random difference.

A hypothesis must have testable implications if it is to have scientific value. If it is not testable, and thus not falsifiable, then it is not science, as Popper contended. The lack of adequate methods often hinders scientific progress, because limited testability restricts what can be of scientific enquiry. Therefore, often, a new, more powerful method propels science ahead. Suddenly, new research areas open up. Outstanding instances include Kary Mullis' development of the polymerase chain reaction (PCR) in molecular biology, recognized by a Nobel Prize in 1993, and the development of the patch clamp in neurobiology by Erwin Nehr and Bert Sakmann, recognized by a Nobel Prize in 1991.

The development of hypotheses is closely associated with the development of models and the planning of experiments. Many hypotheses can be shown to be too imprecise and ambiguous to be rejected and consequently cannot be challenged as

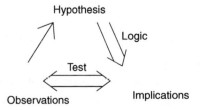

Figure 1.2 The hypotheticodeductive method

Popper requires. Formally, there should be two alternative hypotheses that mutually exclude each other. Then, a decisive experiment should be done to distinguish between them. If a hypothesis is falsified, this may lead to the development of new hypotheses, which in turn can be tested.

Moreover, a hypothesis should have the power to explain. It should relate to existing, generally accepted theoretical basis of the field. There must be good grounds to reject established theories, such as an accepted law of nature. A theory that flounders on the grounds of falsifying experiments may be defended by its remaining adherents who contend that it puts forward ad hoc hypotheses. Whenever newer observations so indicate, it is advisable to modify a hypothesis. In fact, that is part of the scientific process. However, an ad hoc hypothesis differs from a modified hypothesis in that it is not testable and often is more complicated and consequently usually hinders rather than promotes scientific development.

Although falsification has become common ground in empirical biomedical research, its strengths and weaknesses are not always appreciated. According to Popper, a theory or hypothesis should be bold and far-ranging. Its empirical content should be high, that is, it should have great predictable power. Furthermore, the hypothesis should be testable with a radical test. If the results from the empirical test support the hypothesis, it is corroborated (but not verified); if not, it is falsified.

Regardless of how influential Popper's approach has been and still is in empirical research in the biomedical sciences, falsificationism has been severely criticized. Four challenges to it are frequently mentioned:

- First, when we falsify theories, we do not test their prospective robustness. We only test them on past evidence.
- Secondly, a severe test is one that is surprising and unlikely on present evidence. However, to set up a test that is unlikely, we base our knowledge on what is likely, and in so doing we rely on induction. Accordingly, if one really defies induction, there is no reason to act on corroborated theories or hypotheses, because doing so would be induction.
- Thirdly, when we falsify a theory it is on behalf of empirical observations. However, observational statements should also be fallible, and hence the falsification of a theory may be erroneous (if the observational statements are not true).

- Fourthly, Popper's method can lead to falsification of robust and fruitful theories with high empirical content, such as due to errors in the test procedure. In practice we do not falsify a potentially fruitful theory on the basis of only one observation. That is, a theory that is not corroborated is not necessarily falsified. We design new experiments and ad hoc hypotheses to investigate or explain the falsifying observation. Hence, in practice we falsify not single theories, but rather groups or systems of theories.

Aim of science: reducing uncertainty

The primary aim of science is to increase knowledge in order to explain, understand and intervene. We need scientific knowledge to reduce our uncertainty. It is convenient to differentiate among four kinds of 'uncertainty': risk, uncertainty, ignorance and indeterminacy (Table 1.5).

Table 1.5 The modes of uncertainty

PROBABILITY/OUTCOME	KNOWN OUTCOME	UNKNOWN OUTCOME
Known probability	Risk	Ambiguity
Unknown probability	Uncertainty	Ignorance

Risk is when the system behaviour is basically well known, and the chances of different outcomes can be defined and quantified by structured analysis of mechanisms and probabilities. It is a task of science to find the outcomes of a given situation or intervention and its probability; for example, the outcome with respect to survival rate (with respect to cardiovascular disease) when using statins prophylactic for patients with type 2 (non-insulin-dependent) diabetes.

Uncertainty is characterized by knowledge of the important system parameters, but not of the probability distributions. We may know the major outcomes of a certain intervention, but we do not know their respective probabilities. There may be many sources of uncertainty, such as uncertainty in reasoning: how to classify a single case with regards to general categories. There may also be uncertainty in biomedical theory, such as when all mechanisms in a certain field are not known in detail, or because of multifactor causation. Moreover, diseases may be complicated, and it can be difficult to know and understand all of their causes. In the case of uncertainty the main task of science is to provide the probability distributions.

Ignorance is the case when we know neither the outcomes nor their probability distributions. The aim of science is, of course, to find both. However, this is difficult, as we do not know what we do not know.

Even though we would be able to reduce all ignorance to uncertainty and all cases of uncertainty to risk, we still might be subject to *indeterminacy*. It is not always a question of uncertainty due to imprecision (which is assumed to be narrowed by more research), but also a question of how we classify things according to different properties or criteria. When we classify myocardial infarction according to a set of clinical criteria, we will have a different perspective than if we classify it according to the level of troponin in the blood. Likewise, if we investigate pain in terms of neural activity or according to a visual analogue scale (VAS), the risk, uncertainty and ignorance may differ. Processes may not be subject to predictable outcomes from given initial conditions, owing to imprecise classification.

The empirical turn in philosophy of science

Although many of the challenges within traditional philosophy of science (as discussed above) have been addressed, and progress has been made, interesting and fruitful contributions have been fuelled through empirical studies of sciences and scientists. Intimate empirical studies have revealed characteristic social aspects of science. In particular, the norms and activities of scientists have been shown to be basically similar to the norms and activities of other groups in society (Stengers 2000).

The traditional philosophy of science has been theoretical and focused principally on the products of science, that is, knowledge and its conceptual preconditions. The newer approaches are empirical and focus on the social processes of science (and its interaction with material matters). A seminal and famous study of scientific activity (Kuhn 1969) showed that knowledge is not accumulative and that science does not develop in a linear manner. Instead, it evolves in an abrupt way (scientific revolutions) with intervening quiescent periods.

Inspired by Kuhn's paradigmatic conception of scientific progress and by Wittgenstein's theories on rule following and language games (Wittgenstein 2001), a series of science studies, termed the sociology of knowledge (SoK) movement, emerged. The key issue is to show that science is a social activity that follows social patterns in the same manner as do other groups in society. The question of how things are in the world cannot be addressed without the question of how the social group comprising scientists conceives of these things. Things, be it photons or DNA, cannot be attributed a role in our world independent of symbols and meaning.

Hence, while the traditional philosophy of science had procedural criteria of demarcating science from non-science, such as Popper's criteria of refutation, SoK applies social criteria. Whereas the normative aim of traditional science studies was to free science from power inherent in the social structures among scientists and in society, SoK strives to disclose power within the scientific society and to emancipate.

In many respects, the key issue in the classical philosophy of science has been the relationship between scientific theories and nature. In SoK the focus is on the relationship between theory and culture. In what way do scientific theories reflect social structures (instead of structures of the world)? Nevertheless, what appears to be similar in both the traditional philosophy of science and SoK is the focus on epistemological issues: in both cases the key question is *what scientific theories represent*. In the first case they represent patterns in nature, in the second, they represent social structures.

Later studies of science have tried to avoid this *representational* pattern. Their empirical studies of science have investigated not only the relationship between theories and the social processes and structures in science, but *the scientific process* itself, including its material premises. How do scientists behave, and how do they produce the facts of science? This may be called a processual approach (PA), according to which science is the change, restructuring, making new, and stabilizing of things and theories. What characterizes the social process of science is an interaction of methods, material, activities and processes, where negotiations lead to stabilizing and generation of facts. When species of the *Helicobacter pylori* bacteria were found to be associated with gastric and peptic ulcers, scientific debate ensued on the bases of the residing theories, and negotiations on behalf of continued empirical work confirmed that *H. pylori* is a key factor.

There is no question about what the theory represents (either nature or culture), but rather it is a question of negotiation between different scientific groups with regard to what will be considered to compromise facts. Hence, according to the PA the issues are not the relationship between theory and nature/culture (epistemological and representational), but what scientists regard and treat as real (ontological and processual).

1.2 PHILOSOPHY OF THE SOCIAL SCIENCES

A significant part of the overall spectrum of healthcare problems comprises matters that principally are not biological. Should we wish to find why patients do not take prescribed medicines, why wrong medicines are given in hospitals, or why it is difficult to obtain fully informed consent for trials or treatment, we cannot search for answers in human biological research, but instead must turn to the methods of the social sciences.

So, it is essential to know the ways in which the philosophies of the social sciences and the biological sciences differ, so that we do not erroneously use the criteria of one area to judge another. In the social sciences, many different methods are used, and there are various schools of theory. So, the discussion here comprises a brief introduction and does not cover the broad scope of methods and schools of theory.

Interpretation, understanding and explanation

The social sciences differ from the biological sciences in two respects:

- they entail greater elements of interpretation that often enter into compilations of data
- in many cases, a result is an understanding, not an explanation.

Explanation and understanding

The principal goal of inquiry in the biological sciences is to elicit explanations of phenomena studied. One might, for instance, seek the cause of a particular manifestation of a disease.

Some projects in the social sciences also seek causal explanations of social phenomena, but many seek instead an understanding. Understanding is a form of knowledge that enables us to know why a person or a group behaves in a particular way, why and how they experience a specific situation, how they themselves understand their way of life, and so on. We attain understanding through interpretation.

The distinction between explanation and understanding was first expressed by the German philosopher, psychologist and educator Wilhelm Dilthey (1833–1911), who believed that these two ways of understanding the world were characteristic of the natural sciences and the human sciences (*Geisteswissenschaften*), respectively. However, the distinction between explanation and understanding is not as distinct as many believe. Many theories of the social sciences include elements of both causal explanation and non-causal understanding.

Interpretation

All content-bearing objects and statements can be interpreted. People express themselves not just in speech, writing and deeds, but also in architecture, garden design, clothing, etc. If, for example, we enquire into where and why institutions for psychologically ill patients were built, we will find that the history reflects varying understandings of psychological illness. The architecture of the asylum is content bearing. However, here we will focus on the interpretation of texts and other linguistic statements, as it is germane in the discussion of the theory of interpretation, often called hermeneutics.

Interpretation may have many goals, but in general we seek to fathom the information content of the content-bearing material. The various theories of interpretation are based on differing concepts of the nature of content and how it should be located. Is there content in a statement itself, in the thoughts of the person making the statement, in the social structure in which the statement is made, etc.? These differences are germane when analysing the validity of specific methods of the social sciences, but are of lesser importance here in the general discussion of interpretation.

The question of whether one obtains a true interpretation of a text is old. All written religions have sets of hermeneutic rules for interpreting the content of holy texts. For example, in Christian theology, biblical exegesis concerns interpretation of the scriptures.

In modern times, interest arose in the interpretation of secular statements, first as part of literary and historical research, and then as a part of research in the social sciences.

The goal of the various hermeneutic methods that have been developed is to arrive at an understanding of content that can be defended as a valid, intersubjective understanding. That is, it is an understanding that can be substantiated and discussed rationally.

As Popper pointed out, the elements of interpretation enter into all observations and thereby into all forms of science. We lack direct access to the world 'as it is' through our senses. We always view the world through a theoretical filter, and all observations are theoretically loaded. For example, when we say that the sun rises, we reflect the influence of the old geocentric world view in which the sun circled the Earth. And the 'description' that a pathologist gives of a histological preparation seen by microscope is to a large extent an interpretation based on theories of cells, inflammation, etc.

The hermeneutic circle, understanding horizon and 'double hermeneutics'

The hermeneutic interpretation of a text rests on individual parts as well as on the understanding of each individual part related to the whole. Neither an individual part nor the whole text may be interpreted without reference to each other. So, interpretation is circular, the hermeneutic circle. In principle, this circle cannot lead to certain closure, as we will never know whether a deeper analysis of the text may change our interpretation of it. The problem of attaining valid, intersubjective interpretation has long been and still is discussed, and optimistic interpretation theoreticians speak of a hermeneutic spiral that implies that interpretation gets better and better. At the pragmatic level, the problem of the hermeneutic circle is less worthy of attention, as agreement on the meaning of a text usually can be more easily attained.

The concrete interpretation is also influenced by the interpreter's 'horizon of understanding', a concept from *Wahrheit und Methode*, the principal work of German philosopher Hans-Georg Gadamer (1900–2002). Gadamer argues that before I have begun a conversation with another person or begun to interpret a text, I already have bias about them based on my horizon of understanding, a collective term for my world view. My horizon of understanding builds up throughout my life and comprises my understanding of particular words, the connotations that particular words and concepts hold for me, and so on. For a resident of London, the word 'city' connotes financial affairs, while for people elsewhere, it simply connotes an urban concentration of population. Two people engaging in a conversation may believe that they have understood each other without actually having done so. Full understanding is possible only when two conversing people have acquired each other's horizons of understanding ('fused horizons'). Hence there may be a problem of interpretation, as in interviews in the social sciences, which often are too short for the interviewer to understand the interviewee's horizon of understanding. Consequently, a vital part of the interview comprises an effort to find out how the interviewee uses and understands words and concepts in the area being discussed.

Furthermore, English sociologist Anthony Giddens pointed out that within the social sciences, research comprises a 'double hermeneutics' (Giddens 1976, 1990). In reality, the social sciences research interprets interviewees' interpretations of their own understandings, and parts of their understandings arise through concepts that they have acquired from the theories of the social sciences (such as the Marxist concept of class or the incest taboo of psychology). Hence, there is a complex interaction between the interpretations of the researcher and the interviewee, which is why an additional level of interpretation often may be needed to focus on how an interviewee's self-image is affected by the theories of social science. Consequently, an interviewee may be misunderstood if the interviewer does not take such reflections into account.

Power, ideology and interests

Our interpretations of the statements and deeds of others are influenced by aspects in addition to our horizons of understanding. The German philosopher Jürgen Habermas (1929–) pointed out that power, ideology and interests play leading roles. Usually, we are not neutral or objective observers, but interpret according to our power of position, our ideology and the interests we wish to further (Habermas 1986).

In Habermas' view, ideology is not restricted to political ideology. An ideology is simply a set of assumptions that further the interests of a particular group in society. For example, the assertion that 'an extensive hospital system is essential in healthcare' is an apolitical ideology that in addition to safeguarding the interests of patients, furthers the interests of doctors and other healthcare professionals.

A difficulty with ideologies is that they are often concealed, as we neither are aware that we have them nor know where they came from. So, behind our backs, they influence our actions and our interpretations. Consequently, Habermas maintains that the principal task for the critical social sciences is to identify prevailing ideologies so we may be freed from them.

Validity

In the above, we have discussed problems widely recognized, that an interpretation and the understanding that we attain through interpretation can never be 'a final truth' concerning the meaning of a particular statement (unless the statement is extremely simple). So, we are obliged to ask how we can judge the validity of a scientific interpretation. The simple answer is that if a researcher has been aware of these problems and has taken the best possible steps to avoid or avert them (such as by trying to identify which ideologies and interests have influenced the various elements of the research process), there are grounds to rely on the interpretation; not because it is of necessity true, but because it comprises a well-founded hypothesis without significant sources of errors in the research process.

Reductionism and emergence

Some biological researchers contend that there is no need for social scientific interpretation because in the final analysis, all knowledge can be reduced to facts about physical conditions. Social phenomena can be reduced to group psychology, which in turn can be reduced to individual psychology, which in turn can be reduced to neurology, which in turn can be reduced to cellular biology, and so on, until we reach the physical level at which prevailing physical laws provide explanations for all phenomena observed at higher levels. This view, called reductionism, is in strong dispute.

So, here, it is crucial to distinguish between methodological reductionism and general reductionism. In some research projects, methodological reasons may dictate the exploration of one or more factors that can influence the phenom-enon of interest without indicating that other factors are unimportant. We have no methods that can acquire data and at the same time investigate 'the whole'. Of necessity, our attention must be focused on something more specific. Methodological reductionism can be meaningful and necessary, even though we refute general reductionism. If, for instance, we wish to examine a biological relationship, it may be necessary to ignore an ancillary social relationship. Conversely, if we examine a social relationship, it may be necessary to ignore a biological relationship.

Methodological reductionism is itself straightforward, as long as the factors that we examine are sensible. It becomes problematical only when a set of factors is systematically excluded, such as by ignoring the correlation between poverty, social deprivation and disease.

There are many arguments against reducing social phenomena to physics, two of which are summarized here. The first problem confronting the reductionist is that it is doubtful that individual psychology can be reduced to neurophysiologi-cal processes. Dispute persists on the precise description of the relationship between psychological phenomena and cerebral activity, and today we seem no closer to solving the 'mind–brain' riddle than we were a century ago. If this link in the reductionistic chain fails, reductionism as a whole cannot be carried out.

The other problem for the reductionist is that many social phenomena are emergent, that is, they are socially not reducible as they occur at particular social levels and have no meaning when reduced to lower levels (individual psychology, neurology, etc.).

Paper money, for example, is an emergent social phenomenon. A £10 banknote has no value itself (unless you keep it for its portrait of scientist Charles Darwin). It cannot be exchanged for gold or other objects of value at the central bank. But it is integrated in social relationships that enable it to be exchanged for goods or services worth 10 pounds. Otherwise, it is just a small, rectangular scrap of paper.

Emergence at the social level also may be ascribed to a particular set of social conventions or formalized laws. For instance, most societies have the institution of marriage, but the concrete implications of being married and the social effects of it vary from society to society. The human penchant to form pair relationships might be reduced to the biological level, but the concrete institution of marriage in a particular society cannot be similarly reduced. However, it is clear that the concrete, non-reducible institution of marriage affects human actions and considerations, so a full description of these actions and considerations is possible only on the social level.

If the antireductionists are right, scientific effort in the social sphere is useful, and it may employ methods that differ from those applicable at lower levels.

Generalization

Generalizing statistics are often useful in research projects that use quantitative methods. Whenever we take samples from a well-defined population, we express the statistical confidence interval of the results and consequently permit their general extension to other similar populations. In principle, that implies that results from research conducted in the USA may be directly applicable to choices of treatments in Norway. However, it is worth noting that such generalization of results is acceptable only when we have grounds to assume that the populations are in fact similar, as by assuming that there is no biological difference between Americans and Norwegians.

Generalization may be used in much the same way in quantitative social science research, but statistical methods cannot be used in research that is not quantitative. Does this imply that understanding attained in social science research cannot be generalized? Were statistical generalization the only form of generalization available, understanding could not be generalized. Yet there is a form of generalization that is not quantitative and is frequently used across all the sciences. It is theoretical or conceptual generalization, sometimes called transferability. We often generalize, not in exact numbers, such as the cure rate for a particular drug, but rather within a conceptual or a theoretical frame of understanding. For instance, when teleological explanations based on the theory of evolution are used in biology, they rest upon a theoretical generalization of the theory of evolution, not upon a statistical generalization. Social scientific concepts and theories may be generalized in the same manner.

In all forms of generalization, both statistical and conceptual, it is important to keep in mind that conditions change with time. Generalizations that once were valid can be rendered invalid if there are changes in the supporting biological conditions, such as the resistance patterns in bacteria or the structures of families.

REFERENCES

Bhaskar R (1997) A Realist Theory of Science. Verso, London.

Giddens A (1976) New Rules of Sociological Method: A Positive Critique of Interpretative Sociologies. Hutchinson, London.

Giddens A (1990) The Consequences of Modernity. Polity Press, New York.

Habermas J (1986) The Theory of Communicative Action, Reason and the Rationalization of Society. Two Volumes. Polity Press, London.

Hacking I (1983) Representing and Intervening: Introductory Topics in the Philosophy of Natural Science. Cambridge University Press, Cambridge.

Hempel CG (1965) Aspects of scientific explanation. In: Aspects of Scientific Explanation and Other Essays. The Free Press, New York, pp. 331–496.

Kuhn TS (1969) The Structure of Scientific Revolutions. University of Chicago Press, Chicago.

Mackie J (1974) The Cement of the Universe. Clarendon Press, Oxford.

Popper KR (1963) Conjectures and Refutations, The Growth of Scientific Knowledge. Routledge and Kegan Paul, London.

Psillos S (2002) Causation and Explanation. Acumen Publishing, Chesham.

Putnam H (1981) Reason, Truth and History. Cambridge University Press, Cambridge.

Russell B (1959) Mysticism and Logic. Allen & Unwin, London.

Salmon W (1990) Four Decades of Scientific Explanation. University of Minnesota Press, Minneapolis.

Stengers I (2000) The Invention of Modern Science. University of Minnesota Press, Minneapolis.

Wittgenstein L (2001) Philosophical Investigation. Blackwell, Oxford.

FURTHER READING

Boyd R et al. (1991) The Philosophy of Science. MIT Press, Cambridge, MA.

Hacking I (2001) The Social Construction of What? Harvard University Press, Boston.

Hempel CG (1966) Philosophy of Natural Science. Prentice Hall, Englewood Cliffs, NJ.

Hollis MH (1994) The Philosophy of Social Science – An Introduction. Cambridge University Press, Cambridge.

Hugles J (1990) The Philosophy of Social Research (2nd edn). Longman, Harlow.

Lipton R, Odegaard T (2005) Causal thinking and causal language in epidemiology: it's in the details. Epidemiology Perspectives Innovation 2: 8.

Parascandola M, Weed DL (2001) Causation in epidemiology. Journal of Epidemiology and Community Health 55: 905–912.

Popper KR (1959) The Logic of Scientific Discovery. Hutchinson, London.

Williams M, May T (1996) Introduction to the Philosophy of Social Research. Routledge, London.

Wynne B (1992) Uncertainty and environmental learning. Reconceiving science and policy in the preventive paradigm. Global Environmental Change, pp. 111–127.

ETHICS AND SCIENTIFIC CONDUCT

Søren Holm

As for the philosophy of science surveyed in Chapter 1, ethics and scientific conduct comprise a broad subject that is detailed in innumerable published works, many larger than this book. Consequently, this chapter is at most a brief overview of the aspects of ethics applicable to biomedical research. First, we delimit the foundations for ethical assessments of research activities, then move on to discuss the internal ethics of science and research endeavours, and finally look at some specific forms of research misconduct including the fabrication of data and results, false authorship, plagiarism and duplicate publication. The 'external' ethics of science, the obligations towards research participants and the ethics of animal research, are discussed in Chapter 3.

2.1 A BRIEF INTRODUCTION TO ETHICS

Contemplations of ethics have a long history, going back (at least) to the pre-Socratic philosophers in ancient Greece and are an equally venerable component of all major religions. Yet, there still is no agreement on the nature of the correct or best framework or theory for ethical analysis, although there is reasonable agreement concerning the core of the disagreement and the viable contenders for an acceptable ethical framework.

It is, for instance, relatively clear that moral nihilism and moral relativism are not viable options. Cultural relativism about morality is a fairly commonly held position in public debate, essentially claiming that cultures differ in morality, that these differences should be respected, and that morality cannot be discussed across cultures. Although it is clearly a true description of the state of the world that different societies have different moral commitments, cultural relativism is unsustainable as a coherent moral position. If cultural relativism were true, criticism of the moral judgements of people outside your own culture would be nonsensical, universal human rights (even such rights as the right not to be tortured) would be meaningless and it would be impossible to make sense of a notion of moral progress (although most of us, for instance, believe that the abolition of slavery is moral progress wherever it occurs).

Cultural relativism is seen as an attractive position, in part because two distinctions are not made. The first is between the claim that ethical values are universal (e.g. that harming people is wrong wherever and whenever it occurs) and the claim that they are absolute (e.g. that you can never harm a person even to actualize some greater good). The second is between inflexibility and context dependency in the application of values. Many deem cultural relativism to be attractive, because the ostensible alternative is that ethical values are absolute and inflexible. But strictly speaking, a denial of cultural relativism only includes the claim that ethical values are universal; they may well be non-absolute and flexible in their application with sizeable scope for the importance of context. We may, for instance, claim that respect for privacy is a universal value, while still recognizing that the exact contours of such a right will and must vary between societies. For instance, the shape of a right to privacy will depend in part on whether the prevalent types of living accommodation are communal long houses or individual flats.

Moral frameworks hold that values are universal yet permit leeway in application. The principal differences of opinion in ethical theory are about whether the basic level of ethical evaluation is evaluation of acts, states of the world or people, and whether the rightness or goodness of an act is essential in the evaluation of it. The three most often used ethical theories or frameworks are listed in Box 2.1.

> **Box 2.1 Ethical theories and frameworks**
> - Consequentialism
> - Deontological ethics
> - Virtue ethics

Consequentialism

Consequentialism is the simplest possible ethical theory. It holds that goodness is primary and defines the right act as that which maximizes goodness. In thinking about an action, an agent should consider the various possible actions that he or she could perform and then choose the one that maximizes the good consequences. This is equivalent to choosing the act that maximizes the goodness of the state of the world. What consequentialism repudiates is that the type of an act matters in itself. That an act comprises lying should not matter in thinking about whether to perform it, the only thing that matters is whether it has good consequences. Different consequentialist theories differ in what they count as good consequences, whether they claim that good consequences should be maximized, and what class of entities the calculation should include. The two first differences between consequentialist theories often do not matter for practical purposes, but the last difference can have significant implications. For instance, it matters hugely for the ethics of animal research whether (some) animals are included in the class for whom good consequences should be maximized.

The main criticism of consequentialist theories is that in some situations, they can justify actions that most people think are wrong, for instance the sacrifice of the interests (perhaps even lives) of a few research participants in order to gain important scientific knowledge. This criticism has led to the development of a variant of consequentialist theory called rule consequentialism. Rule consequentialism claims that we should not consider individual acts in isolation but instead should focus on the rules that will have the best consequences if followed. A rule consequentialist might, for instance, contend that it makes sense to have a rule against lying, because following the rule maximizes good consequences over time.

Deontological ethics

Deontological theories are in some ways the opposite of consequentialist theories. According to deontological theories of ethics, the primary concern is whether an act is right, not whether it has good consequences. An act of lying should not be performed even if it has good consequences. Various deontological theories differ in how actions are determined to be right and in how one should choose which of two wrong acts to perform in a situation in which there is no right action. Deontological theories fit well with our prereflective commitments to the belief that there are some acts that are wrong in themselves (e.g. torturing newborn children). The principal criticism of them is that they fail to explain why, for instance, 'white lies' should be considered as seriously wrong as implied.

Virtue ethics

Virtue theory differs from consequentialism and deontology in that it focuses on the person performing an act instead of on the act itself. According to virtue theory it is possible to identify the set of character traits and motives that a morally good person should possess (these are the character traits that are designated as 'virtues'). The morally right action is the action that a virtuous person performs and that flows from his or her virtues. The main criticisms of virtue theory are that there is no consensus on a list of virtues and that the theory has difficulty accounting for the fact that even morally evil people seem able to perform the occasional good act.

2.2 SCIENTIFIC CONDUCT AND MISCONDUCT

Let us now consider the ethical obligations of scientists in their research activities. Clearly, we can derive sundry ethical obligations from the general framework discussed above. All systems of ethics can, for instance, explain why lying and other forms of dishonesty are problematic in most circumstances, and that deceiving others for your own gain is prickly. Similarly, all forms of ethical theory find the exploitation of the powerless by the powerful problematic. This clearly supports rules against well-known types of scientific misconduct, including fabrication of data, plagiarism and false or gift authorship.

But perhaps we may derive better or more specific guidance on proper scientific conduct from an analysis of the purpose and goals of the research enterprise. It is clear from the analysis presented in Chapter 1 that there is no univocal and uncontested definition of science or research. But most would accept that a core feature of science is that it aims at producing publicly available, well-justified knowledge achieved only through a long-term effort involving many different researchers and research groups. Indeed, that view is corroborated in the famous remark that 'we see further because we stand on the shoulders of giants' (although, as it has been mischievously pointed out, we would also see further if we stood on the shoulders of midgets); the original remark often is attributed to Isaac Newton, but seems to have originated with Bernard of Chartres in the twelfth century.

If we accept this categorization of science as a goal-driven activity, it makes sense to ask how the participants must act in order to achieve the goals of the activity.

This question has been analysed extensively by philosopher Knut Erik Tranøy and by many sociologists, most notably Robert Merton.

Tranøy argues that scientific work is characterized by and requires three different kinds of norms (Tranøy 1988, 1996), as listed in Box 2.2.

Box 2.2 Requirements of norms
- Internal norms
- Linkage norms
- External norms

All three guide scientific work, but in different ways. If scientific activity did not interact with society at large, only the internal norms would be guiding. But because there are varieties of interaction with society, linkage and external norms also come into play.

Internal norms

Within internal norms, we distinguish between epistemic norms that guide the activity of each individual researcher, and social norms that guide the collaboration between researchers and research groups in a scientific endeavour. In epistemic norms, Tranøy includes truth seeking, testability, consistency, coherency and simplicity, and in the social norms he includes openness, open-mindedness and honesty. He argues convincingly that unless individual researchers and research communities accept these norms and act in accordance with them, science as an activity aimed at generating knowledge cannot succeed. Consequently, these norms are mandatory. But they are not imposed from the outside; they arise through the nature of scientific activity.

Linkage and external norms

Linkage norms include utility, fruitfulness and relevance and explain why society permits scientists freedom of research, whereas ultimately external norms are the limits society places on scientific conduct (e.g. in relation to research participants). Evidently, there are significant similarities between the results of Tranøy's analysis and those obtained by American sociologist Robert Merton. Based on studies of scientists, Merton claimed that the scientific community was committed to a set of norms denoted by the acronym CUDOS, for 'Communism of knowledge' (subsequently changed to 'Communalism' when 'Communism' became contentious), 'Universality', 'Disinterestedness' and 'Organized scepticism' (Merton 1968).

Tranøy's analysis is significant because it provides a reason for labelling certain activities as problematic scientific misconduct, even if we believe that the general ethical frameworks do not apply to science or scientific activity. Tranøy shows that there are purely internal reasons related to the epistemic claims that science makes.

2.3 MISCONDUCT AND WHY IT OCCURS

Scientific misconduct is generally regarded to be illicit, so it is difficult to assess its general prevalence and to probe its specific instances. That said, available evidence suggests that scientific misconduct is not rare.

A recent American survey of 1645 clinical trial research coordinators found that 21.5% had first hand knowledge of incidents of scientific misconduct during the previous year (scientific misconduct was defined in accordance with the US Office of Research Integrity (ORI) definition as: 'Fabrication, falsification, plagiarism, and other practices that seriously deviate from accepted standards when proposing, conducting and reporting research' (Office of Research Integrity 2006, Broome et al. 2005, p. 264)).

A review in the UK conducted in 2000 looks at a variety of studies that seem to indicate a frequency of serious misconduct in 0.5–1% of clinical research projects (Evans 2000).

What triggers misconduct in science? In most cases, self-interest seems the most plausible explanation. By operating outside the rules, it is easier to get results and papers on one's list of publications. In some cases, the psychological basis apparently is more complex, but there is no cause to believe that the perpetrators are mentally ill or psychologically deranged.

An archetypal justification for misconduct was given by American researcher Eric T. Poehlman before he was sentenced to 366 days in prison 'because his actions led to a loss to the government, obstruction of justice, and abuse of a position of trust' (Office of Research Integrity 2006, p. 1). On investigation by the ORI he had been found to have falsified or fabricated data in at least 12 publications and 19 grant applications. Poehlman's explanation of his conduct is extracted in Box 2.3.

Box 2.3 Poehlman's explanation

In a letter to Judge William Sessions, III, US District Court for the District of Vermont, Eric T. Poehlman said he had convinced himself that it was acceptable to falsify data for the following reasons:

'First, I believed that because the research questions I had framed were legitimate and worthy of study, it was okay to misrepresent "minor" pieces of data to increase the odds that the grant would be awarded to UVM* and the work I proposed could be done.'

'Second, the structure at UVM created pressures which I should have, but was not able to stand up to. Being an academic in a medical school setting, I saw my job and my laboratory as expendable if I were not able to produce. Many aspects of my laboratory, including salaries of the technicians and lab workers, depended on my ability to obtain grants for the university. I convinced myself that the responsibility I felt for these individuals, the stress associated with that responsibility, and my passion and personal ambition justified "cutting corners".'

'Third, I cannot deny that I was also motivated by my own desire to advance as a respected scientist because I wanted to be recognized as an important contributor in a field I was committed to.' (Office of Research Integrity 2006, p. 5)

*UVM is the abbreviation for the University of Vermont at Burlington.

In addition to his prison sentence Mr Poehlman was debarred for life from applying for or receiving federal research grants.

2.4 FABRICATION AND OTHER FORMS OF MISCONDUCT AFFECTING THE TRUTH CLAIMS OF SCIENTIFIC FINDINGS

Scientific misconduct is most serious when it affects the truth claims of scientific findings, as it then undermines the cumulative nature of scientific work and development and may lead to practical applications that are harmful to patients.

Fabrication – the invention of data – arguably is the most blatant form of misconduct affecting truth claims. It ranges from the invention of all data reported to the invention of some of it (for instance because of recruitment problems and time constraints). It is not common, but it is not exceptionally rare either, especially when cases in which all data are fabricated are taken into account. Each year, ORI records several cases of fabrication.

A recent high-profile case illustrates the perils of fabrication and falsification. Early in 2006, in Norway it was discovered that John Sudbø, a young yet prominent researcher in the diagnosis, development and treatment of oral premalignant lesions and carcinomas, had fabricated all data in an article published in October 2005 in the Lancet by himself and co-authors. Later investigations found evidence of fabrication and falsification in a number of Sudbø's previously published papers, including some published in the New England Journal of Medicine (University of Oslo 2006). Lancet editor Richard Horton called the case 'the worst the research world has seen' (Aftenposten 2006), but that may be overstatement due to the Lancet having been one of the journals that published Sudbø's fraudulent research. All of the papers were retracted. But because of the original and high-profile nature of Sudbø's 'research' it is likely that some of his 'findings' have led to suboptimal treatment of patients.

Other types of misconduct in this category are suppression of unwanted results or intentionally biased analysis of the data to obtain 'desired' results. Both lead to misleading information in scientific records.

The same is true whenever researchers publish false or misleading accounts of their methodology in order to slow down competing research groups. Even if the research findings themselves are not affected, the deliberate introduction of falsehoods into the scientific record is tantamount to misconduct.

Plagiarism

Plagiarism is claiming the work of another to be one's own (see Box 2.4). In publications, plagiarism may be total, such as the submitting of copies or translations of papers previously published by others elsewhere (actually not rare), or more limited, such as copying and pasting other people's work into your own papers either in their original form or slightly paraphrased. Ubiquitous word processing and the Internet have made plagiarism easy and consequently commonplace. It is generally recognized that there is more plagiarism in university student essays than there used to be. However, the ease of Internet searches has simplified the detection of plagiarism, and several dedicated professional plagiarism detection packages are now available.

Box 2.4 Tom Lehrer on plagiarism

A verse of *Lobachevsky*, a song by Tom Lehrer (1928–):

I can never forget the day I first meet the great Lobachevsky.
In one word he told me secret of success in mathematics:
Plagiarize!

Plagiarize,
Let no one else's work evade your eyes,
Remember why the good Lord made your eyes,
So don't shade your eyes,
But plagiarize, plagiarize, plagiarize –
Only be sure always to call it please 'research'.

The most prolific plagiarist of all times in biomedical research probably was E.A.K. Alsabti, who worked in the USA, the UK and South Africa in the late 1970s. The complete extent of his plagiarism is not known, but he is suspected of having plagiarized about 60 full articles (Lock 1996).

Another form of plagiarism is the theft of ideas. For instance, there have been incidents when referees for journals have seen an interesting idea in a paper they are refereeing and have set their laboratory to work on the idea while holding up the refereeing process, so that they themselves can submit a paper based on the idea before it is published.

One of the most prominent cases concerning scientific priority and potential plagiarism was the dispute between Luc Montagnier and Robert Gallo over who first isolated and identified the virus now known as HIV. Montagnier published and submitted patent applications on tests for the virus first, but Gallo nonetheless claimed priority and managed to get the valuable US patent. Both researchers denied the claim of the other and the dispute was only resolved when the then presidents of the USA and France intervened to broker an agreement whereby the two researchers agreed to be named as co-discoverers of the virus and to share the patents. The case is well documented and there is little doubt that Gallo used a virus sample that he had obtained from Montagnier and essentially isolated the same virus; see the papers by Prusiner, Montagnier and Gallo in a Science special collection (Science 2002, Cohen 1993).

2.5 AUTHORSHIP ISSUES

Disputes about authorship are probably the most common of conflicts within research groups. Bylines on papers have several functions within the scientific community. They designate who was involved in a published work and accordingly who should share the honour related to the findings reported. Because of this significance, being an author plays a prominent role in employment, promotion and grant-awarding decisions. Authorship thus functions as a sort of currency that can be cashed later in the researcher's career. Hence, the frequency of disputes is understandable.

The International Committee of Medical Journal Editors promulgates authorship rules aimed to minimize the number of disputes and to ensure that problematic forms of authorship can be clearly identified (reproduced in Appendix 1). Note that these rules primarily concern the biomedical sector; other sectors may have different rules. The February 2006 version of the rules is summarized in Box 2.5.

Box 2.5 International Committee of Medical Journal Editors rules of authorship

1. Substantial contributions to conception and design, or acquisition of data, or analysis and interpretation of data.
2. Drafting the article or revising it critically for important intellectual content.
3. Final approval of the version to be published.
4. Authors should meet conditions 1, 2, and 3.

While

5. Acquisition of funding, collection of data, or general supervision of the research group, alone, does not justify authorship.

It is generally recognized that there are five main types of misconduct related to authorship (see Box 2.6).

> ## Box 2.6 Authorship issues
> - Exclusion from authorship
> - Gift authorship
> - Authorship achieved by coercion
> - Unsolicited authorship
> - Refusal to accept responsibility as an author when other misconduct is detected

Exclusion from authorship

Exclusion from authorship happens when someone who has contributed significantly to a project and fulfils the criteria for authorship is not named in the byline, although he or she so wishes (there is no requirement of listing contributors against their wishes). This happens most often to junior researchers, but can also happen where a research group has split before publication. Unjustified exclusion from authorship is tantamount to theft.

Gift authorship

Gift authorship is the case where someone who has not fulfilled the criteria for authorship nonetheless is offered authorship. There are different scenarios in which gift authorship might occur: it may be a swap: 'I'll give you author status on my paper if you give me author status on yours', it may be to gain the endorsement of a famous name on a paper to help it through the peer-review process; it may be a way to 'improve' the CV of junior researchers in a laboratory; or it may be a way for a pharmaceutical firm to get a prestigious name on a review essentially written by the company.

Authorship by coercion

Authorship achieved by coercion commonly occurs when a senior researcher, often the head of a laboratory, demands to be an author on all publications from the laboratory, regardless of whether or not he or she has fulfilled the criteria for authorship.

Unsolicited authorship

Unsolicited authorship is where someone is listed as an author without their knowledge or consent. An example from the anonymous case records of the Committee on Publication Ethics is given in Box 2.7.

Box 2.7 Paper submitted by a PR company without the knowledge of the authors

'A paper was submitted for which there were seven contributors, but no corresponding author. The only identification of who had sent the paper was an accompanying e-mail from a public relations company.

When contacted by the editorial office, the PR company confirmed that the paper was to be considered for possible publication. The named contributors were then contacted and asked whether they had given permission for their name to be attached to the paper, asked who was the corresponding author, and also if they wished to declare any conflict of interest.

This produced a very interesting flurry. One author said the paper had been produced as a result of a seminar to which he and the other contributing authors had been invited. He himself believed that he was simply giving advice to the drug company concerned, for which he had received a fee. He believed that a misunderstanding had led the PR company to send the paper for review, but that he had no knowledge that they had done so, and suggested that the paper be shredded.

Another author telephoned to say he could remember very little about it and certainly hadn't seen the final document. A third author telephoned in some distress, anxious that he might be accused of some form of misconduct and had never thought that his involvement would lead to a paper being submitted to a journal.

The most interesting letter of all was from the first named author who had subsequently written an editorial for the journal that was fairly critical of the drug concerned. The PR company who was acting for the drug company, she said, had submitted the paper on her behalf without her knowledge. (...)

The same company had previously published another article to which they had put her name, but which she had not written. This author feels very abused, particularly as she wrote to the PR company requesting that they did not use her name again.'

(COPE case 00/06)

Unsolicited authorship almost always also involves ghost authorship; that is, the person who really wrote the paper is not listed as an author.

Refusal to accept responsibility

In accepting that one's name appears in the byline of a paper, a person also accepts responsibility for at least a part of its content. Yet, in many cases where fabrication or some other forms of serious misconduct has been revealed in a jointly authored paper, people who gladly were listed as co-authors suddenly renounce responsibility for the paper. This is either in itself a form of misconduct, or it points to earlier misconduct in accepting authorship without due care, as in the revising of a manuscript for publication.

2.6 SALAMI, IMALAS AND DUPLICATE PUBLICATION

A final type of misconduct worth noting is the phenomena of salami and imalas publication. Both terms were coined by Professor Povl Riis, the first chairman of the Danish national ethics committee, and were taken from salami, the highly salted, flavoured Italian sausage that is usually served in thin slices, imalas being salami spelled backwards (Riis 1995). Salami and imalas publication seek to maximize the number of papers published from a given work done by reducing each to the 'least publishable unit'. Salami publication involves carving up the results of work done into the thinnest possible slices that can still be published. Imalas publication is the sequential publishing of what are essentially the same results, but with a few new data included in the analysis each time (e.g. publishing results of a study with a planned recruitment of 100 people, after data on the first 30 and 60 people have been generated).

Salami publication constitutes misconduct, because it makes it more difficult for the users of the research results to gain an overview of the complete project. An especially problematic type of salami publication is where a large trial is published at the same time as parts of the trial are published (for instance, reporting on the patients recruited in one of the participating institutions or countries). If the link between the complete trial and the part is not made clear in the publications this may lead to double counting of the evidence in later reviews or meta-analyses. Imalas publication leads to the literature being cluttered with interim results, which again makes it more difficult to gain an overview of the definitive results of a project.

The limiting case of imalas publication is duplicate or multiple publication of the same research results as if they were new. In addition to the general effects of imalas publication, this involves a direct deception of the second journal, as most journals prohibit double publication. Duplicate publication is generally only acceptable if the first publication is in an international journal and language and the second publication is in a national language and journal, and the relationship between the two papers is made clear.

2.7 THE INVESTIGATION AND PUNISHMENT OF SCIENTIFIC MISCONDUCT

Scientific misconduct undermines the internal value system of science and tarnishes the external validity of scientific claims, so it should be minimized, like other forms of crime.

This means that allegations of scientific misconduct should be investigated and that those found guilty of it should be punished. But who should investigate and what punishment should be meted out?

Traditionally, scientific misconduct has been investigated by the institution employing the researcher against whom allegations are raised. But such an approach is problematic, because the institution is not a neutral or disinterested party. It has interests in maintaining its reputation and in maintaining good relations with grant-awarding bodies. Previously such interests often led to the suppression of allegations of scientific misconduct and the persecution of whistle blowers, whereas more recently they have sometimes led to overreactions and severe sanctions against researchers accused of misconduct before a case has been fully and impartially investigated. Consequently, several countries have set up non-institutional systems for investigating allegations of scientific misconduct.

An example is the Danish Committees on Scientific Dishonesty (DCSD), upon which full information is available at http://danmark.dk/portal/page/pr04/ FIST/FORSIDE/UDVALGENE_VIDENSKABELIG_UREDELIGHED in Danish and English. The DCSD is established by law as part of the Danish Research Agency and consists of three separate committees for health sciences, natural and technical sciences, and social and human sciences, respectively. An allegation of misconduct can be made to these committees against researchers working at Danish public institutions including universities or at organizations supported by public funds. An initial investigation will be made and if the relevant committee thinks that there might be a case of misconduct an ad hoc investigative committee will be established consisting of members of the DCSD and experts in the relevant area of research. At the end of the investigation, a determination is made as to whether misconduct has occurred and if so its nature. This determination is sent to the researcher's employer(s) and to the journals in which the research has been published. All investigated cases are reported anonymously in the annual reports. The DCSD will also investigate cases where researchers who have been publicly accused of misconduct seek to have their names cleared.

The DCSD has no formal punishments at its disposal when it has determined that misconduct has taken place, but the American ORI has a range of sanctions that it can impose on researchers or institutions when misconduct has been proved. It can debar them from federal funding, most commonly for one to three years, but there have been lifetime debarments, and it can require that their employer implements strict supervision even of senior researchers.

Journals are sometimes involved in the investigation of allegations of scientific misconduct, especially whenever the journal is well resourced and where no other organization is willing to undertake the investigation. The range of sanctions available to journals is obviously more limited than those available to a national body, but they include official retraction of papers, which will then also be listed as retracted on databases such as Medline, refusal to publish more papers by the same author and publication of the finding of misconduct.

REFERENCES

Aftenposten (2006) Research scam makes waves. Aftenposten, 17 January.

Broome ME et al. (2005) The Scientific Misconduct Questionnaire – Revised (SMQ-R): validation and psychometric testing. Accountability in Research: Policies and Quality Assurance 12(4): 263–280.

Cohen J (1993) HH: Gallo guilty of misconduct. Science 259: 168–170.

Evans SJW (2000) How common is it? Journal of the Royal College of Physicians of Edinburgh 30(1): 9–12.

Lehrer T (1953) Lobachevsky. In: Songs by Tom Lehrer (originally an LP recording).

Lock S (1996) Research misconduct: a résumé of recent events. In: Lock S, Wells F (eds) Fraud and Misconduct in Medical Research (2nd edn). BMJ Publishing Group, London.

Merton R (1968) Science and democratic social structure. In: Social Theory and Social Structure (enlarged edn). The Free Press, New York. (The article was first published in 1942.)

Office of Research Integrity (2006) Newsletter 14: 4.

Riis P (1995) Authorship and scientific dishonesty. Annual Report 1994, Committee on Scientific Dishonesty, Copenhagen, pp. 33–39.

Science (2002) AIDS and HIV: historical essays. Science 298: 1726–1731.

Tranøy KE (1988) Science and ethics. Some of the main principles and problems. In: Jones AKI (ed.) The Moral Import of Science. Essays on Normative Theory, Scientific Activity and Wittengenstein. Sigma, Bergen, pp. 111–136.

Tranøy KE (1996) Ethical problems of scientific research: an action-theoretic approach. The Monist 79: 183–196.

University of Oslo (2006) Investigative Report on J. Sudbø and co-authors [Rapport fra granskningskommisjon oppnevnt av Rikshospitalet – Radiumhospitalet HF og Universitetet i Oslo]. English translation available at: http://radium.no/general/docs/ekbom/Inquiry_report_2006.doc

FURTHER READING

Lock S, Wells F, Farthing M (eds) (2002) Fraud and Misconduct in Medical Research (3rd edn). BMJ Publishing Group, London.

Macrina FL (2005) Scientific Integrity (3rd edn). ASM Press, Washington, DC. (This book presents many good case studies.)

APPENDIX 1

Authorship rules, in: International Committee of Medical Journal Editors. Uniform Requirements for Manuscripts Submitted to Biomedical Journals: Writing and Editing for Biomedical Publication, February 2006 version.

II.A Authorship and contributorship

II.A.1 *Byline authors*

An 'author' is generally considered to be someone who has made substantive intellectual contributions to a published study, and biomedical authorship continues to have important academic, social, and financial implications. (1) In the past, readers were rarely provided with information about contributions to studies from those listed as authors and in acknowledgements. (2) Some journals now request and publish information about the contributions of each person named as having participated in a submitted study, at least for original research. Editors are strongly encouraged to develop and implement a contributorship policy, as well as a policy on identifying who is responsible for the integrity of the work as a whole.

While contributorship and guarantorship policies obviously remove much of the ambiguity surrounding contributions, it leaves unresolved the question of the quantity and quality of contribution that qualify for authorship. The International Committee of Medical Journal Editors has recommended the following criteria for authorship; these criteria are still appropriate for those journals that distinguish authors from other contributors.

- Authorship credit should be based on 1) substantial contributions to conception and design, or acquisition of data, or analysis and interpretation of data;

2) drafting the article or revising it critically for important intellectual content; and 3) final approval of the version to be published. Authors should meet conditions 1, 2, and 3.

- When a large, multi-centre group has conducted the work, the group should identify the individuals who accept direct responsibility for the manuscript (3). These individuals should fully meet the criteria for authorship defined above and editors will ask these individuals to complete journal-specific author and conflict of interest disclosure forms. When submitting a group author manuscript, the corresponding author should clearly indicate the preferred citation and should clearly identify all individual authors as well as the group name. Journals will generally list other members of the group in the acknowledgements. The National Library of Medicine indexes the group name and the names of individuals the group has identified as being directly responsible for the manuscript.
- Acquisition of funding, collection of data, or general supervision of the research group, alone, does not justify authorship.
- All persons designated as authors should qualify for authorship, and all those who qualify should be listed.
- Each author should have participated sufficiently in the work to take public responsibility for appropriate portions of the content.

Some journals now also request that one or more authors, referred to as 'guarantors', be identified as the persons who take responsibility for the integrity of the work as a whole, from inception to published article, and publish that information.

Increasingly, authorship of multi-centre trials is attributed to a group. All members of the group who are named as authors should fully meet the above criteria for authorship.

The order of authorship on the byline should be a joint decision of the co-authors. Authors should be prepared to explain the order in which authors are listed.

II.A.2 Contributors listed in acknowledgements
All contributors who do not meet the criteria for authorship should be listed in an acknowledgments section. Examples of those who might be acknowledged

include a person who provided purely technical help, writing assistance, or a department chair who provided only general support. Editors should ask authors to disclose whether they had writing assistance and to identify the entity that paid for this assistance. Financial and material support should also be acknowledged.

Groups of persons who have contributed materially to the paper but whose contributions do not justify authorship may be listed under a heading such as 'clinical investigators' or 'participating investigators', and their function or contribution should be described – for example, 'served as scientific advisors', 'critically reviewed the study proposal', 'collected data', or 'provided and cared for study patients'.

Because readers may infer their endorsement of the data and conclusions, all persons must give written permission to be acknowledged.

CHAPTER 3

ETHICS IN HUMAN AND ANIMAL STUDIES

Søren Holm and Bjørn Reino Olsen

3.1 INTRODUCTION

The focus of this chapter is on the basic principles of research ethics as it applies to biomedical research with humans and on animals. It builds on the more general introduction to goodness and ethics in science presented in Chapter 2.

Section 3.2 is concerned with human research ethics. It introduces the basic principles of human research ethics, takes a closer look at two influential international regulatory protocols on human research ethics and discusses specific problems raised (1) when research participants are unable to consent, (2) by the use of placebos in randomized controlled trials (RCTs), and (3) by the use of vulnerable research populations. It also considers the role of research ethics committees (RECs). The controversial problems raised by research on embryos or foetuses are not discussed, since they will be relevant to very few readers of this book, and the discussion would require an in-depth analysis of the moral status of embryos and foetuses (Harris & Holm 2003a). Human biomedical research ethics is regulated by law or statutory regulations in most countries, as well as at the international level, as discussed in Section 3.3. Although regulatory details differ from country to country, the existence of international documents with long histories has resulted in significant harmonization of the national regulations and an almost universal acceptance of certain research ethical principles at the regulatory level.

Section 3.4 is concerned with animal research ethics and considers the justification of animal research, the obligations to minimize animal suffering and the question of whether certain species require specific protections.

3.2 BASIC PRINCIPLES OF HUMAN RESEARCH ETHICS

The basic principles of human biomedical research ethics are based on the sociological view that the typical research project is characterized by the following features listed in Box 3.1 (please note that this is *not* a claim that all projects share these features, just that the typical project exhibits them).

Box 3.1 Characterization of the typical research project

- The research is aimed primarily not at the benefit of the actual research participants but at gaining knowledge that will be of benefit in the future.
- The participants possibly may be harmed by participation.
- The researchers are socially powerful, or at least more powerful than the average participant.
- The researchers are those who know most about the project.
- The researchers have a personal interest in the success of the project.
- Participants may, in clinical research, depend on the researchers for their continued clinical care in cases where the researchers also may be the participants' physicians.

This means that there is an asymmetry in power and knowledge and that the participants are in a certain sense 'used' so that others (i.e. the researchers and future patients) may benefit. Using people in this way is wrong unless they understand that they are being so used and have agreed to participate (otherwise they are truly research 'subjects'). This follows from the more general ethical principles of respect for self-determination and respect for bodily integrity. There may be circumstances where conscription for the common good is accepted, but biomedical research is not usually seen as one of those circumstances (for arguments that it should be or that there is at least a moral obligation to be a research participant, see Evans 2004 and Harris 2005).

This analysis of the research context leads to three main principles of research ethics listed in Box 3.2.

Box 3.2 The main principles of research ethics

- The potential harm to research participants should be minimized.
- Participation should be voluntary and based on an adequate understanding of the project.
- Participants should have an absolute right to withdraw from the study.

The second of these principles is then operationalized as a requirement for voluntary, informed consent, which can be parsed into four discrete elements listed in Box 3.3.

Box 3.3 The elements of voluntary, informed consent

Information elements:

- full information given
- full information understood

Consent elements:

- consent ability (legal and actual)
- voluntarism (absence of coercion)

Exactly what constitutes full information is controversial, but as discussed in Section 3.3, it is specified in some detail in international documents. Because of the centrality of informed consent in biomedical research ethics, all circumstances where participants cannot give informed consent become 'problem cases' even if they are not ethically problematic in any other way.

Like other societal activities, biomedical research is subject to general considerations concerning social justice that impinge on the choice of how we should select and recruit research participants.

3.3 INTERNATIONAL REGULATION

Let us look more closely at two significant and influential examples of international regulatory documents to substantiate the commonalities mentioned above, to understand better the basic principles and to discuss some specific issues. These documents are the Helsinki Declaration of the World Medical Association (WMA) (Appendix 1) and the Oviedo Convention of the Council of Europe and its additional protocol on biomedical research (Appendices 2 and 3; Appendix 2 contains the text of the relevant chapter in the Convention itself, and Appendix 3 is the more detailed protocol).

Although the Helsinki Declaration is technically binding only on researchers who are members of a medical association that is a member of the WMA, it has nevertheless attained the status of an authoritative human rights document, because it was one of the first documents in the area and because it has had significant influence on how human biomedical research is regulated in most countries.

The Oviedo Convention is technically binding on those states that have signed and ratified it, but it also will be used by the European Court of Human Rights as an aid in interpretation in specific human rights cases even against European states that have not ratified the Convention (Plomer 2005).

Consent

Both documents contain a specification of the information that should be given to potential research participants (Table 3.1), and it is evident that both require significant amounts of information to be given to prospective participants and furthermore require that this information is understood. Hence, obtaining informed consent to research participation is not a simple matter and is best understood as a

Table 3.1 Comparison of Helsinki Declaration and Oviedo Convention

TOPIC	HELSINKI DECLARATION	OVIEDO CONVENTION AND ADDITIONAL PROTOCOL ON BIOMEDICAL RESEARCH
Consent and information – basic rule	Article 22 In any research on human beings, each potential subject must be adequately informed of the aims, methods, sources of funding, any possible conflicts of interest, institutional affiliations of the researcher, the anticipated benefits and potential risks of the study and the discomfort it may entail. The subject should be informed of the right to abstain from participation in the study or to withdraw consent to participate at any time without reprisal. After ensuring that the subject has understood the information, the physician should then obtain the subject's freely given informed consent, preferably in writing. If the consent cannot be obtained in writing, the non-written consent must be formally documented and witnessed.	Article 13 – Information for research participants 1 The persons being asked to participate in a research project shall be given adequate information in a comprehensible form. This information shall be documented. 2 The information shall cover the purpose, the overall plan and the possible risks and benefits of the research project, and include the opinion of the ethics committee. Before being asked to consent to participate in a research project, the persons concerned shall be specifically informed, according to the nature and purpose of the research: i of the nature, extent and duration of the procedures involved, in particular, details of any burden imposed by the research project; ii of available preventive, diagnostic and therapeutic procedures; iii of the arrangements for responding to adverse events or the concerns of research participants; iv of arrangements to ensure respect for private life and ensure the confidentiality of personal data; v of arrangements for access to information relevant to the participant arising from the research and to its overall results; *(Continued)*

Table 3.1 (*Continued*)

TOPIC	HELSINKI DECLARATION	OVIEDO CONVENTION AND ADDITIONAL PROTOCOL ON BIOMEDICAL RESEARCH
		vi of the arrangements for fair compensation in the case of damage; vii of any foreseen potential further uses, including commercial uses, of the research results, data or biological materials; viii of the source of funding of the research project. 3 In addition, the persons being asked to participate in a research project shall be informed of the rights and safeguards prescribed by law for their protection, and specifically of their right to refuse consent or to withdraw consent at any time without being subject to any form of discrimination, in particular regarding the right to medical care. Article 14 – Consent 1 No research on a person may be carried out, subject to the provisions of both Chapter V and Article 19, without the informed, free, express, specific and documented consent of the person. Such consent may be freely withdrawn by the person at any phase of the research. 2 Refusal to give consent or the withdrawal of consent to participation in research shall not lead to any form of discrimination against the person concerned, in particular regarding the right to medical care.

Persons unable to consent

24. For a research subject who is legally incompetent, physically or mentally incapable of giving consent or is a legally incompetent minor, the investigator must obtain informed consent from the legally authorised representative in accordance with applicable law. These groups should not be included in research unless the research is necessary to promote the health of the population represented and this research cannot instead be performed on legally competent persons.

25. When a subject deemed legally incompetent, such as a minor child, is able to give assent to decisions about participation in research, the investigator must obtain that assent in addition to the consent of the legally authorised representative.

26. Research on individuals from whom it is not possible to obtain consent, including proxy or advance consent, should be done only if the physical/mental condition that prevents obtaining informed consent is a necessary characteristic of the research population. The specific reasons for involving research subjects with a condition that

3 Where the capacity of the person to give informed consent is in doubt, arrangements shall be in place to verify whether or not the person has such capacity.

Article 15 – Protection of persons not able to consent to research

1 Research on a person without the capacity to consent to research may be undertaken only if all the following specific conditions are met:

i the results of the research have the potential to produce real and direct benefit to his or her health;

ii research of comparable effectiveness cannot be carried out on individuals capable of giving consent;

iii the person undergoing research has been informed of his or her rights and the safeguards prescribed by law for his or her protection, unless this person is not in a state to receive the information;

iv the necessary authorisation has been given specifically and in writing by the legal representative or an authority, person or body provided for by law, and after having received the information required by Article 16, taking into account the person's previously expressed wishes or objections. An adult not able to consent shall as far as possible take part in the authorisation procedure. The opinion of a minor shall be taken into consideration as an increasingly determining factor in proportion to age and degree of maturity;

v the person concerned does not object.

(Continued)

Table 3.1 (*Continued*)

TOPIC	HELSINKI DECLARATION	OVIEDO CONVENTION AND ADDITIONAL PROTOCOL ON BIOMEDICAL RESEARCH
	renders them unable to give informed consent should be stated in the experimental protocol for consideration and approval of the review committee. The protocol should state that consent to remain in the research should be obtained as soon as possible from the individual or a legally authorised surrogate.	2 Exceptionally and under the protective conditions prescribed by law, where the research has not the potential to produce results of direct benefit to the health of the person concerned, such research may be authorised subject to the conditions laid down in paragraph 1, sub-paragraphs ii, iii, iv, and v above, and to the following additional conditions: i the research has the aim of contributing, through significant improvement in the scientific understanding of the individual's condition, disease or disorder, to the ultimate attainment of results capable of conferring benefit to the person concerned or to other persons in the same age category or afflicted with the same disease or disorder or having the same condition; ii the research entails only minimal risk and minimal burden for the individual concerned; and any consideration of additional potential benefits of the research shall not be used to justify an increased level of risk or burden. 3 Objection to participation, refusal to give authorisation or the withdrawal of authorisation to participate in research shall not lead to any form of discrimination against the person concerned, in particular regarding the right to medical care.

Emergency research	26.	Research on individuals from whom it is not possible to obtain consent, including proxy or advance consent, should be done only if the physical/mental condition that prevents obtaining informed consent is a necessary characteristic of the research population. The specific reasons for involving research subjects with a condition that renders them unable to give informed consent should be stated in the experimental protocol for consideration and approval of the review committee. The protocol should state that consent to remain in the research should be obtained as soon as possible from the individual or a legally authorised surrogate.

Article 19 – Research on persons in emergency clinical situations

1 The law shall determine whether, and under which protective additional conditions, research in emergency situations may take place when:

  i a person is not in a state to give consent, and

  ii because of the urgency of the situation, it is impossible to obtain in a sufficiently timely manner, authorisation from his or her representative or an authority or a person or body which would in the absence of an emergency situation be called upon to give authorisation.

2 The law shall include the following specific conditions:

  i research of comparable effectiveness cannot be carried out on persons in non-emergency situations;

  ii the research project may only be undertaken if it has been approved specifically for emergency situations by the competent body;

  iii any relevant previously expressed objections of the person known to the researcher shall be respected;

  iv where the research has not the potential to produce results of direct benefit to the health of the person concerned, it has the aim of contributing, through significant improvement in the scientific understanding of the individual's condition, disease or disorder, to the ultimate attainment of results capable of conferring benefit to the person concerned or to other

(Continued)

Table 3.1 (*Continued*)

TOPIC	HELSINKI DECLARATION	OVIEDO CONVENTION AND ADDITIONAL PROTOCOL ON BIOMEDICAL RESEARCH
		persons in the same category or afflicted with the same disease or disorder or having the same condition, and entails only minimal risk and minimal burden. 3 Persons participating in the emergency research project or, if applicable, their representatives shall be provided with all the relevant information concerning their participation in the research project as soon as possible. Consent or authorisation for continued participation shall be requested as soon as reasonably possible.
Use of placebo control	29. The benefits, risks, burdens and effectiveness of a new method should be tested against those of the best current prophylactic, diagnostic, and therapeutic methods. This does not exclude the use of placebo, or no treatment, in studies where no proven prophylactic, diagnostic or therapeutic method exists. Note of clarification on paragraph 29 of the WMA Declaration of Helsinki The WMA hereby reaffirms its position that extreme care must be taken in making use of a placebo-controlled trial and that in general this methodology should only be used in the absence of existing proven	Article 23 – Non-interference with necessary clinical interventions 1 Research shall not delay nor deprive participants of medically necessary preventive, diagnostic or therapeutic procedures. 2 In research associated with prevention, diagnosis or treatment, participants assigned to control groups shall be assured of proven methods of prevention, diagnosis or treatment. 3 The use of placebo is permissible where there are no methods of proven effectiveness, or where withdrawal or withholding of such methods does not present an unacceptable risk or burden.

therapy. However, a placebo-controlled trial may be ethically acceptable, even if proven therapy is available, under the following circumstances:

– Where for compelling and scientifically sound methodological reasons its use is necessary to determine the efficacy or safety of a prophylactic, diagnostic or therapeutic method; or

– Where a prophylactic, diagnostic or therapeutic method is being investigated for a minor condition and the patients who receive placebo will not be subject to any additional risk of serious or irreversible harm.

All other provisions of the Declaration of Helsinki must be adhered to, especially the need for appropriate ethical and scientific review.

process that involves presenting both written and oral information, engaging the person in reflection concerning the project, and allowing the person sufficient time to make up his or her mind as to whether they want to participate. Only if this process is completed successfully is the signature on the consent form a valid documentation of consent. In the design of the recruitment procedures for a specific process, it is necessary to consider how much time this process will take, and allow adequate time for it. Unless there is some medical necessity (i.e. treatment has to start now and not tomorrow), potential participants should be given ample time to make their decision, and should not be coerced into deciding here and now.

It is well known that informational materials often are written in complicated language and are thus difficult to understand. In order for the average person to understand these documents, they should be no more difficult to read than the sports pages of a popular newspaper. Only when participants are exclusively recruited among academics should academic-style informational materials be used.

Persons unable to consent

There are many situations in which prospective research participants are legally or factually unable to consent, but here only three of these will be considered: research on children, research on permanently incapacitated adults and research in the emergency setting where the disease process has rendered a person temporarily unable to consent (e.g. research on stroke or on cardiopulmonary resuscitation).

With regard to children who are not legally able to consent, the general rule is that their parents or guardians can consent to research participation, if this is not against the interests of the child. But then, no researcher should ask for a child's participation if it is against the interest of the child. Because children are seen as a vulnerable and exploitable group, both Helsinki and Oviedo have restrictions on what kind of research children can participate in (Table 3.1). It has been argued that these restrictions are too narrow, since children are already moral agents and we have no reason to believe that they do not want to help other people (Harris & Holm 2003b). The counter-argument is that if the restrictions were less narrowly defined, children will be exploited by more or less unscrupulous researchers.

Although children are legally unable to consent they can clearly have valid views concerning whether or not they want to participate in a specific research project, and even if they are so young that we think they cannot form a valid opinion on such a complex matter, they nevertheless need to know what is going to happen to them during the research. It is therefore generally accepted that children should be informed using age-appropriate informational materials and techniques. The assent of older children should be sought and any refusal respected, even if the parents want the child to participate.

Permanently incapacitated adults differ from children in that they often have no legally defined proxy decision maker. They fall into two groups: persons who have previously been competent and persons who have never been competent. With regard to consent both groups are essentially in the same legal position as children (we do not imply that they are actually comparable to children in any other way). They cannot consent to research personally, but they may in some instances be able to assent or refuse. Inclusion in the research can only take place if there is a proxy who is legally authorized to give proxy consent to research, and this proxy must make decisions based on an understanding of the best interests of the incapacitated person. If the person has previously been competent, the proxy can take account of the previously expressed wishes and values of the person, and decide in accordance with these. Some countries allow competent persons to execute advance directives concerning their treatment that become operative when the person becomes incompetent; these are usually limited to treatment but a few countries allow advance directives for research as well.

In the acute case, it is very unlikely that there is a legally recognized proxy because many legal systems only allow proxies to be appointed through a legal process after the person has become incompetent, and even if there is an automatic legal presumption that for instance the spouse becomes the proxy, there may be no time to contact the spouse before the patient has to be entered into the research protocol. Requiring proxy consent may therefore completely block this kind of emergency research or make it very, very difficult. This was for instance the case in the USA for some years during the 1990s when the consent requirement was interpreted very restrictively. But it is clearly important that research into, for instance, improvement of cardiopulmonary resuscitation takes place. It is important that research ethics does not become so protective of a specific group that it makes it impossible to do research on their condition and its treatment and leaves them in what the Danish research ethicist and co-drafter of the 1975 version of Helsinki, Professor Povl Riis, has called a 'golden ghetto'. This has been increasingly recognized in recent years and exemptions from consent requirements for emergency research can now be found both in Helsinki and Oviedo. As can be seen in Table 3.1, the exceptions in Oviedo are considerably more narrowly drafted than the exceptions in Helsinki, especially with regard to what kind of research that can take place without consent.

Randomized controlled trials

Specific problems arise in RCTs, especially with regard to the use of placebo in the control group. If there is an effective treatment for the condition in question, if this is not given to the participants in the control group, and if they would have received this treatment if they were not in the trial, they are worse off than they would otherwise have been by being trial participants (and we would know in advance that everyone in the control group would be worse off).

A small detriment in well-being is probably something a research participant can consent to (i.e. whether my ingrown toe nail is treated this week, or in two weeks' time if the randomization places me in the control group), but whether it should be possible to consent to a potentially large detriment in well-being is more controversial.

Helsinki contains what looks like an absolute prohibition of the use of placebo if there is effective treatment available (Table 3.1), but this has been modified by the explanatory footnote which seems to be more a change in position than an explanation. The footnote's second clause brings Helsinki in alignment with Oviedo and allows what we could call low-risk or low-harm use of placebo, but the first clause seems to allow even high-risk uses of placebo if it is necessary 'for compelling and scientifically sound methodological reasons'. It is difficult to see how this can be ethically justified, unless the scientific question is so important that we truly believe that participants should be allowed to sacrifice their own interests (and maybe even themselves) in the interest of science.

A somewhat similar issue arises in some kinds of psychiatric research where it is methodologically advantageous to have a long 'washout period' where the patient's ordinary therapy is stopped before the patient commences the trial medication. This practice is ethically problematic if there is a significant risk of deterioration in the washout period.

A considerably more charged situation arises in those cases where placebo-controlled research is carried out in a developing country by a research sponsor from a developed country. Here the situation may occur that there is an effective treatment for the condition in question, that everyone gets this treatment in the sponsoring country, but that it is so expensive that no one gets it in the country where the research takes place (this has been the case in a number of HIV/AIDS trials). Would using placebo in the control group entail a breach of Helsinki or Oviedo? (Note that Oviedo applies to this research if the sponsor is European according to Article 29 of the additional protocol.) The answer to this question hinges on the interpretation of whether 'there are no methods of proven effectiveness', to use the Oviedo formulation, is given a global or a local interpretation. Is this condition satisfied if the method is not used here, or is it only satisfied if the method is not used anywhere?

The reason that this question gets complicated and murky and the discussion often heated is that it intersects with other controversial questions concerning the relations between rich and poor countries (e.g. Do rich countries exploit the poor? Do rich countries have an obligation to help the poor? Is sponsoring in another country research that you would not allow at home, hypocritical or expressive of double standards?) We may note that no one in the control group is actually worse off in this scenario. Patients in the control group do not get effective treatment, but they would not have received it anyway if they had not been research participants. It has been suggested that this makes the research acceptable either on its own, or if the research results furthermore will benefit the community in which the research is carried out. There is, however, still no agreement on this issue.

Vulnerable research participants

There are groups of research participants that are vulnerable to exploitation, not because they are unable to give consent, but because they are institutionalized, socially powerless or dependent on the researchers in some way.

For these groups the voluntarism of their consent becomes an issue and RECs will often require specific justification of why such a group is chosen as the research population and/or assurance that it will be made even more clear than usual that participation is purely voluntary and that it is acceptable to refuse and that this will have no negative consequences.

Epidemiological and biobank research

It is evident that the model of research covered by biomedical research ethics is derived from typical clinical research, but large-scale epidemiological research is gaining in importance and it is increasingly combined with the analysis of biological samples from large, structured sample collections ('biobanks'). These projects are often of very long duration, they may use data and samples collected for unrelated purposes, and the focus of the research may change during the life of the project. This means that the ideal situation of full informed consent at the beginning of the project becomes problematic. Suggestions for how this problem can be solved include 'broad' or 'unspecific' consent and some form of recurrent consent. A requirement for recurrent or renewed consent to new uses of old data and samples clearly creates a significant burden on researchers, and such consents may be impossible to obtain if the participants are untraceable or dead. This idea has thus been strongly resisted by the research community. However, broad consent is not unproblematic. The broader the consent is, the less valid it is. Have I really consented to a given use of my samples, in the ethical sense of the term 'consent', if I have consented to 'any future use' and the use is one that neither I nor the researchers could envisage at the time of consent? In research ethics there is great resistance to the idea that consent creates some form of contract between researcher and research participant, but considering consent as constituting a contract is probably the only way to justify relying on broad consent.

It has also been suggested that epidemiological research on non-sensitive topics may not require consent, because there is no possibility of harming the research participants. However, this argument has not been universally accepted.

The role of research ethics committees

To ensure that the research ethics regulations are adhered to, researchers are required to submit a research protocol including the relevant patient information and consent forms to an REC [in some countries known as research ethics board (REB) or institutional review board (IRB)] for approval before the project commences. The REC will typically have both professional and lay members, and will assess the scientific validity of the proposal, any ethical problems that the project contains and the information and consent documents, and will decide whether the project should be allowed to begin, or whether elements of it need to be modified. REC approval is a legal requirement in many countries and it is reinforced by the requirement of most journals that they will only publish research that has REC approval.

In some countries, RECs also have control functions concerning the actual conduct of the research, that is, whether it is conducted in accordance with the approved protocol.

Data protection, good clinical practice and other regulations influencing biomedical research

Specific research ethics regulation is not the only kind of regulation that influences the ethical conduct of biomedical research and the work of RECs. Most countries have data protection laws that independently require consent for the collection and processing of person-identifiable health data, except for official, statistical purposes. It may also be necessary to obtain permission from the national data protection agency to store and process person-identifiable health data.

Some countries also have specific rules concerning the storage and use of human tissue, and these again require explicit consent for research use in many contexts.

The international Good Clinical Practice guidelines that apply to all research performed to support registration of drugs or medical devices in the USA, the European Union and Japan also require explicit, informed consent from all research participants and approval by an independent REC.

This means that even if Helsinki and Oviedo suddenly disappeared there would not be much difference in the basic ethical requirements for the conduct of biomedical research on adult, competent human beings. Only the more specific rules concerning specific cases would be affected.

3.4 THE ETHICS OF ANIMAL RESEARCH

The two basic problems in the ethical analysis of animal research are (1) whether it is at all justifiable to use animals in research, and (2) the conditions under which it is justifiable. It is immediately obvious that the standard justification for human research will not work in the case of animals. Animals cannot consent to be research subjects, and there is no way in which the animal can voluntarily take upon itself the burdens that being a research subject entails (even in veterinary research aimed at benefiting a specific species). Some argue that the use of animals in research cannot be justified at all, since research infringes on the animal's right to life or causes animal suffering. It is highly controversial whether animals can have a right to life, and if so who would have the obligation to protect that right. It is, however, generally accepted that animals can suffer, that their well-being can be negatively affected; it is very difficult to argue that animal suffering morally should not matter at all. The main reason it is wrong to kick your dog is simply that it causes the dog pain, not as Immanuel Kant seemed to believe, that it desensitizes you. To the degree that animal research causes pain and other kinds of suffering, we therefore need a justification for causing this pain. What is the good that outweighs the suffering?

The answer is that the good is the scientific knowledge and the future medical benefits that flow from animal research. Although it is sometimes denied by animal rights activists, there is no doubt that animal research has been and continues to be a necessary component of the long and complicated research processes leading to medical progress. But this is still not sufficient to provide a compelling justification for the practice of animal research. A further argument is needed to show that the amount of animal suffering caused by research is justified by the beneficial outcome (medical or veterinary progress). To make that assessment is extremely complicated, partly because we have no good handle on estimating the magnitude of the various kinds of suffering involved, that is, the animal suffering caused and the human and animal suffering averted, and partly because it is difficult to estimate the exact contribution of animal research to medical benefits.

The argument has so far been on the general level of biomedical research as one, comprehensive activity. This means that even if we accept that animal research is justified in general, we do not necessarily have to accept that every individual project involving animals is justified. There may well be individual projects where the expected benefits do not outweigh the predicted suffering of the experimental animals.

The official regulation of animal research therefore has two aims. First, to ensure that the general level of animal suffering caused by animal research is minimized; this is for instance the justification for requiring training and certification of researchers and laboratory technicians and good housing conditions for all laboratory animals. Secondly, to evaluate individual projects or broader research plans to ensure that there is an acceptable balance between suffering and benefit in each case.

Over the years, a consensus has emerged that a useful tool for reaching these aims is the 'Three Rs' approach. Based on concepts initially developed by Russell and Burch (1959), current definitions of the Three Rs are listed in Box 3.4.

Box 3.4 The 'Three Rs'

- *Refinement*: Improvement of all aspects of the lifetime experience of animals to reduce suffering and improve welfare.
- *Reduction*: The use of fewer animals in each experiment without compromising scientific output and the quality of biomedical research and testing, and without compromising animal welfare.
- *Replacement*: The use of methods that permit a given scientific purpose to be achieved without conducting experiments or other scientific procedures on living animals.'

(Nuffield Council on Bioethics 2005)

Some countries now require researchers to specify how they have taken the three Rs into account in their project planning.

Because considerations of animal suffering are so central in the standard approach to animal research ethics, the most problematic category of research is usually taken to be experiments where the animal survives for a long time in a state involving some suffering. In contrast, experiments where the animal is anaesthetized, experimented upon and then painlessly killed are deemed to be less problematic. Here it is important to note that a significant proportion of short- and long-term toxicological studies that fall into the most problematic category are legally mandated as part of premarketing testing of new chemicals and pharmaceuticals.

A question that still causes controversy is whether there should be different kinds of protection for different kinds of animals, and what the rational basis for that increased protection is. For instance, some countries have had special categories for animals that are traditionally kept as pets, so that you could carry out in pigs experiments that you could not carry out in dogs. Such special protection for animals that we humans happen to like does not seem rationally warranted. Whatever evidence there is suggests that pigs can suffer to the same degree as dogs, and that they are probably more intelligent than dogs. A more rational approach is to consider what kind of suffering an animal can experience based on its cognitive abilities and species-specific lifestyle. Sentience is the basic criterion for any kind of suffering, but there are other kinds of suffering than pain. Animals that usually live in social groupings may suffer if housed in single cages, and animals that have higher cognitive capacities may (like humans) suffer from boredom if housed in an unstimulating environment, or may suffer because of the anticipation of pain to come. Different countries have operationalized these considerations differently in their regulations, but a typical classification of increasing protection would be:

Not protected:

- non-vertebrates (no specific protection because not believed to be sentient)

Increasingly protected:

- vertebrates (a few countries also protect cephalopods)
- mammals
- non-human primates
- great apes (many countries now prohibit or strongly discourage research on great apes).

A hierarchical list like this is clearly not sufficiently sensitive to all the many differences between animal species, and it is therefore important that both researchers and laboratory technicians are knowledgeable about the specific specie(s) they are working with so that they can minimize species-specific suffering (e.g. animals that have rooting as a normal behaviour should have rooting materials in their cages, and animals that live in social groups should not be housed individually).

Animal research and other uses of animals

In most industrialized countries there are quite stringent rules concerning the use of animals in biomedical research and about the breeding and housing conditions of these animals. There is often a glaring inconsistency between these rules and the rules, or absence of rules, concerning other human uses of animals. Whereas all countries prohibit cruelty to animals, there is often very little enforcement of these rules. Exotic vertebrates can be kept as pets without knowing anything about their requirements for well-being. Mammals are used in dangerous sports where many of them will suffer. One mammal can be used to hunt and kill another mammal (no, we are not thinking about fox hunting in the UK, but about the activities of domestic cats all over the world). And, perhaps most importantly in terms of numbers, farm mammals are kept under conditions that are much worse than the conditions required for laboratory animals. It is worth considering whether animal rights activists should not first target these other, much more problematic uses of animals, before targeting animal experimentation. Even vegans may want to benefit from progress 'bought' through the suffering of animals in biomedical research.

REFERENCES

Evans HM (2004) Should patients be allowed to veto their participation in clinical research? Journal of Medical Ethics 30: 198–203.

Harris J (2005) Scientific research is a moral duty. Journal of Medical Ethics 31: 242–248.

Harris J, Holm S (2003a) Abortion. In: La Follette H (ed.) The Oxford Handbook of Practical Ethics. Oxford University Press, Oxford, pp. 112–135.

Harris J, Holm S (2003b) Should we presume moral turpitude in our children? – Small children and consent to medical research. Theoretical Medicine 24: 121–129.

Nuffield Council on Bioethics (2005) The ethics of research involving animals – a guide to the report. Nuffield Council on Bioethics, London.

Plomer A (2005) The Law and Ethics of Medical Research: International Bioethics and Human Rights. Cavendish Publishing, London.

Russell WMS, Burch RL (1959) The Principles of Humane Experimental Technique. Methuen, London, available at: http://altweb.jhsph.edu/publications/humane_exp/hettoc.htm

FURTHER READING

Brody BA (1998) The Ethics of Biomedical Research – An International Perspective. Oxford University Press, New York.

Emanuel EJ et al. (2003) Ethical and Regulatory Aspects of Clinical Research – Readings and Commentary. Johns Hopkins University Press, Baltimore.

APPENDIX 1
WORLD MEDICAL ASSOCIATION DECLARATION OF HELSINKI
ETHICAL PRINCIPLES FOR MEDICAL RESEARCH INVOLVING HUMAN SUBJECTS

A. Introduction

1. The World Medical Association has developed the Declaration of Helsinki as a statement of ethical principles to provide guidance to physicians and other participants in medical research involving human subjects. Medical research involving human subjects includes research on identifiable human material or identifiable data.

2. It is the duty of the physician to promote and safeguard the health of the people. The physician's knowledge and conscience are dedicated to the fulfilment of this duty.

3. The Declaration of Geneva of the World Medical Association binds the physician with the words, 'The health of my patient will be my first consideration', and the International Code of Medical Ethics declares that, 'A physician shall act only in the patient's interest when providing medical care which might have the effect of weakening the physical and mental condition of the patient.'

4. Medical progress is based on research which ultimately must rest in part on experimentation involving human subjects.

5. In medical research on human subjects, considerations related to the well-being of the human subject should take precedence over the interests of science and society.

6. The primary purpose of medical research involving human subjects is to improve prophylactic, diagnostic and therapeutic procedures and the understanding of the aetiology and pathogenesis of disease. Even the best proven prophylactic, diagnostic, and therapeutic methods must continuously be challenged through research for their effectiveness, efficiency, accessibility and quality.

7. In current medical practice and in medical research, most prophylactic, diagnostic and therapeutic procedures involve risks and burdens.

8. Medical research is subject to ethical standards that promote respect for all human beings and protect their health and rights. Some research populations are vulnerable and need special protection. The particular needs of the economically and medically disadvantaged must be recognized. Special attention is also required for those who cannot give or refuse consent for themselves, for those who may be subject to giving consent under duress, for those who will not benefit personally from the research and for those for whom the research is combined with care.

9. Research Investigators should be aware of the ethical, legal and regulatory requirements for research on human subjects in their own countries as well as applicable international requirements. No national ethical, legal or regulatory requirement should be allowed to reduce or eliminate any of the protections for human subjects set forth in this Declaration.

B. Basic principles for all medical research

10. It is the duty of the physician in medical research to protect the life, health, privacy, and dignity of the human subject.

11. Medical research involving human subjects must conform to generally accepted scientific principles, be based on a thorough knowledge of the scientific literature, other relevant sources of information, and on adequate laboratory and, where appropriate, animal experimentation.

12. Appropriate caution must be exercised in the conduct of research which may affect the environment, and the welfare of animals used for research must be respected.

13. The design and performance of each experimental procedure involving human subjects should be clearly formulated in an experimental protocol. This protocol should be submitted for consideration, comment, guidance, and where appropriate, approval to a specially appointed ethical review committee, which must be independent of the investigator, the sponsor or any other kind of undue influence. This independent committee should be in conformity with the laws and regulations of the country in which the research experiment is performed.

    The committee has the right to monitor ongoing trials. The researcher has the obligation to provide monitoring information to the committee, especially any serious adverse events. The researcher should also submit to the committee, for review, information regarding funding, sponsors, institutional affiliations, other potential conflicts of interest and incentives for subjects.

14. The research protocol should always contain a statement of the ethical considerations involved and should indicate that there is compliance with the principles enunciated in this Declaration.

15. Medical research involving human subjects should be conducted only by scientifically qualified persons and under the supervision of a clinically competent medical person. The responsibility for the human subject must always rest with a medically qualified person and never rest on the subject of the research, even though the subject has given consent.

16. Every medical research project involving human subjects should be preceded by careful assessment of predictable risks and burdens in comparison with foreseeable benefits to the subject or to others. This does not preclude the participation of healthy volunteers in medical research. The design of all studies should be publicly available.

17. Physicians should abstain from engaging in research projects involving human subjects unless they are confident that the risks involved have been adequately assessed and can be satisfactorily managed. Physicians should cease any investigation if the risks are found to outweigh the potential benefits or if there is conclusive proof of positive and beneficial results.

18. Medical research involving human subjects should only be conducted if the importance of the objective outweighs the inherent risks and burdens to the subject. This is especially important when the human subjects are healthy volunteers.

19. Medical research is only justified if there is a reasonable likelihood that the populations in which the research is carried out stand to benefit from the results of the research.

20. The subjects must be volunteers and informed participants in the research project.

21. The right of research subjects to safeguard their integrity must always be respected. Every precaution should be taken to respect the privacy of the subject, the confidentiality of the patient's information and to minimize the impact of the study on the subject's physical and mental integrity and on the personality of the subject.

22. In any research on human beings, each potential subject must be adequately informed of the aims, methods, sources of funding, any possible conflicts of interest, institutional affiliations of the researcher, the anticipated benefits and potential risks of the study and the discomfort it may entail. The subject should be informed of the right to abstain from participation in the study or to withdraw consent to participate at any time without reprisal. After ensuring that the subject has understood the information, the physician should then obtain the subject's freely given informed consent, preferably in writing. If the consent cannot be obtained in writing, the non-written consent must be formally documented and witnessed.

23. When obtaining informed consent for the research project the physician should be particularly cautious if the subject is in a dependent relationship with the physician or may consent under duress. In that case the informed consent should be obtained by a well-informed physician who is not engaged in the investigation and who is completely independent of this relationship.

24. For a research subject who is legally incompetent, physically or mentally incapable of giving consent or is a legally incompetent minor, the investigator must obtain informed consent from the legally authorised representative in accordance with applicable law. These groups should not be included in research unless the research is necessary to promote the health of the population

represented and this research cannot instead be performed on legally competent persons.

25. When a subject deemed legally incompetent, such as a minor child, is able to give assent to decisions about participation in research, the investigator must obtain that assent in addition to the consent of the legally authorised representative.

26. Research on individuals from whom it is not possible to obtain consent, including proxy or advance consent, should be done only if the physical/mental condition that prevents obtaining informed consent is a necessary characteristic of the research population. The specific reasons for involving research subjects with a condition that renders them unable to give informed consent should be stated in the experimental protocol for consideration and approval of the review committee. The protocol should state that consent to remain in the research should be obtained as soon as possible from the individual or a legally authorised surrogate.

27. Both authors and publishers have ethical obligations. In publication of the results of research, the investigators are obliged to preserve the accuracy of the results.

Negative as well as positive results should be published or otherwise publicly available. Sources of funding, institutional affiliations and any possible conflicts of interest should be declared in the publication. Reports of experimentation not in accordance with the principles laid down in this Declaration should not be accepted for publication.

C. Additional principles for medical research combined with medical care

28. The physician may combine medical research with medical care, only to the extent that the research is justified by its potential prophylactic, diagnostic or therapeutic value. When medical research is combined with medical care, additional standards apply to protect the patients who are research subjects.

29. The benefits, risks, burdens and effectiveness of a new method should be tested against those of the best current prophylactic, diagnostic, and therapeutic methods. This does not exclude the use of placebo, or no treatment, in

studies where no proven prophylactic, diagnostic or therapeutic method exists.[1]

30. At the conclusion of the study, every patient entered into the study should be assured of access to the best proven prophylactic, diagnostic and therapeutic methods identified by the study.[2]

31. The physician should fully inform the patient which aspects of the care are related to the research. The refusal of a patient to participate in a study must never interfere with the patient–physician relationship.

32. In the treatment of a patient, where proven prophylactic, diagnostic and therapeutic methods do not exist or have been ineffective, the physician, with informed consent from the patient, must be free to use unproven or new prophylactic, diagnostic and therapeutic measures, if in the physician's judgment it offers hope of saving life, re-establishing health or alleviating suffering. Where possible, these measures should be made the object of research, designed to evaluate their safety and efficacy. In all cases, new information should be recorded and, where appropriate, published. The other relevant guidelines of this Declaration should be followed.

The WMA hereby reaffirms its position that extreme care must be taken in making use of a placebo-controlled trial and that in general this methodology should only be used in the absence of existing proven therapy. However, a placebo-controlled trial may be ethically acceptable, even if proven therapy is available, under the following circumstances:

- Where for compelling and scientifically sound methodological reasons its use is necessary to determine the efficacy or safety of a prophylactic, diagnostic or therapeutic method; or
- Where a prophylactic, diagnostic or therapeutic method is being investigated for a minor condition and the patients who receive placebo will not be subject to any additional risk of serious or irreversible harm.

---

1. Note of clarification on paragraph 29 of the WMA Declaration of Helsinki.
2. Note of clarification on paragraph 30 of the WMA Declaration of Helsinki.

All other provisions of the Declaration of Helsinki must be adhered to, especially the need for appropriate ethical and scientific review.

The WMA hereby reaffirms its position that it is necessary during the study planning process to identify post-trial access by study participants to prophylactic, diagnostic and therapeutic procedures identified as beneficial in the study or access to other appropriate care. Post-trial access arrangements or other care must be described in the study protocol so the ethical review committee may consider such arrangements during its review.

APPENDIX 2
OVIEDO CONVENTION (COUNCIL OF EUROPE, EUROPEAN TREATY SERIES 164)

Chapter V – Scientific research

Article 15 – General rule
Scientific research in the field of biology and medicine shall be carried out freely, subject to the provisions of this Convention and the other legal provisions ensuring the protection of the human being.

Article 16 – Protection of persons undergoing research
Research on a person may only be undertaken if all the following conditions are met:

i there is no alternative of comparable effectiveness to research on humans;

ii the risks which may be incurred by that person are not disproportionate to the potential benefits of the research;

iii the research project has been approved by the competent body after independent examination of its scientific merit, including assessment of the importance of the aim of the research, and multidisciplinary review of its ethical acceptability;

iv the persons undergoing research have been informed of their rights and the safeguards prescribed by law for their protection;

v the necessary consent as provided for under Article 5 has been given expressly, specifically and is documented. Such consent may be freely withdrawn at any time.

Article 17 – Protection of persons not able to consent to research

1 Research on a person without the capacity to consent as stipulated in Article 5 may be undertaken only if all the following conditions are met:

   i the conditions laid down in Article 16, sub-paragraphs i to iv, are fulfilled;

   ii the results of the research have the potential to produce real and direct benefit to his or her health;

   iii research of comparable effectiveness cannot be carried out on individuals capable of giving consent;

   iv the necessary authorisation provided for under Article 6 has been given specifically and in writing; and

   v the person concerned does not object.

2 Exceptionally and under the protective conditions prescribed by law, where the research has not the potential to produce results of direct benefit to the health of the person concerned, such research may be authorised subject to the conditions laid down in paragraph 1, sub-paragraphs i, iii, iv and v above, and to the following additional conditions:

   i the research has the aim of contributing, through significant improvement in the scientific understanding of the individual's condition, disease or disorder, to the ultimate attainment of results capable of conferring benefit to the person concerned or to other persons in the same age category or afflicted with the same disease or disorder or having the same condition;

   ii the research entails only minimal risk and minimal burden for the individual concerned.

Article 18 – Research on embryos in vitro

1 Where the law allows research on embryos in vitro, it shall ensure adequate protection of the embryo.

2 The creation of human embryos for research purposes is prohibited.

APPENDIX 3
OVIEDO CONVENTION – ADDITIONAL PROTOCOL CONCERNING BIOMEDICAL RESEARCH, CHAPTER I–IX (COUNCIL OF EUROPE, EUROPEAN TREATY SERIES 195)

Chapter I – Object and scope

Article 1 – Object and purpose
Parties to this Protocol shall protect the dignity and identity of all human beings and guarantee everyone, without discrimination, respect for their integrity and other rights and fundamental freedoms with regard to any research involving interventions on human beings in the field of biomedicine.

Article 2 – Scope
1 This Protocol covers the full range of research activities in the health field involving interventions on human beings.
2 This Protocol does not apply to research on embryos in vitro. It does apply to research on foetuses and embryos in vivo.
3 For the purposes of this Protocol, the term 'intervention' includes:
   i   a physical intervention, and
   ii any other intervention in so far as it involves a risk to the psychological health of the person concerned.

Chapter II – General provisions

Article 3 – Primacy of the human being
The interests and welfare of the human being participating in research shall prevail over the sole interest of society or science.

Article 4 – General rule
Research shall be carried out freely, subject to the provisions of this Protocol and the other legal provisions ensuring the protection of the human being.

Article 5 – Absence of alternatives
Research on human beings may only be undertaken if there is no alternative of comparable effectiveness.

Article 6 – Risks and benefits

1 Research shall not involve risks and burdens to the human being dispropor-tionate to its potential benefits.

2 In addition, where the research does not have the potential to produce results of direct benefit to the health of the research participant, such research may only be undertaken if the research entails no more than acceptable risk and acceptable burden for the research participant. This shall be without prejudice to the provision contained in Article 15 paragraph 2, sub-paragraph ii for the protection of persons not able to consent to research.

Article 7 – Approval

Research may only be undertaken if the research project has been approved by the competent body after independent examination of its scientific merit, including assessment of the importance of the aim of research, and multidisciplinary review of its ethical acceptability.

Article 8 – Scientific quality

Any research must be scientifically justified, meet generally accepted criteria of scientific quality and be carried out in accordance with relevant professional obligations and standards under the supervision of an appropriately qualified researcher.

Chapter III – Ethics committee

Article 9 – Independent examination by an ethics committee

1 Every research project shall be submitted for independent examination of its ethical acceptability to an ethics committee. Such projects shall be submitted to independent examination in each State in which any research activity is to take place.

2 The purpose of the multidisciplinary examination of the ethical acceptability of the research project shall be to protect the dignity, rights, safety and well-being of research participants. The assessment of the ethical acceptability shall draw on an appropriate range of expertise and experience adequately reflecting professional and lay views.

3   The ethics committee shall produce an opinion containing reasons for its conclusion.

Article 10 – Independence of the ethics committee

1   Parties to this Protocol shall take measures to assure the independence of the ethics committee. That body shall not be subject to undue external influences.
2   Members of the ethics committee shall declare all circumstances that might lead to a conflict of interest. Should such conflicts arise, those involved shall not participate in that review.

Article 11 – Information for the ethics committee

1   All information which is necessary for the ethical assessment of the research project shall be given in written form to the ethics committee.
2   In particular, information on items contained in the appendix to this Protocol shall be provided, in so far as it is relevant for the research project. The appendix may be amended by the Committee set up by Article 32 of the Convention by a two-thirds majority of the votes cast.

Article 12 – Undue influence

The ethics committee must be satisfied that no undue influence, including that of a financial nature, will be exerted on persons to participate in research. In this respect, particular attention must be given to vulnerable or dependent persons.

Chapter IV – Information and consent

Article 13 – Information for research participants

1   The persons being asked to participate in a research project shall be given adequate information in a comprehensible form. This information shall be documented.
2   The information shall cover the purpose, the overall plan and the possible risks and benefits of the research project, and include the opinion of the ethics

committee. Before being asked to consent to participate in a research project, the persons concerned shall be specifically informed, according to the nature and purpose of the research:

    i of the nature, extent and duration of the procedures involved, in particular, details of any burden imposed by the research project;

   ii of available preventive, diagnostic and therapeutic procedures;

  iii of the arrangements for responding to adverse events or the concerns of research participants;

  iv of arrangements to ensure respect for private life and ensure the confidentiality of personal data;

   v of arrangements for access to information relevant to the participant arising from the research and to its overall results;

  vi of the arrangements for fair compensation in the case of damage;

 vii of any foreseen potential further uses, including commercial uses, of the research results, data or biological materials;

viii of the source of funding of the research project.

3   In addition, the persons being asked to participate in a research project shall be informed of the rights and safeguards prescribed by law for their protection, and specifically of their right to refuse consent or to withdraw consent at any time without being subject to any form of discrimination, in particular regarding the right to medical care.

Article 14 – Consent

1   No research on a person may be carried out, subject to the provisions of both Chapter V and Article 19, without the informed, free, express, specific and documented consent of the person. Such consent may be freely withdrawn by the person at any phase of the research.

2   Refusal to give consent or the withdrawal of consent to participation in research shall not lead to any form of discrimination against the person concerned, in particular regarding the right to medical care.

3   Where the capacity of the person to give informed consent is in doubt, arrangements shall be in place to verify whether or not the person has such capacity.

Chapter V – Protection of persons not able to consent to research

Article 15 – Protection of persons not able to consent to research

1   Research on a person without the capacity to consent to research may be undertaken only if all the following specific conditions are met:

    i   the results of the research have the potential to produce real and direct benefit to his or her health;

    ii   research of comparable effectiveness cannot be carried out on individuals capable of giving consent;

    iii the person undergoing research has been informed of his or her rights and the safeguards prescribed by law for his or her protection, unless this person is not in a state to receive the information;

    iv   the necessary authorisation has been given specifically and in writing by the legal representative or an authority, person or body provided for by law, and after having received the information required by Article 16, taking into account the person's previously expressed wishes or objections. An adult not able to consent shall as far as possible take part in the authorisation procedure. The opinion of a minor shall be taken into consideration as an increasingly determining factor in proportion to age and degree of maturity;

    v   the person concerned does not object.

2   Exceptionally and under the protective conditions prescribed by law, where the research has not the potential to produce results of direct benefit to the health of the person concerned, such research may be authorised subject to the conditions laid down in paragraph 1, sub-paragraphs ii, iii, iv, and v above, and to the following additional conditions:

    i   the research has the aim of contributing, through significant improvement in the scientific understanding of the individual's condition, disease or disorder, to the ultimate attainment of results capable of conferring benefit to the person concerned or to other persons in the same age category or afflicted with the same disease or disorder or having the same condition;

    ii   the research entails only minimal risk and minimal burden for the individual concerned; and any consideration of additional potential benefits of the research shall not be used to justify an increased level of risk or burden.

3 Objection to participation, refusal to give authorisation or the withdrawal of authorisation to participate in research shall not lead to any form of discrimination against the person concerned, in particular regarding the right to medical care.

Article 16 – Information prior to authorisation

1 Those being asked to authorise participation of a person in a research project shall be given adequate information in a comprehensible form. This information shall be documented.

2 The information shall cover the purpose, the overall plan and the possible risks and benefits of the research project, and include the opinion of the ethics committee. They shall further be informed of the rights and safeguards prescribed by law for the protection of those not able to consent to research and specifically of the right to refuse or to withdraw authorisation at any time, without the person concerned being subject to any form of discrimination, in particular regarding the right to medical care. They shall be specifically informed according to the nature and purpose of the research of the items of information listed in Article 13.

3 The information shall also be provided to the individual concerned, unless this person is not in a state to receive the information.

Article 17 – Research with minimal risk and minimal burden

1 For the purposes of this Protocol it is deemed that the research bears a minimal risk if, having regard to the nature and scale of the intervention, it is to be expected that it will result, at the most, in a very slight and temporary negative impact on the health of the person concerned.

2 It is deemed that it bears a minimal burden if it is to be expected that the discomfort will be, at the most, temporary and very slight for the person concerned. In assessing the burden for an individual, a person enjoying the special confidence of the person concerned shall assess the burden where appropriate.

Chapter VI – Specific situations

Article 18 – Research during pregnancy or breastfeeding

1 Research on a pregnant woman which does not have the potential to produce results of direct benefit to her health, or to that of her embryo, foetus or child after birth, may only be undertaken if the following additional conditions are met:
   i the research has the aim of contributing to the ultimate attainment of results capable of conferring benefit to other women in relation to reproduction or to other embryos, foetuses or children;
   ii research of comparable effectiveness cannot be carried out on women who are not pregnant;
   iii the research entails only minimal risk and minimal burden.
2 Where research is undertaken on a breastfeeding woman, particular care shall be taken to avoid any adverse impact on the health of the child.

Article 19 – Research on persons in emergency clinical situations

1 The law shall determine whether, and under which protective additional conditions, research in emergency situations may take place when:
   i a person is not in a state to give consent, and
   ii because of the urgency of the situation, it is impossible to obtain in a sufficiently timely manner, authorisation from his or her representative or an authority or a person or body which would in the absence of an emergency situation be called upon to give authorisation.
2 The law shall include the following specific conditions:
   i research of comparable effectiveness cannot be carried out on persons in non-emergency situations;
   ii the research project may only be undertaken if it has been approved specifically for emergency situations by the competent body;
   iii any relevant previously expressed objections of the person known to the researcher shall be respected;
   iv where the research has not the potential to produce results of direct benefit to the health of the person concerned, it has the aim of contributing,

through significant improvement in the scientific understanding of the individual's condition, disease or disorder, to the ultimate attainment of results capable of conferring benefit to the person concerned or to other persons in the same category or afflicted with the same disease or disorder or having the same condition, and entails only minimal risk and minimal burden.

3   Persons participating in the emergency research project or, if applicable, their representatives shall be provided with all the relevant information concerning their participation in the research project as soon as possible. Consent or authorisation for continued participation shall be requested as soon as reasonably possible.

Article 20 – Research on persons deprived of liberty

Where the law allows research on persons deprived of liberty, such persons may participate in a research project in which the results do not have the potential to produce direct benefit to their health only if the following additional conditions are met:

i   research of comparable effectiveness cannot be carried out without the participation of persons deprived of liberty;

ii the research has the aim of contributing to the ultimate attainment of results capable of conferring benefit to persons deprived of liberty;

iii the research entails only minimal risk and minimal burden.

Chapter VII – Safety and supervision

Article 21 – Minimisation of risk and burden

1   All reasonable measures shall be taken to ensure safety and to minimize risk and burden for the research participants.

2   Research may only be carried out under the supervision of a clinical professional who possesses the necessary qualifications and experience.

Article 22 – Assessment of health status

1 The researcher shall take all necessary steps to assess the state of health of human beings prior to their inclusion in research, to ensure that those at increased risk in relation to participation in a specific project be excluded.

2 Where research is undertaken on persons in the reproductive stage of their lives, particular consideration shall be given to the possible adverse impact on a current or future pregnancy and the health of an embryo, foetus or child.

Article 23 – Non-interference with necessary clinical interventions

1 Research shall not delay nor deprive participants of medically necessary preventive, diagnostic or therapeutic procedures.

2 In research associated with prevention, diagnosis or treatment, participants assigned to control groups shall be assured of proven methods of prevention, diagnosis or treatment.

3 The use of placebo is permissible where there are no methods of proven effectiveness, or where withdrawal or withholding of such methods does not present an unacceptable risk or burden.

Article 24 – New developments

1 Parties to this Protocol shall take measures to ensure that the research project is re-examined if this is justified in the light of scientific developments or events arising in the course of the research.

2 The purpose of the re-examination is to establish whether:
   i   the research needs to be discontinued or if changes to the research project are necessary for the research to continue;
   ii research participants, or if applicable their representatives, need to be informed of the developments or events;
   iii additional consent or authorisation for participation is required.

3 Any new information relevant to their participation shall be conveyed to the research participants, or, if applicable, to their representatives, in a timely manner.

4 The competent body shall be informed of the reasons for any premature termination of a research project.

Chapter VIII – Confidentiality and right to information

Article 25 – Confidentiality

1 Any information of a personal nature collected during biomedical research shall be considered as confidential and treated according to the rules relating to the protection of private life.

2 The law shall protect against inappropriate disclosure of any other information related to a research project that has been submitted to an ethics committee in compliance with this Protocol.

Article 26 – Right to information

1 Research participants shall be entitled to know any information collected on their health in conformity with the provisions of Article 10 of the Convention.

2 Other personal information collected for a research project will be accessible to them in conformity with the law on the protection of individuals with regard to processing of personal data.

Article 27 – Duty of care

If research gives rise to information of relevance to the current or future health or quality of life of research participants, this information must be offered to them. That shall be done within a framework of health care or counselling. In communication of such information, due care must be taken in order to protect confidentiality and to respect any wish of a participant not to receive such information.

Article 28 – Availability of results

1 On completion of the research, a report or summary shall be submitted to the ethics committee or the competent body.

2 The conclusions of the research shall be made available to participants in reasonable time, on request.

3 The researcher shall take appropriate measures to make public the results of research in reasonable time.

Chapter IX – Research in States not parties to this Protocol

Article 29 – Research in States not parties to this Protocol

Sponsors or researchers within the jurisdiction of a Party to this Protocol that plan to undertake or direct a research project in a State not party to this Protocol shall ensure that, without prejudice to the provisions applicable in that State, the research project complies with the principles on which the provisions of this Protocol are based. Where necessary, the Party shall take appropriate measures to that end.

CHAPTER 4

RESEARCH METHODOLOGY: STRATEGIES, PLANNING AND ANALYSIS

Haakon Breien Benestad and Petter Laake

4.1 INTRODUCTION

The achievements of individuals drive progress in science. That is most famously recognized by the Nobel Prizes, awarded each year since 1901 for outstanding achievements in chemistry, physics, physiology or medicine, for literature and for peace. Just what constitutes achievement has been debated since ancient times. In the sciences, an achievement arguably might be classed as a discovery or as an improvement. A discovery is the finding of something previously unknown, such as Roentgen's discovery of X-rays or the Curies' discovery of radium. An improvement is the producing of something better. Alfred Nobel's achievement of 1866, of mixing kieselguhr, an absorbent, in highly unstable nitroglycerine to make dynamite, a stable explosive, was an improvement. With today's level of scientific knowledge, spectacular discoveries are few. Likewise, few improvements can have consequences as rewarding as Alfred Nobel's invention of dynamite, which underpinned the business on which he built the fortune that now funds the Nobel Prizes. Most scientists achieve through contributing to the greater body of scientific knowledge, such as by providing support for a view or hypothesis that is generally accepted but not proven. In concert, contributions can be significant, as illustrated by an example within the fields covered by this book. Early in the twentieth century, the hypothesis that signals propagated between nerve cells by chemical, not electrical means was accepted, though unproven; yet decades passed before the first neurotransmitter was identified.

Scientific developments are classed neither as discoveries nor as improvements, but nonetheless bring about progress. Mathematical modelling can clarify weaknesses in, or insufficient databases for, prevailing opinion. Compiling, improving or testing laboratory methods may be significant contributions. The special skills of entrepreneurship are a prerequisite for coordinating the work of a large number of staff, in basic research, applied research and other big science, as well as for successfully appealing to those who provide funding. 'It takes all sorts to make the [scientific] world'; see Box 4.1 for other reflections.

Box 4.1 On scientific problems

Unfortunately, solving problems is not all there is to the scientific endeavour. Even more important than solving problems is finding relevant problems – formulating questions that really matter. The world offers us an infinite number of problems to solve, of which we select some and disregard others. Much of the art of doing science is then deciding which problems to concentrate on and which to ignore.

... The ability to solve problems requires logical thinking, and hence a rational mind, whereas the ability to identify consequential problems is only in part based on logic; mostly it is based on instinct, intuition, subconscious perception, a sixth sense, inborn proclivity, talent, irrational impulse or whatever you might want to call it.

There is plenty of evidence from the history of science for an independent assortment of these two traits.

(Klein 1985)

In all cases, it helps to have thorough knowledge, to be prepared for the unexpected and to see analogies between dissimilar scientific fields with intuition and imagination. Pasteur believed that 'chance favours the prepared mind'. So serendipity counts. Indeed, Horace Walpole's entertaining story about the three princes from Serendip, now Sri Lanka, is often quoted in scientific writings, as by Austin (1978).

4.2 YOUR SCIENTIFIC PROBLEM

First and foremost, a problem must have significance. Its solution must have value, not just fascination for you. Attaining the goal may require time and effort, so avoid trivialities. Plan thoroughly in advance!

Many problems have long been known but not tackled, because the means to solve them were not available. Elucidations of greater portions of molecular biology, such as the structure of genetic material, the principles of protein synthesis and the regulation of gene activity, were required before the problems of developmental biology could be attacked. How does a fertilized egg develop into a complete, multicellular organism? Technology is often said to build on basic research, but information can flow the other way. A new technique, method or apparatus often enables the solution of an older, 'overripe' problem in biology. That is worthy. But learning a technique and exploiting it repetitively for all it is worth constitutes technically orientated, not problem-orientated research, which is hardly praiseworthy and can spawn many publications (Box 4.2).

Box 4.2 On untrodden paths and unexpected finds

In basic research, everything is just the opposite. What you need at the outset is a high degree of uncertainty; otherwise it isn't likely to be an important problem. You start with an incomplete roster of facts, characterised by their ambiguity; often the problem consists of discovering the connections between unrelated pieces of information. You must plan experiments on the basis of probability, even bare possibility, rather than certainty. If an experiment turns out precisely as predicted, this can be very nice, but it is only a great event if at the same time it is a surprise. You can measure the quality of the work by the intensity of astonishment. The surprise can be because it did turn out as predicted (in some lines of research, 1% is accepted as a high yield), or it can be confoundment because the prediction was wrong and something totally unexpected turned up, changing the look of the problem and requiring a new kind of protocol. Either way, you win.

(Thomas 1974)

Sir Peter Medawar, the philosopher of science who shared the 1960 Nobel Prize in Physiology or Medicine with Sir Frank MacFarlane Burnet, maintains that 'science is the art of the soluble'. Accordingly, you should conduct definitive enquiry that either supports or disproves your working hypothesis. Be bold: attack your favoured hypothesis from several aspects; try to falsify it so that you do not end up with a 'so-what' paper. It must be possible to disprove the hypothesis.

Heed the principle of 'Occam's (Ockham's) razor', that is, select the simple. Choose preferably the simplest hypothesis or explanation that is compatible with the facts you seek to interpret. Complex assumptions with many reservations and thereby exceptions that require modifications of the hypothesis are uninspiring and difficult to falsify. Also be a bit lazy, in the sense that your set-up is not overly involved, lest you increase the risk of error on the way. And proceed unhurriedly, step by step, so that you always have an overview of the potential outcomes and consequences of your project.

Your experiment or intervention must have the requisite controls. Often, researchers show their skills best in the method checks and other controls included in their experimental design. Appropriate controls are vital. A control should differ from the test in only one aspect, that is, the variable being investigated, be it a chemical, a training scenario, whatever. The control should be congruent with the test. The control patients should be in accord with the test patients, the control animals in accord with the test animals and the control phase in accord with the test phase in trials involving experimental animals that serve as their own controls (baseline versus experimental phase). As an aside, you may advantageously eliminate interindividual variations by arranging for patients, experimental animals, organ preparations or cell cultures to be their own controls, such as by successively applying test and control conditions.

Reproducibility is as vital as relevance, testability and congruence. A potentially publishable finding should not be released before it is replicated in one or preferably two new, independent experiments. Likewise, you should be sceptical of sensational published developments until they have been confirmed by independent research groups.

4.3 MORE ON SCIENTIFIC PROBLEMS

All science builds on observations, and causal explanation of observed phenomena is routine in the natural sciences. Causality can be straightforward whenever there is only one cause. Yet most often, several causal factors may act simultaneously to produce an observed effect. Causal explanations represent a natural starting point for understanding or explaining an observed phenomenon. However, studies of interesting associations or findings should not be ignored if a relationship cannot be explained by cause and effect. Regardless of the starting point taken in the philosophy of science, causal thought enters into the explanation of the empirical associations essential to the hypothetical–deductive method. Understanding is enhanced through hypotheses that can be falsified by empirical trials. (For further discussion, see Section 1.1 of Chapter 1.)

Research distinguishes between variables connected to the starting points for studies or experiments and the variables used for explanatory purposes. The variable to be studied is called the outcome variable, the response variable or the effect variable, whereas the variable used for explanatory purposes is called the explanatory variable, the predictor variable or the exposure variable. The research hypothesis includes a description of the association between the outcome variable and one or more explanatory variables. The hypothesis that we seek to falsify, the null hypothesis, is that there is no association between the outcome variable and at least one of the explanatory variables.

4.4 LITERATURE, METHODS AND TECHNIQUES

The body of scientific literature is extensive and overwhelming. You cannot know it all, but you should have a solid and not overly narrow basic knowledge of it, so that you can think straight and sense what is and what is not biologically plausible. Moreover, you must be sure that your work does not simply duplicate that already carried out by others. Through the alerting services offered by the various data-bases (e.g. MyNCBI in PubMed), it is possible to set up search strategies that include important keywords and the names of important researchers within your subject field (see Section 5.9 of Chapter 5).

That said, getting started with the actual practical work also is essential, lest too much time be spent in theoretical preparation for it. It is tough being the Lone Ranger, but beneficial if the group you are in can divide reading among members and hold regular literature seminars.

Choosing the optimal materials and methods is, of course, decisive. For the experimenter, the question can be whether to approach the scientific problem using test subjects, experimental animals, an organ culture, a cell culture or a sub-cellular system. Often, the best procedure is to think through and possibly conduct experiments at several levels, preferably in the reverse order, with the simplest and most controllable set-up first. The higher in the test hierarchy, the greater the chance for physiologically meaningful results, but also for misleading results. For example, if you inject a new hormone in an experimental animal, there is a non-negligible risk that the concentration at the locus of action and the time–effect profile are above and beyond the physiological. Many years ago, a member of an academic appraisal committee wrote that 'the oestrogen doses given to mice would have been more suitable for an elephant'. Conversely, it is possible that the agent you inject metabolizes or in some other way deactivates so rapidly that you observe no effects at all. Or a compensatory mechanism might take over, or the mechanism you seek manifests itself only under special conditions, as often has been the disap-pointing experience with transgenic animals that otherwise seem to have no flaws. An injected substance A can provoke the secretion of hormone B, which in turn gives measured response C. Asserting that A causes C is then slightly sketchy. Maybe response C is not observed at all, because physiological, homoeostatic reac-tions (negative feedback) contribute to maintaining steady state and thereby reduce C to an unmeasurable level. Removing one cause of depression does not guarantee improvement in the patient's self-esteem.

Misleading results or artefacts can also deceive whenever one works with sys-tems that are simpler than intact organisms. Perhaps the best precaution is meticulously to chart the experimental set-up, as outlined in Box 4.3. But it is easy to be deceived. Many new cancer treatments, held promising following

Box 4.3 Experimenter's checklist

METHODS

- Precision: quartile interval, coefficient of variation?
- Accuracy: compare to a 'gold standard'
- Dose–response?
- Time lapse?
- Physical–chemical conditions: pH, osmolarity, etc.?
- Positive controls? Sensitivity controls? Controls with known answers?
- Negative controls: specificity controls?
- Milieu control?
- Statistics: Sample size? Power? Uncertainty measure: 95% confidence interval, standard error of the mean? P-value? Parametric or non-parametric test? Regression or correlation?

STUDY SET-UP

- Test = control plus one factor?
- Scoring with a blind or double-blind arrangement?
- Selection criteria, exclusion criteria, randomization (sample, sequence), representative population?
- Validity? Large gap between observation and interpretation? Several routes to the point?
- Repeatability? How many replicate studies? Reproducibility?

INTERPRETATION

- Biologically likely?
- Statistically as well as biologically significant? (The interocular impact test.)

encouraging results on experimental animals or in vitro on cancer cells, have disappointed upon finally being tested on patients.

Often we are most interested in conclusions valid for people. Nonetheless, experiments are often conducted elsewhere. For instance, the early experiments of the 1940s on nerve cells were conducted on the giant axon of a squid, because sufficiently precise instruments were not yet available to conduct similar experiments on human nerve fibres. Indeed, controlled experiments on animals can reveal mechanisms and connections not as readily observable in human patients. In any event, the lack of understanding of some phenomena, such as tickling, stitches caused by strenuous exercise, migraine and fibromyalgia, is due in part to an absence of sufficiently comparable animal models and in part to their not being amenable to in vitro study. Consequently, the newer transgenic animals hold great promise. Examples include mouse anaemia that is similar in many respects to thalassaemia, Mediterranean anaemia, and another model that ostensibly simulates schizophrenia. It is less expensive and less emotionally stressful to use small animals, such as mice and rats, rather than larger mammals. Moreover, small rodents may be selectively bred to be as alike as identical twins. Momentous decisions must be made before apes, livestock, cats or dogs are used. But in studies of cardiophysiology or pathophysiology, where rodents cannot be used, dogs are preferable to cats, because the neural regulation (vagus tonus at rest) of the dog heart is the closer to that of the human heart.

In any case, it is best, as well as often convincing, to reach conclusions in many ways, such as by addressing a scientific problem at several levels. However, this is an ideal that may be audacious in practice, particularly if you are new in a research group and attempt to change its ways. Time available is your most limited resource, and as an old adage advises: 'Never change a winning team!' The experimental set-up is certainly thorough, documented and proven, and many data are available for further experimentation. That said, a reluctance to bring in new procedures and models may easily result in stagnation. So it is advantageous if the group as a whole occasionally innovates, and you can take part in that effort. It is wise first to gain experience through pursuing the literature on materials and methods, but if possible, do not postpone going to laboratories where procedures are routine. In that, you save effort and may gain new contacts.

4.5 RESEARCH CONDITIONS

Research hypotheses are studied empirically, through confronting collected data. Medical studies comprise experimental studies and observational studies. The experimental studies include clinical trials, which are studies of interventions on humans. The goal of all empirical studies is to arrive at valid, reproducible conclusions concerning associations. This is achieved through well-planned studies, with valid designs and valid analyses.

The experiment, which involves an intervention, is typified by a study of the association between the outcome variable and a single explanatory variable. For instance, this is the case in a study of the effect of a new drug, in other words, an assessment of whether there exists a difference in the outcome variable between a treatment group and a placebo group. The equivalent situation exists in laboratory trials exploring, for example, the effect of a hormone on the growth rate in a cell culture.

Causal relationships are often multifactorial. Even then, the effect of a single explanatory variable can be studied by randomizing test units into two groups that differ by that variable alone. This is the case when test subjects are randomized into treatment and placebo groups. Randomization ensures that the two groups are 'identical', save for the explanatory/cause variable to which the treated group is subjected. An experiment that permits testing a hypothesis on a single causal factor also simplifies the statistical analysis in that only the difference between groups needs to be statistically tested.

Observational studies differ from experimental studies in that they are multifactorial, because a randomized arrangement in test or control groups cannot be implemented. Nonetheless, the influence of a particular causal factor on the difference between the test group and the control group is often of interest. It is not directly possible to assess in an observational study, as controls of other causal factors are not included in the design, that is, through randomization. Consequently, the causal factor of interest must be and can be studied through statistical analysis using multivariate statistical methods.

Confounding arises when the effect of interest is intermixed with the effects of other variables. The effects of confounding may be reduced through appropriate design or through special statistical analysis. In experiments or clinical trials, randomization is meant to eliminate the effects of confounding. In observational studies, confounding is more severe, so multivariate methods are used to reduce its effects. In multivariate statistical methods, control of causal factors other than the principal explanatory variable is accomplished by the use of a statistical model that includes all the principal causal factors. The effect of the causal factor of interest may then be estimated. Examples of such analyses include linear regression analysis, logistic regression and Cox analysis.

Examples of observational studies in the medical and biological sciences are listed in Box 4.4.

Box 4.4 Observational studies
- Cross-sectional studies and surveys
- Cohort studies and longitudinal studies
- Case–control studies

Cross-sectional studies and *surveys* are conducted on a study population at a particular time. At that time, the study population may include people of various ages, at various phases of an illness or with various causal factors of the illness. The outcome variable and the causal factors are not tracked over time, so a cross-sectional study is not well suited to exploring the relationship between cause and effect. A cross-sectional study is suited to studying the prevalence but not the incidence of an illness (see also Section 9.7 of Chapter 9).

Cohort studies are prospective studies or follow-up studies of a study population. In studying the association between causal factors and an illness, we start with a healthy population and study test subjects until the illness occurs or the study is stopped. Key examples of such studies include progress studies and survival studies that are often of long duration. The information on the association between the causal factors and the illness lies in the number of cases and the length of the follow-up time before the illness occurs. Age, other causal factors and the outcome variable are monitored over time, so this sort of study is well suited to examining the relationship between cause and effect, and both the incidence and prevalence of a disease may be studied. Typical follow-up studies include those of the association between the incidence of cancer and nutrition or diet.

Longitudinal studies are conducted on test subjects or experimental animals through repeated monitoring over time. At the same time, the progress of the outcome variable and that of the causal factors are studied. Data are acquired through repeated measurements, so the parameters do not include follow-up time but are limited to observations that the disease and the changes in the causal factors occurred before a particular monitoring point. In cohort studies, the study population is usually followed until an event occurs and then the follow-up time is recorded. Longitudinal studies are useful whenever there are relatively rapid changes in disease or causal factors. They are well suited for studying the associations between causal factors and effect variables. Examples of longitudinal studies include investigations of the effects of hormones upon growth in animals or humans, as with monthly measurements of growth. Such studies may be conducted over months or years.

A drawback of cohort and longitudinal studies is that they are of long duration and are costly. Moreover, a long-term study often suffers from the loss of some test subjects. The challenge in conducting a study then is to ensure that a decline in the number of test subjects does not threaten its validity or conclusions.

Case–control studies are characterized by the test subjects being selected with the outcome variable as the starting point. In studies of disease, the outcome variable is characterized by ill/healthy, which in turn become the case (test) and the control, respectively. The causal factors may be analysed retrospectively, back in time, to examine time lapse. Case–control studies may be conducted by matching to variables other than the main explanatory variable(s), such as age and gender. In this manner, the sample of test subjects is equalized with respect to the matching variables, and the process may be viewed as a randomization in an effort to re-create the situation of the classic experiment. In cancer epidemiology associations with diet or sun exposure are often studied in case–control studies.

4.6 DATA TYPES

Observations provide data, and data must be analysed to provide the basis of valid and reproducible conclusions. In the natural sciences, statistics such as mean, median, variation and measures of association are of interest. These magnitudes are computed in ways that depend on the types of data included in an analysis, and they are expressed on scales suiting their nature. The computations are numerical, so the scales accordingly are represented as numbers. In turn, the scales affect the presentation of the data, the choice of the effect measure and the statistical analysis. As listed in Box 4.5, there are two principal types of data, categorical and numerical.

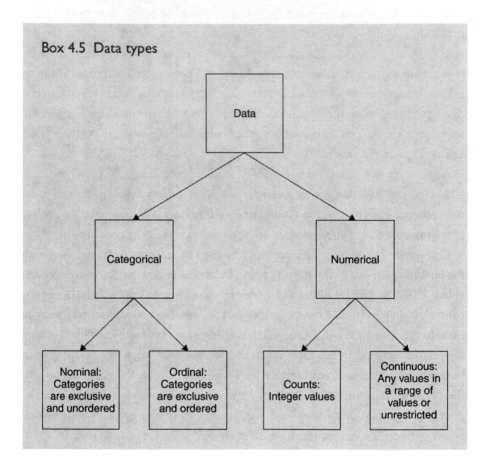

Box 4.5 Data types

Categorical data are classified into nominal or ordinal data. Smoking is an example of a nominal variable, with a division into non-smoker (category value 0) and smoker (category value 1). Ordinal data are in dedicated categories in a way such that there is underlying ranking of the categories. Smoking habits comprise an example of ordinal data, with the categories of non-smoker (category value 0), former smoker (category value 1), occasional smoker (category value 2) and daily smoker (category value 3). Numerical data are count data or continuous data.

Data acquired over time or space may be expressed in discrete numbers, such as count data. The number of times an event occurs, for instance, cases per year, is a typical example. Continuous data are characterized by the measurement of ranges of variables. Typical examples include weight, concentration, blood pressure and age, for which the data comprise the actual observed values.

4.7 TECHNIQUES

Technical proficiency is essential, as it enables the experimenter to rely upon results and reproduce findings. As in other endeavours, proficiency comes from practice, practice and more practice. You must develop a feel for how methods work in your hands, perform a series of replicate measurements and register the spread of the results, preferably expressed as the coefficient of variation, CV = (standard deviation)/(arithmetic mean). You might precisely measure blood haemoglobin level with an analytical CV = 1.3%, while your serum folate analysis has a CV = 16%. This determines the number of replications in the test and control groups. The more precise the analysis, the fewer parallel measurements are needed to find a mean value with a low degree of uncertainty, or a significant difference between two groups of data. At this point, it often may be wise to consult a statistician to perform a power calculation, if you cannot perform it yourself. That is, you should calculate the amount of data required so that you can, for instance, assume with 80% probability that you can reveal a 30% difference between two groups of data, at the 5% level, when the analytical plus the biological CV (which permits estimation of the variance) are known.

Avoiding systematic error is essential. That is, your work must be accurate. Precision and accuracy are often confused, but may be distinguished by example. If you shoot at a target with a rifle on which the sights are skewed, your shots may fall close to each other (high precision), but systematically be up to the right of the bull's eye (low accuracy). Accuracy requires that you calibrate your measuring instruments against a recognized standard, which may or may not be commercially available. The nature of your procedure will determine how often you (or your instrumentation engineer) must recalibrate the instruments.

The principal aspects of quality control, as performed at a large hospital laboratory, may be a paradigm for a research laboratory. They are as follows. Two types of references are used: calibrators with known, stated values, and control material used daily.

Calibrators are used to establish working characteristics or standard curves (the functional connection between instrument response and analytical concentration). Some calibrators are offered by independent companies, but for many analyses, the only calibrators available are those supplied by the instrument manufacturer as part of the reagent kit.

Controls are used to monitor analytical quality, as in the daily, internal quality control in which control material in various concentrations is included in all analyses. Repetitions are often included in long series of analyses, usually at the start, in the middle and at the end, to monitor for instrument drift. The controls should be at least at two levels, one low and one high, when the working characteristic or standard curve is linear, and at three levels when the characteristic is non-linear. For diagnostic applications, the levels should bracket the range over which the exact measurement data are decisive for determining whether clinical intervention is indicated. For pure technical quality assurance, as in a research project, there should be a control in the middle as well as at either end of the working range where standard curves deflect, such as at ED_{20} and ED_{80} in a radioimmunoassay.

In principle, controls other than those provided by the manufacturer of the instruments are preferable, although in practice this is often not possible, as in haematological analyses. For many analyses, controls may be made in-house, by freezing portions that later can be thawed just before use. Controls should have the same matrix, that is, the same composition of protein content, ion concentration, etc., as that of the test material.

Moreover, at intervals of two weeks to six times a year, external organizations (LabQuality, Murex, etc.) send out control materials for analysis, so that individual laboratories may report back their findings. This procedure is called external quality assessment.

Some set-ups may require exceptional vigilance to spot systematic errors. Might recorded observations change if an experimental animal moves, or might its breathing have an effect? Do subjective choices enter the picture? For example, do you record from the largest nerve cells, take blood from the most accessible arteries, count cells in areas of a preparation in which you expect to find many specimens of the type of cell that you hope are numerous? In such cases, the recording of data should be blinded.

Establishing the course of dose–response for an analysis is a rule of craftsmanship and may be essential for proper performance of an experiment. Moreover, you may profit in future experiments, in which a particular dose can be selected for continuation of an investigation. Either saturation doses or half-maximum effect doses (ED_{50}) may be chosen; in such cases you need to know the strength of influence of the dose–response relation.

Time lapse also is significant. For example, in a subsequent trial you may wish to record data at a time when you are certain that the maximum response will occur. Perhaps your experimental apparatus has a long response time that you may wish to alter or at least be aware of. A lethargic mechanical part might be replaced by an electronic component, or other causes of unacceptable latency may be removed. (When non-invasive ultrasonic blood-flow measurements became available and replaced invasive measurements on experimental animals, it became possible to measure blood flow in humans and, by recording changes in the course of a couple of heart beats, corroborate that changes in the bloodstream were triggered by nerves, not hormones, which act more slowly.) It is also essential to know whether measurement results are stable over a period of time, so reducing the need to recalibrate instruments or compensate for drift. An experimental set-up is splendid if you can document a stable base level, then observe an outcome resulting from the experimental intervention, and finally return to base level (baseline) when the intervention ceases. In such cases, it is advantageous to compare the observations of the measured variables with the means of the two nearly identical control phases.

For in vitro trials, the researcher must preside over physicochemical conditions, such as pH, osmolarity, temperature and gas pressure.

Positive controls are also essential, particularly with negative findings. Positive controls are related to calibrators, or control material in various concentrations, and to dose–response trials, but are most often used at higher, more complex levels. For example, for a particular intervention, an observed effect was not as expected. This may be interpreted in many ways: you may have (1) falsified your working hypothesis, (2) been misled by a compensatory reaction in a preparation, experimental animal or test subject, that more or less masked the response you expected, (3) not actually intervened in the way you thought you would, such as because a chemical agent no longer was active (past its 'best by' date), or (4) used cells that have lost their ability to respond. Then you are in a position where you must document that the intervention has proceeded as it should. For example, if you find that a graft versus host (GvH) reaction does not affect blood formation as assumed, then the occurrence of a genuine GvH reaction must be confirmed, such as by demonstrating enormous spleen growth.

Sensitivity control goes a step farther in that you must show not only that an intervention acts but also document a strength–effect relation, as discussed above for individual analyses. If the stimulus strength required to elicit the desired effect or outcome is physiologically unreasonable, you are on thin ice. Supraphysiological stimuli can cause various artefacts. In intact animals, unspecific stress reactions or alerting responses may have triggered the measured effects. Large drug doses or other factors may have unspecific effects. Sensitive cells and tissue that are more or less roughly handled in vitro can react otherwise in situ. Cell lines comprise transformed cells that do not necessarily respond in culture as do normal varieties in vivo. Cell cultures also can change apparently spontaneously; they can be infected by mycoplasma or be subjected to genetic or epigenetic transformation.

Examples such as these remind us of the importance of specificity controls or *negative controls*. Before telling the media that you have discovered a remarkable substance that kills cancer cells, it is best to check that it does not also kill normal cells. (Moreover, you should definitely not publish in the general media before your scientific paper is published!) Perhaps a more realistic example might be that you have isolated a factor that believably specifically stimulates a type of cell to secrete a particular hormone. You must then show that you have observed an actual secretion process and not the result of cell damage (as indicated by release of a cytoplasmic enzyme such as lactate dehydrogenase). Furthermore, specificity should be proven by confirming that the effect occurs via a particular receptor. This can be done by blocking the effect with an antagonist or antibody to the receptor. Alternatively, you may use a ribozyme or inhibitory RNA molecule that cleaves or inactivates messenger RNA for the receptor protein, or create a 'knock-out' mouse that lacks the receptor and hence also the hormone secretion that you have shown in wild-type cells. Specificity can also occur in an intact organism, for 'geographical reasons', even though you cannot demonstrate it in vitro: autocrine, juxtacrine, neurocrine and paracrine agents are site specific in an organism and have no remote effects as they cannot diffuse at a distance.

Finally, you should have further controls of your study conditions; these may be called milieu controls. Test subjects usually should be at rest, balanced, fasting, comfortable and properly informed on the nature of the study (see Section 3.2 of Chapter 3). Experimental animals should be handled according to regulations, adapted to their new environment in the local experimental animal centre. You must give proper consideration to diurnal physiological variation of variables, preferably by recording data at the same time each day. Anaesthesia, analgesia, circulatory and respiratory parameters, fluid and electrolyte concentrations should all be controlled and standardized. These precautions are parallel to those of the physical and chemical conditions for in vitro tests.

4.8 REPEATABILITY, REPRODUCIBILITY AND RELIABILITY

In the design of an analysis, structure must be distinguished from variation. Structure means that which is sought about a problem, such as if there is a difference between a new and an old drug or if there is an association between the concentration and the effect of a drug. Variation arises because test entities are observed with measuring instruments that never are completely precise and because we observe with other sources of random error. If variation is small, structure is readily observable. Large variation can complicate the task.

In assessing the variation or the random error of observations, a distinction is made between repeatability and reproducibility. Repeatability is understood as the degree to which the same results are seen when measurements are repeated under identical experimental conditions. In turn, identical experimental conditions denote identical methods and observers. Repeatability is expressed by the variation coefficient $CV = SD/\bar{X}$, where $\bar{X}$ is the mean and SD the standard deviation of the observations, defined by the expressions

$$\bar{X} = \frac{1}{n} \sum_{i=1}^{n} X_i \quad \text{and} \quad SD = \sqrt{\frac{1}{n-1} \sum_{i=1}^{n} (\bar{X}_i - \bar{X})^2}$$

To understand the concept of repeatability, note also that whenever data follow a normal distribution, about 95% of the observations will fall within $\bar{X} \pm 1.96SD$.

Reproducibility is understood as the degree of variation when the study conditions change. It may, for example, depend on the measurement method or on the observer. Hence, in planning a study, it is wise to consider the factors that may cause variation. For instance, in a study of the changes in blood pressure, we must assess the aspects of the measurements that may cause variation. They may include considerations of whether the subjects are in a supine or sitting position, of the measurement apparatus used and of the time of day when the measurements are made. After all such matters have been considered, we can assess the aspects that are vital and create variation and perhaps control them in the design of the experiment.

Whenever the variation of the experimental method is unknown, it is wise to start with a preliminary study that enables assessment of the repeatability and the reproducibility of the measurements to be made. Measures of reliability are meant as means to identify and estimate the reproducibility, say, over measurement apparatus or positions. For instance, an assessment of the degree to which a supine or sitting position affects blood pressure measurement should be made before a blood pressure study starts, by performing measurements on test subjects. In this way, a measure is attained that permits studying the variations between measurement methods as well as between test subjects. The measure of reliability is calculated as the intraclass correlation coefficient (ICC) for continuous data and as Cohen's kappa coefficient for categorical data. Cohen's kappa coefficient is a variety of ICC for categorical data, so it is sufficient to describe ICC briefly.

Consider an assessment of the reliability of an experimental method. A reproducibility study is conducted by repeating measurements on the same test subjects using different methods of measurement. The concept of the computation of reliability measures such as ICC or Cohen's kappa coefficient is as follows. Let S_B^2 represent the variations (variance) among test subjects. This is called the between-subjects variation. Further, let S_W^2 be the variation associated with test subjects, which is called the within-subjects variation. It arises because several measurements are performed on the same test subject. ICC is the ratio of the between-subjects variation and the total variation. Accordingly, ICC is a measure that varies between 0 and 1, and the closer to 1, the higher the reliability. High reliability is attained when the within-subjects variation and hence the variation between measurement methods is small.

Two measurements are performed on each test subject, one with each of two test methods. This gives the following set-up with n test units. Here, X_1 and X_2 are the measurements using the two methods, $\overline{X}$ is the mean of the two measurements, and S^2 is the variance among the test subjects and thereby between the test methods. The results of the study are given by:

$$
\begin{array}{cccc}
X_1 & X_2 & \overline{X} & S^2 \\
X_{11} & X_{21} & \overline{X}_1 & S_1^2 \\
X_{12} & X_{22} & \overline{X}_2 & S_2^2 \\
\vdots & \vdots & \vdots & \vdots \\
X_{1n} & X_{2n} & \overline{X}_n & S_n^2
\end{array}
$$

Then $S_i^2 = 1/2(X_{1i} - X_{2i})^2$ for the within-subjects variation for test person i, and the total within-subjects variation is $S_W^2 = 1/n \sum_{i=1}^{n} S_i^2$. Further, $S_B^2 = 2/(n-1) \sum_{i=1}^{n} (\overline{X}_i - \overline{X})^2$ is the variation between the subjects. Then we have an estimate of the intraclass correlation as $(S_B^2 - S_W^2)/(S_B^2 + S_W^2)$.

Example: A study aims to assess the reproducibility of blood pressure measurements with respect to a supine or sitting position. This is done by measuring the blood pressure of all test subjects in both a supine (measurement 2) and a sitting (measurement 1) position. All test subjects are measured in both positions so that variations both within and between people can be estimated. The measurements are listed in Table 4.1.

In this study, $S_W^2 = 2.25$, and $S_B^2 = 101.03$. The variation between subjects is far greater than that within subjects, so high reliability is anticipated. In addition, ICC $= (101.03 - 2.25)/(101.03 + 2.25) = 0.96$.

Of course, there is no one answer to the question of whether a reliability check is required to ensure agreement between two sets of measurements, but a guideline is given in Box 4.6.

Box 4.6 shows that there is nearly perfect agreement between the measurements of the example above for the supine and sitting positions.

Table 4.1 Blood pressure measurements of 10 test participants in two positions

PERSON	POSITION 1	POSITION 2	MEAN	VARIANCE
1	93	91	92	2.0
2	78	79	78.5	0.5
3	84	81	82.5	4.5
4	72	69	70.5	4.5
5	83	85	84	2.0
6	90	91	90.5	0.5
7	80	80	80	0.0
8	91	89	90	2.0
9	82	79	80.5	4.5
10	73	75	74	2.0

Box 4.6 Reliability: degree of agreement between two measurement methods or situations
- 0.00–0.20: poor
- 0.20–0.40: moderate
- 0.40–0.60: good
- 0.60–0.90: very good

Reliability computations are particularly requisite in tests of the reproducibility of differing measurement methods or in tests with repeated measurements on the same test subjects. Another example is analysis of the interrater reliability of the agreement of data recorded on the same test participants by different observers.

4.9 VALIDITY, EFFECT MEASURE AND CHOICE OF STATISTICAL TEST

There are three types of validity, as listed in Box 4.7.

As its name implies, concept validity is associated with the concept being studied, while internal and external validity concern the conclusions drawn from a study. Validity entails consideration of whether the associations or differences

> **Box 4.7 Validity**
> - Concept validity
> - Internal validity
> - External validity

observed are real which, of course, is germane to all research. Validity also applies to conclusions; a large gap between conclusions drawn and data acquired may indicate poor validity. Even though a chemotherapeutic agent kills cold virus in a cell culture, you cannot conclude that it is a good medicine for the common cold, and even though professors report an average working week of 55 hours, it is not certain that they actually work that much.

Concept validity assesses the degree to which the data reflect the variables that we wish to study but cannot register directly. The problem studied should be operationalized to be suitable and adequate, and accordingly, several variables or tests may be used. This is particularly the case in social medicine and psychiatry. A variable is said to be valid whenever association is strong, such that the data are relevant to the approach.

Internal validity is associated with valid inference in a population studied and consequently has potential weaknesses, as listed in Box 4.8.

> **Box 4.8 Internal validity may be threatened by:**
> - Sample selection bias
> - Information bias
> - Statistical confounding

Bias can occur when the sample is not representative of the population we want to study. Non-response or exclusion from a study may induce bias. Loss of subjects seriously threatens the validity of a study and is a particular problem in long-duration studies and in study designs in which defection is different in the groups being compared. Results from clinical trials may be biased when the treatment group and the placebo group differ.

Information bias occurs whenever test subjects report incorrectly or when the information registered in a study is flawed in other ways. This results in misclassification of variables, which is particularly serious whenever misclassification differs between groups studied, such as in clinical trials with treatment and placebo groups or in case–control studies.

Research in the medical and biological sciences uses statistical methods and statistical testing of hypotheses. Internal validity is weakened by improper use of methods. For example, statistical validity depends on using the right effect measures and statistical tests, so as to avoid type I and type II errors (see Section 8.3 of Chapter 8).

The effect measure is derived from the outcome variable. It is the quantity that provides the statistical description of the circumstance of interest in a study. For example, there may be a mean difference in the outcome variable between a treatment group and a placebo group. The effect estimate along with its uncertainty, presented in the form of a confidence interval, enables us to assess the significance of the results of a study. The P-value provides information on the degree to which the result of a sample can be generalized to apply to a larger study population. Many choose 5% as a threshold for the statistical significance of a result and consequently present only results having P-values less than 0.05. Usually, there is no reason to reduce the information contained in the P-value to a statement of significant or not significant, so presenting the P-value is recommended. That said, P-values higher than 0.05 may contain considerable information.

A valid statistical analysis comprises the parameters listed in Box 4.9.

Box 4.9 Presentation of main statistical results
- Effect estimate, possibly adjusted for confounding.
- Uncertainty of the effect estimate, expressed as a confidence interval (preferably 90–95%).
- Results of one or more hypothesis tests, expressed in P-values.

Most studies have more than one effect measure and entail more than one statistical test. If the limit for rejecting a test is set at 0.05, the probability of rejecting at least one of several tests will be greater than 0.05. So the significance levels of the individual tests must be adjusted to preclude excessive testing, such as by applying the Bonferroni multiple comparison correction.

The choice of a statistical method is closely associated with the effect measure used, which in turn is closely related to the type of data for the variables. Table 4.2 comprises an overview of effect measures and statistical tests used with various types of data. Regardless of the method used, a valid statistical analysis will contain the parameters listed in Box 4.9.

External validity concerns generalization, such as by indicating the population to which a conclusion may be generalized. It is a complex challenge, as conclusions

Table 4.2 Statistical tests for independent and dependent samples

NUMBER OF GROUPS	EFFECT MEASURE	INDEPENDENT SAMPLE	DEPENDENT SAMPLE (REPEATED MEASUREMENTS)
Nominal data			
Two or more groups	Relative risk/ odds ratio	Chi-squared test, logistic regression	McNemar's test
Ordinal data			
Two or more groups	Difference in medians	Wilcoxon–Mann– Whitney test, Kruskall–Wallis test	Wilcoxon sign test, Friedman's test
Count data			
Two or more groups	Incidence rate ratio	Poisson regression	
Continuous data			
Two or more groups	Difference in means, difference in medians	t-Test, ANOVA, Wilcoxon–Mann– Whitney test, Kruskall–Wallis test	Paired t-test, Wilcoxon sign test, Friedman's test
Continuous explanatory variable	Regression coefficient	Linear regression	Repeated measurements

depend on design, population and statistical model. The relation between the sample and the population is essential. Yet, regardless of how meticulous we have been in planning and analysis, a discussion of external validity is to a great degree marked by assessments and speculations and not clear criteria. That said, conclusions should be explicit, such that the reader may assess the external validity.

4.10 EXPERIMENTAL PROTOCOL

An experimental protocol is usually required as part of an application for research funding. It takes perseverance to think meticulously through all the possibilities and consequences of a choice of procedure, instead of rushing to the laboratory and starting work. Yet you have to. In any case, ethics committees, experimental animal panels, database authorities and the like will require a project description before you are allowed to start even preliminary research. Even if risk assessment is not imposed, it is wise to think ahead, not least to minimize the chance of failure in practical work (or of replicating work already done), as well as minimizing the number of unproductive pilot trials. (Count on the project taking three times as much time as you anticipate at the outset.)

You may supplement the basic experimental protocol of Box 4.10 with aspects discussed elsewhere in this chapter, as in Box 4.3 on the checklist for an experimenter. Consult others, perhaps by holding a seminar in your research group and including in it technicians who may put in a word or two. Medawar (1979) believed that: '... technicians are colleagues in collaborative research: they must be kept fully in the picture about what an experiment is intended to evaluate and about the way in which the procedures decided upon by mutual consultation might "conduce to the sum of the business" (Bacon).'

> **Box 4.10 Experiment protocol: a checklist**
> - Project title?
> - Problem and goal: the what and why of the project.
> - Background: what is now known? With references to published work.

- The experimental set-up: the how of the project.
- Material: what will be subjected to the trial (patients, animals, organs, cells, subcellular systems, chemicals)?
- Variables: what will be measured – and when?
- Analysis: how will data be processed?
- Presentation of results: scientific journal or report or perhaps also a popular press article? Order of authors' names?
- Financing and administration: who is paying; who is responsible for what; what are the deadlines?
- Time plan: when will the project finish?
- If a grant application is involved, this list might be supplemented with:
  - Why the project should be conducted, why it is important.
  - Indication of expertise: are all procedures to be used routine in your group or cooperating groups, or do you intend to devote time and resources to establishing special procedures?

You should describe the scientific problem or the goal of your project, clarify how you intend to find answers, explain what has been done previously in the field and analyse the consequences and possible results of your suggested set-up. After describing the experimental design, material and methods, you can substantiate the study's reliability and validity (discussed above and in Box 4.11). In any event, you should think through validity and reliability, even though they are not specifically included in your experimental protocol.

Box 4.11 On scientific conclusions

'... to establish firmly the logical relationship between the hypothesis and the specific prediction following from the hypothesis. It is no use spending a great deal of time, effort and money for the purposes of designing and executing a series of experiments to answer a particular question if, at the end of all this, answering that question is not a decisive test of the hypothesis. For example, suppose the hypothesis is that a high-fat diet can result in premature death. Suppose also that it is known that diets rich in fat can cause obesity. To show that obese subjects have a decreased life expectancy in comparison with non-obese controls does not necessarily support the original hypothesis. Obese individuals might differ from controls not only in the size of their fat stores but also in the intake of some other foodstuffs. A decisive test of the hypothesis would require the two groups of individuals to be identical in all factors (including degree of obesity and dietary habits) except for fat intake ...'

(Scott & Waterhouse 1986)

How extensive or numerous should your experiments or trials be? Answering that question often entails compromising your goals, as dictated by the time, funds and resources available. In any case, the research material must be as large as the power calculation implies, to preclude the risk of type II error, that is, the amount of data is inadequate to reveal a real and significant difference between test and control.

Think of how you may avoid systematic error. Will the essential variables change simply because of time elapsed? Will they, for instance, change with diurnal cycles, the seasons of the year, apparatus wear, lapse of biological material and chemicals, or the effects of repeated measurements on biological preparations? Observer bias is to be avoided, as discussed above and in Box 4.12. Subjective evaluations, such as differential counts of white blood cells, must be blind, that is, with hidden identification of preparations. Trials with human subjects should be double-blind so that neither the test subjects nor the research leaders are aware of whether a subject is in the test group or the control group.

Box 4.12 Example of observer bias ('experimenter effects')

The experimenter effect is particularly prominent in behavioural research, where people exchange signals unintentionally, without speaking. These signals may be transmitted by gestures, by auditory or visual channels, by touch or even by smell. In every experimental situation the experimenter may thus convey to the subjects his (or her) feelings without even knowing that he has done so. ...

Rosenthal and his collaborators carried out experiments designed to detect the experimenter effect on rats. ... They had to train rats in seven different tasks. The experimenters were deliberately biased by having been provided with false information that some of the rats were 'bright' and the others were 'dull' while in fact all the rats were from the same colony, were of the same age and sex and had performed similarly. The 'intelligence' of the rats was said to have been determined in previous maze running experiments. Eight teams were given rats described as 'bright', and six teams were told their rats were 'dull'. At the end of the experiment, the experimenters had to rate themselves, as well as the rats. It turned out that the experimenters believing their subjects to be generally 'bright' observed better performance on the part of the rats and rated themselves as more 'enthusiastic, friendly, encouraging, pleasant and interested' in connection with the performance of their rats, than the experimenters working with 'dull' rats. The differences between the two groups were statistically significant. The explanation given to the experimenter effect in rats was that the rats defined as 'bright' and supposedly performing better, were liked better by their experimenters and were therefore touched more. Indeed, Bernstein showed in 1957 that rats learned better when they were handled more by the experimenters. If mere physical contact could affect the learning behaviour of rats surely more dramatic effects may be expected in human experimentation.

(Kohn 1986)

You should state how you will process the results. Which statistical test will you use (see Table 4.2)? Will your methods be parametric or non-parametric? Will you use regression analysis? Will you control for multiple significance testing using the Bonferroni multiple comparison correction or another correction? Will you use a special test that permits a check of whether you have attained statistical significance on the way in an experimental series, before you continue to accumulate new data?

It is wise for collaborators to agree in advance on responsibilities and deadlines, on who will compile the first draft to a scientific paper, on who is entitled to be a co-author according to the Vancouver guidelines at www.icmje.org (also see Section 2.5 of Chapter 2) and on the order of authors named. You should also allow for unanticipated changes on the way that could result in rearrangement of these matters.

Trials on human subjects should be cleared by the relevant ethics committee, and often the test subjects must be insured. Likewise, experiments on animals must be authorized by the relevant agency (see Section 3.4 of Chapter 3).

All concerned should have approved the protocol and have a copy of it.

As the research protocol is constructed, it is also wise to compile a register form for the individual experiments. The form should include spaces for trial number, title, data, goal of the trial and reference to detailed description of the method. Moreover, it should include routine data that are recorded on each day of the experiment, such as the time for the various partial procedures, the cell concentrations, the room temperature, etc. Finally, there should be space for conclusions, so that you are obliged to learn from each experiment. You can staple instrument printouts to the register form. The form and its attachments should be collected in a protocol with fixed pages. The protocol should have a table of contents in which trial numbers and titles are entered sequentially. Increasingly, however, electronic notebooks are taking over for these protocols, in combination with digital storage of research data. The following guidelines, based on ones prepared by the University of Newcastle upon Tyne, illustrate how strict the rules for protocols may be.

1. Records of primary experimental data and results should always be made using indelible materials. Pencils or other easily erasable materials must not be used. Where primary research data and results are recorded on audio or videotape (e.g. interviews), the tape housing should be labelled as set out in paragraph 4.
2. Complete and accurate records of experimental data and results should be made on the day they are obtained and the date should be indicated clearly in the record. When possible, records should be made in a hard-backed, bound notebook in which the pages have been numbered consecutively.

3. Pages should never be removed from notebooks containing records of research data. If any alterations are made to records at a later date they should be noted clearly as such, the date of the alteration stated and the alteration signed by the person making it.

4. Machine printouts, photographs, tapes and other such records should always be labelled with the date and with an identifying reference number. This reference number should be clearly recorded in the notebook referred to above, along with other relevant details, on the day the record is obtained. If possible, printouts, photographs, tapes and other such records should be affixed to the notebook. When this is not possible (e.g. for reasons of size or bulk), such records should be maintained in a secure location in the university for future reference. When a 'hard copy' of computer-generated primary data is not practicable, the data should be maintained in two separate locations within the university, on disk, tape or other format.

5. When photographs and other such records have been affixed to the notebook, their removal at a later date for the purpose of preparing copies or figures for a thesis or other publication should be avoided. If likely to be needed, two copies of such records should be made on the day the record is generated. If this is not practicable, then the reason for removing the original copy and the date on which this is done should be recorded in the notebook, together with a replacement copy or the original if this can be reaffixed to the notebook.

6. Custody of all original records of primary research data must be retained by the principal investigator, who will normally be the supervisor of the research group, laboratory or other forum in which the research is conducted, and who shall follow any instructions on confidentiality issued by an appropriate ethics committee. An investigator may make copies of the primary records for his or her own use, but the original records should not be removed from the custody of the principal investigator. The principal investigator is responsible for the preservation of these records for as long as there is any reasonable need to refer to them, and in any event for a minimum period of 10 years.

Some scientists recommend that you always carry a notebook, in which you can jot ideas that come at unexpected moments. It is said that Loewi dreamed of how he could prove the existence of the postulated 'Vagus stuff' (that we now know is acetylcholine) that was believed to be exuded from the vagus nerve to the heart, causing it to beat more slowly. He woke, made a note of a research plan, and fell asleep again. The next morning, he could not understand what he had written. The dream came again. He woke, went to his laboratory and set up two frog heart preparations. Both hearts beat in a Ringer saline solution, one with and one without the vagus nerve. He electrically stimulated the vagus nerve of one heart for a couple of minutes, so that its beat slowed. He then transferred the saline solution from it to the other heart, which then also began to beat more slowly, as if it had been vagus stimulated. This proved the existence of a Vagus stuff (Mazarello 2000).

4.11 EXPERIMENTAL ROUTINE

You are now conducting laboratory experiments, collecting data. Try to establish workable routines, which further efficient work, and promote cleanliness and a physically and socially pleasant working environment. Laws, regulations and common sense should regulate the storage, use and disposal of drugs, poisons, radioactive materials, microbes and other chemicals. In this it pays to be meticulous. An otherwise worthy experiment is worthless if you use unstable chemicals or biological materials that have lost their potency because they have been improperly stored or handled, such as storage at high temperature, repeated freezing and thawing of proteins, failure to use desiccants or the premature opening of deep-frozen containers so that moisture condenses on their contents. If you doubt that your reagents are fully active, buy new. Your working time probably costs more.

Cleaning can be critical. Traces of aluminium or detergents can ruin the contents of a reagent tube or Petri dish. Some plastics adsorb proteins and peptides, as well as cells if the medium has no proteins. So, you should check for abnormal loss of key ingredients during cleaning, concentrating solutions or separation procedures. Biologically active substances such as mycoplasma or endotoxin can cause misleading results in many types of trials. So you should routinely test for the absence of these 'robbers'. If you must sterilize the experimental material, will it tolerate autoclaving, dry heat, X-ray radiation, boiling or ethylene-oxide gas treatment? Or might you be obliged to filtrate solutions that cannot be bought sterile?

The responsibility for routine controls or calibration, as discussed in Section 4.7 above, can be distributed among the staff of your group. For instance, even electronic scales should be checked twice a year, preferably with old-fashioned physical weights.

Ensure proper documentation of your results, raw data as well as detailed procedure descriptions, to enable you accurately to re-establish the experiment at some later date. Finally, do not believe that you can remember everything you have done up to the time you write the final paper, not least because that may be later than you think!

REFERENCES

Austin JH (1978) Chase, Chance and Creativity, The Lucky Art of Novelty. Columbia University Press, New York.

Klein J (1985) Hegemony of mediocrity in contemporary sciences, particularly in immunology. Lymphology 18: 122–131.

Kohn A (1986) False Prophets. Basil Blackwell, Oxford.

Mazarello P (2000) What dreams may come? Nature 408: 523.

Medawar PB (1979) Advice to a Young Scientist. Human Action Wisely Undertaken. Harper & Row, New York.

Scott EM, Waterhouse JM (1986) Physiology and the Scientific Method. Manchester University Press, Manchester.

Thomas L (1974) The Lives of a Cell: Notes of a Biology Watcher. Viking Press, New York.

FURTHER READING

Altman DG (1991) Practical Statistics for Medical Research. Chapman & Hall, London.

Beveridge WIB (1974) The Art of Scientific Investigation. Heinemann, London.

Medawar PB (1969) Induction and Intuition in Scientific Thought. Jayne Lectures for 1968. American Philosophical Society, Philadelphia.

Strike PW (1996) Measurement in Laboratory Medicine. Butterworth-Heinemann, Oxford.

White VP (1988) Handbook of Research Laboratory Management. ISI Press, Philadelphia.

LITERATURE SEARCH AND PERSONAL REFERENCE DATABASES

Anne-Marie B. Haraldstad and Ellen Christophersen

'If I have seen further it is by standing on the shoulders of giants'
Isaac Newton (*The Oxford Dictionary of Quotations*)

Research does not occur in a vacuum, but is part of a professional reality and builds upon previously acquired knowledge. It requires reliability, originality, professional relevance and usability. If the results are to be credible, it is essential that the research is based on methods and statistics appropriate to the approach or hypothesis under discussion and testing. In other words, it emphasizes the need for a systematic approach, systematized quality-controlled implementation and, not least, solid documentation. What about literature searches? Should equally strict demands be made here or are literature searches the weakest link in the research chain? If the same requirements are not demanded of the literature search as of the research protocol itself, in the worst case the work will have a weak foundation and its credibility will be in doubt since valuable information and background material may have been overlooked.

5.1 INFORMATION LITERACY

The body of biomedical information is complex, voluminous and growing all the time. In today's flood of information, knowledge and expertise are required to sort out that which is relevant and reliable. Handling information is therefore a skill that should be developed along with professional competence. Information literacy entails identifying the information required, the location of suitable sources, the critical assessment of information and its organization and efficient use. In summary, information literacy is comprised of the ability to:

- recognize a need for information
- identify and locate relevant sources of information
- know how to access sources
- evaluate the quality of information
- organize information
- use information efficiently.

It deals with a carefully considered relationship to professional information and a deliberate choice of sources of information. The aim is to use the best sources first, be they books, journals, newspapers, radio and television, communications with colleagues, etc. Information literacy concerns the use of all available sources of information. It concerns establishing a plan and a structure to avoid random and fortuitous results.

Searching by content or by methodology: two approaches to a literature search

Most literature searches are conducted on specific topics, e.g. on biological subjects, diseases or drugs, and the terms used during the search are put together in various logical combinations. Such searches do not differentiate between the various types of publications. The results are displayed at random, and make no discrimination between studies concerning human relationships, animal experiments, laboratory tests, cases, protocols, primary and secondary sources, and basic and clinical research. This may be a workable approach if the aim is to find 'everything'. However, if the intention is to find appropriate literature as a basis for clinical practice, it is essential that the literature fulfils the criteria for clinical research. As the majority of the references in Medline, for instance, concern basic research, specific search strategies must be used to sift out clinically relevant literature.

Hence, 'methodology filters' have been developed for various categories of clinical queries, e.g. diagnosis, treatment/prevention, prognosis, aetiology and qualitative studies. The filters comprise lists of key concepts of methods and statistics that together reflect the research design corresponding to the clinical query involved.

5.2 SYSTEMATIC LITERATURE SEARCH

This section describes a method of carrying out a systematic literature search, i.e. a systematic approach to literature retrieval that is both methodical and reasoned. As with other research procedures, literature searches should be documented and verifiable. There are no set ways of going about this, so documentary procedure is germane. The procedure described in the first part of this chapter is not dependent on databases and is transferable; thereafter, the chapter describes the principal databases used in medicine.

Identify the best source

Literature searches may be conducted using electronic, print sources or both. No one single database provides answers to all medical queries. Medline is the best known of the principal biomedical databases, but it should not be used to the exclusion of other potentially better sources of information which, if ignored, can introduce systematic distortion in research.

A structure based on principal concepts is essential to establishing a systematic search strategy.

Primary studies and primary resources

Primary resources are databases or other information sources in which the bulk of the content comprises individual studies, so-called 'primary studies'. A primary study may present skilled, methodical research, but its conclusions apply to that specific study only. Examples of primary resources are Medline, EMBASE and other general databases.

Secondary studies and secondary resources

Secondary studies are based on research carried out on existing research. Here, the approach is based on existing studies, both published and unpublished, and the data are coalesced for analysis and to draw conclusions. The results comprise systematic reviews, meta-analyses, clinical guidelines or economic evaluations. The conclusions drawn from such secondary studies will have greater significance and will embody the understanding of a topic in a manner beyond that possible in a single study.

Secondary resources are databases or other sources of information which provide access to comparisons of primary studies. The Cochrane Library and Clinical Evidence, as well as secondary journals such as ACP Journal Club and several evidence-based journals (Evidence-Based Medicine/Dentistry/Nursing, etc.), are examples of secondary resources.

Reviews and systematic reviews

A review article may or may not be systematic, as it depends on the method of collating primary studies. A review article presents the results and conclusions of two or more publications on a particular topic. If the authors fail to explain their inclusion criteria and procedure, the reader should question whether essential studies have been omitted, and treat the conclusions accordingly. The method is open to manipulation as the authors may promote their own theories in an impressive manner, having deliberately selected studies supporting these theories. However, the results are never better than those of the primary studies on which they are based. On the other hand, a conventional review article can offer a good, broad introduction to a new topic.

Systematic reviews and meta-analysis

Systematic reviews often cover a more limited subject area. They are a product of work which aims to provide an overview of the collated literature on specific topics and to draw conclusions from these articles. The method's section of a systematic review should include a description of the procedures and search strategies used to collate the individual studies of a specific, relevant topic. Often, search strategies are listed from Medline and other databases. Documented, systematic procedure is requisite in the location, evaluation and summary of the individual studies.

Meta-analyses use statistical methods to analyse the results of several studies as though they were one single study. This method requires that these studies have comparable research designs and figures. If this is not the case, the publication will be a systematic review, not a meta-analysis. Not all meta-analyses build on systematic literature searches. Again, an examination of the method's section will reveal the breadth and depth of the initial literature search. The relationship between the different types of study can be illustrated in Figure 5.1, where the best studies are those within the shaded area.

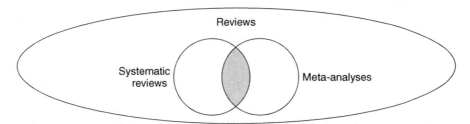

Figure 5.1 Types of review article

5.3 HOW TO FORMULATE A QUERY: PICO

Each and every search begins with a query. Formulating a concise query is one of the most important steps in the search process. A literature search, begun with only a vague concept of what to look for, may well produce vague results. It is therefore time well spent to structure a search in advance. New ideas on search strategy often come to mind during a search, but that does not preclude the clarification of concepts in advance. Literature searches are conducted on each database's own premises, and a search strategy comprising several components will improve the conduct of the search and provide a better perspective over the strategy used.

Begin by dividing the query into groups of keywords and synonyms and use a thesaurus to find relevant search terms. In the example here, MeSH (Medical

Subject Headings) terms are used to search in Medline. Be aware that other databases use other subject terms. For example, EMBASE has its own thesaurus, EMTREE.

PICO: a focused approach

PICO (the acronym for population/patient/problem, intervention, comparison, outcome) is a useful tool in formulating a search strategy. Although searches may not have all four constituent components of PICO, most will have two or three of them.

The details of an exhaustive literature search – not all of which are involved in all searches – are listed in Boxes 5.1 and 5.2. The search sequence of Box 5.1 concerns the details common to basic and clinical research, while ancillary details relevant to clinical research are given in Box 5.2.

Example of a focused query: Does urban air pollution cause an increase in the mortality of a population?

Box 5.1 Systematic literature search in basic and clinical research
- What is the question? An approach which focuses on the problem will define the search elements.
- The subject determines which databases and sources of information are the most suitable, in order of relevance.
- Time-frame: how far back is far enough?
- Listing of relevant print-based sources.
- Fully exploit the sources: knowledge and search skills.
- Assess the quality of the search results. Try a variety of approaches to the subject.
- Both during and after finishing work, repeat the search to catch the latest information. Identify possible retracted studies.
- Store or file literature references using bibliographic software or compile a bibliography.

Box 5.2 Ancillary details of literature searches in clinical research

- What is the principal clinical question (diagnosis, treatment, prognosis, aetiology, lived experience, healthcare costs)?
  - Which research design answers this question best?
- The research design determines which databases best address the problem.
- List in prioritized order, best source first.
- Assess the potentials of primary and secondary studies, that is, review articles and individual studies.
- Use Table 5.4 as an aid to draw up a list of the databases in order of priority whenever a search on the research design is essential for the query.

The query can be broken down and entered in a PICO form. The search concepts and synonyms are listed in the columns of the form, both in free text and from the database thesaurus, as shown in Table 5.1. Thereafter, affiliated search terms are collated with OR (vertical list). Each column is an independent search element, as shown. Finally, the searches are combined with AND across all columns.

Table 5.1 PICO form

PATIENT/ POPULATION/ PROBLEM	INTERVENTION/ EXPOSURE	COMPARISON	OUTCOME
Who? Urban populations ↓ City population Inner-city population (MeSH): Urban population	What? Air pollution ↓ Poor air quality Polluted air (MeSH): Exp air pollution Exp air pollutants	Alternatives? Rural population ↓ (MeSH): Rural population	Result? Increased mortality ↓ Death Death rate (MeSH): Mortality Survival rate Life expectancy

OR ↕ (left)   OR ↕ (right)

← AND →

The PICO form is then transferred to the search strategy, as shown. Although not all search terms are included, the structure of the search is evident, as illustrated in Figure 5.2.

Number	Parameter	Hits	Coverage
1	Urban population (kw)	28 310	Collect all concepts that
2	Urban population (tw)	1 705	characterize the patient
3	1 OR 2	29 002	group/population/problem
4	Exp air pollution (kw)	35 114	Collect all concepts that
5	Exp air pollutants (kw)	31 757	characterize exposure or
6	Poor air quality (tw)	26	intervention
7	4 OR 5 OR 6	58 760	
8	Mortality (kw)	28 799	Collect all concepts that
9	Life expectancy\(kw)	8 130	characterize the
10	Death rate (tw)	4 890	outcome/result
11	8 OR 9 OR 10	38 940	
12	3 AND 7 AND 11	75	Final combination

kw: keyword; tw: textword; exp: exploded/expanded term

Figure 5.2 Example of a search structure

Boolean operators

The AND operator narrows the search. The OR operator is used to broaden the search to collect for instance synonymous concepts. Use the NOT operator cautiously. It is a powerful operator that can eliminate valuable references.

As shown in Figure 5.3, the various combinations of results from a search can be represented by circles. The dark shadings indicate the search results using the AND, OR and NOT operators.

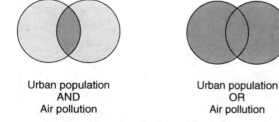
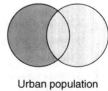

Urban population
AND
Air pollution

Urban population
OR
Air pollution

Urban population
NOT
Air pollution

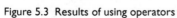

Figure 5.3 Results of using operators

The AND, OR and NOT operators are the most used. The choice of database vendor and the applications offered determine the search functionality supported by that base. The search strategy can be set up in the form: 'term *operator* term' (Table 5.2).

Table 5.2 Examples of uses of search terms and operators

TERM AND OPERATOR	RESULT
birth	Finds all documents containing that term
birth OR labour	Finds all documents containing either or both of the terms
birth AND forceps	Finds all documents containing both terms, but not necessarily adjoining
forceps ADJ delivery	The terms are adjacent to each other in the document
delivery NEARn birth	The terms are located near, within n word(s) of each other, independent of order
birth WITH pain	The terms are in the same field, e.g. both are in the abstract
back NEXT pain	The terms are adjacent to each other in the order entered
injur*	Searches the word stem and gives injury, injuries, injured, etc.

The databases with the greatest number of available operators support the greatest flexibility. For example, a Medline search via Dialog, with its register of operators, affords greater opportunity for refining a literature search than if the PubMed interface is used to search Medline.

The operators supported by the most central databases are listed in Table 5.3.

Table 5.3 Operators supported by the principal database vendors

OPERATOR	OVID	PUBMED	COCHRANE	SILVERPLATTER	DIALOG
AND	x	x	x	x	x
OR	x	x	x	x	x
NOT	x	x	x	x	x
ADJ	x			x	x
NEXT			x		x
NEAR			x	x	x
WITH				x	x

5.4 SEARCH TECHNIQUE

Print out and keep the search history as part of the documentation. State the databases used and time spans covered.

- In planning a search strategy, remember that the indexers making the entries in the databases use the most specific subject terms to describe the content of each article. So, use a thesaurus (such as MeSH/EMTREE) to find subject terms and their interrelations to each other.
- Remember that subject terms can have their own 'histories', that is, they may have been changed in the course of time. Check Scope Note (Medline Ovid)/MeSH Browser in PubMed, and formulate the search strategy accordingly.
- As no indexing is completely correct in all contexts, you should consider using combinations of thesaurus search and textword search.
- Some database vendors support searching several bases in one operation (Ovid: 'Select more than one database to search'). It can be a tempting short-cut. However, such a search across databases cannot take into consideration their differing structures, thesauri, types of publication, and so on. Surveys have shown that a large number of references are indexed in different ways, depending on the base, such as in Medline and EMBASE. Hence, cross-searches should be avoided whenever the goal is to find all available information on a particular topic. However, at the beginning of a work a cross-search will identify interesting resources to be exploited one by one.
- Think of alternative terms and synonyms. Do not rely on only one term.
- In addition, perform an author search.

Selection

- Use several bibliographic databases.
- Use various types of databases or secondary resources (PubMed's related articles function, US National Library of Medicine's Locator Plus, etc.).

Subject heading versus publication type

A subject heading (term) designates content, whereas a publication type delineates a category of reference or the manner in which the information is compiled. Consequently, subject heading and publication type must not be confused with each other, as their results differ. Publication type is a criterion for limiting a search, as stated under *Limits* for the individual databases.

For example, in Medline the subject heading Randomized Controlled Trials (plural) applies to articles that describe methods or discuss applications, while Randomized Controlled Trial (singular), limited as publication type, retrieves articles that actually are Randomized Controlled Trials.

The same applies for Meta-analysis (MeSH) and Meta-analysis (publication type), and Review Literature (MeSH) and Review Literature (publication type). Whenever the subject heading and publication type are identical as in these examples, Meta-analysis and Review Literature are searched in the same way as any other subjects. A search may be limited using the database's *Limits/Limit to*.

Expanding the search

- The members of a research team should search independently, as each then may find relevant, unique references.
- Check the subject headings assigned to relevant articles and re-enter them to initiate new searches.
- Expand the search using related terms (mortality/survival rate).
- Sometimes you may augment a search using contrasting terms, e.g. entering 'rural population' in a search concerned with 'urban population'.
- Consider broadening the timescale of the search, and check to see whether the subject headings entered have changed with time.
- Be aware of how various databases handle expanded search statements (Explode). In PubMed, all search statements are automatically expanded whenever possible, but in Medline Ovid, search statements are expanded only upon command.

Narrowing the search

- Consider whether the term should restrict hits only to articles for which it describes principal content. Limit by using *Focus of the article* (Ovid Medline)/ *Major topic* (PubMed). Consider: in literature searches on a single topic, limit to the major topic. In literature searches on more than one topic, do not limit to the major topic.
- Be cautious in limiting to subheadings. Try to limit the search in other ways. For example, by 'human', or publication type, methodology filters, etc.

- For studies concerning humans only, use the 'human' limit. Limiting also with 'not animal' further reduces the number of hits, because many articles are assigned indexing tags of both 'human' and 'animal'.
- Be cautious in limiting to a specific language, to preclude language bias, as valuable information may be overlooked.
- The delineations of age groups differ between databases, so be sure to use the designation applicable to the database in use.
- Limiting to age groups, human or animal studies, language and publication type is carried out at the end of a search, once the formulation of the subject search is completed.

Free text search

Free text search entails search in available text for individual words, as in titles and abstracts. Hence, the search strategy must take the following into consideration:

- Synonyms, such as those arising from differing terminology between countries or disciplines (e.g. bed sores/decubitus ulcer; SIDS/cot death).
- Truncation (wild card): includes singular and plural forms, and other orthographic endings; for instance, inju$ will search injury, injuries or injured. Remember that truncation symbols may differ from database to database, such as * $: ?
- British versus American spelling (tumour/tumor). Search both: tumo?r.
- Abbreviations of terms (AIDS/acquired immunodeficiency syndrome).
- Trademarked or generic name (Tamiflu/oseltamivir phosphate).
- A free text search also elicits false drops, as words may be used in other connotations than those relevant to the query at hand.

5.5 METHODOLOGY FILTERS

Each year, millions of medical publications are released. Their quality varies. The basis for work of a high quality lies in research work with a thorough basis, carried out to a high standard. Consequently, the higher quality publications must be sifted out by formulating a search strategy that in addition to the topic in

> **Box 5.3 Methodology filters in brief**
> - One filter for each study design
> - Devised and tested to retrieve high-quality research papers
> - Reduce the number of hits in a search result
> - Used to elicit quality rather than quantity
> - Methodology filters are adapted to the structures of individual databases

question focuses on the methods described. This is particularly the case in clinical research and practice, known as evidence-based medicine (EBM).

Methodology filters (see Box 5.3) are based on previous systematic literature searches for a specific time span, in which the goal is to find search strategies that yield optimal results from each of the various research designs. Methodology filters are so formulated that the search result closely approaches the 'gold standard', i.e. a completely successful search. Each methodology filter comprises the words and concepts that together delineate the research design.

There is no standard structure for databases, so methodology filters reflect the existing diversity of databases. Because of this, methodology filters must be constructed for each individual database and must suit the search terms and publication types used by that base. Collections of methodology filters are available on the Internet, such as:

PubMed filter table: www.ncbi.nlm.nih.gov/entrez/query/static/clinicaltable. html

Centre for Evidence-Based Medicine: www.cebm.net

Methodology filters: shortcuts

Methodology filters comprise general search strategies that describe research designs and consequently can suitably be combined with any subject search seeking a specific research design. The filters may consist of comprehensive search strategies, so shortcut filter access should be used whenever possible, so as to avoid having to key in an entire methodology search. In databases, methodology filters are available via clinical queries.

PubMed filters

In PubMed, methodology filters are implemented as automatic searches accessible in Clinical queries. The filter tables list the search parameters of the individual filters. PubMed has filters for:

- aetiology
- diagnosis
- therapy
- prognosis
- clinical prediction guides
- systematic reviews.

In addition to the choice of search filter, *Sensitivity/Broad* or *Specificity/Narrow* may be checked. A literature search that focuses on sensitivity will elicit a greater number of references than a corresponding search focused on specificity.

- *Sensitivity*: a broad search that captures as many relevant references as possible.
- *Specificity*: a narrow search that weeds out as many irrelevant references as possible.

The *Systematic Reviews* filter is matched to systematic reviews and meta-analyses. The *Medical Genetics Searches* entry finds citations related to various topics in medical genetics.

5.6 QUALITY: CRITICAL APPRAISAL

Upon completion of the basic literature search, work continues by sifting information from the list of references to decide which papers, books or other media should be read in detail. At this point, it is natural to contact the local library to order publications not in your own collection. There is no simple way to determine whether a work fulfils quality requirements, but there are a few checkpoints to keep in mind in deciding whether a more thorough evaluation is required. Some concern your own work methods, the extent to which your own search technique may be said to be quality controlled and is able, for instance, to discover that some studies may have been retracted. It is also important to check whether your information comes from abstracts or from full texts. Other checkpoints relate to the study at hand: is it of sufficient quality to be included in the dossier? Critical appraisal must be used in this process. For example, visit www.phru.nhs. uk/casp/casp.htm for a description of relevant procedures (see also Chapter 15).

Abstract or full text?

Most references in bibliographic databases, such as Medline and EMBASE, include abstracts. Ordering a document in full text may be costly and take time, so it is tempting to make do with abstracts. However, as shown by Pitkin et al. (1999), it is inadvisable to rely only upon abstracts. Pitkin's study aimed to assess the degree to which the content of abstracts was consistent with the content of the articles abstracted. He found that 'the proportion of deficient abstracts varied widely (18–68%) and to a significant degree … '. Even for the renowned medical journals included in the survey (BMJ, Lancet, JAMA, New England Journal of Medicine, Annals of Internal Medicine, Canadian Medical Association Journal), the conclusion was that the content of an abstract often was inconsistent with or even in contrast to the content of the article abstracted. This indicates that basing work on information gleaned from abstracts alone is risky (Pitkin et al. 1999). Moreover, many people may maintain that it constitutes unethical research.

Errata/retracted publications

Scientists are concerned with keeping abreast of developments in their respective fields. There are several ways to keep informed, among them by using a current awareness service such as the Selective Dissemination of Information (SDI) (see Section 5.9). Another way is to subscribe to a journal's contents alerting service to receive tables of contents via e-mail. By acquiring references in this manner only, one may miss articles subsequently retracted. There can be many reasons for retracting a publication. The authors may have reasons for retracting a publication. Employers may not accept responsibility for the publication. In the worst case, the work may have been exposed as research misconduct or fraud. Regardless of the severity of the reason, research studies are retracted every year.

Yet they remain listed as references in databases, marked in various ways: retracted publication, retraction of publication, erratum.

Searches have shown that databases differ in the speed in which they trace and register such changes. Consequently, a paper may be listed in Medline as a valid reference but simultaneously be tagged Erratum in EMBASE. This is yet another reason for searching in more than one database.

The individual scientist is obliged to ensure that such incomplete or flawed studies are not included in accessed background material. Studies have shown that even though a paper has been retracted and marked accordingly in databases, it may appear in citations. Scientists who glean citations from reference lists published by others, not from their own well-formulated literature searches in databases, can overlook retracted papers. Moreover, the practice is unethical. To be on the safe side, consider finishing all searches with 'limit to retracted publications' (Medline).

Quality of reference lists/bibliographies

It is wise to note references as soon as they are accessed. Using bibliographic management software eases registration. It is better to register too many references than too few, as in the early stages of work it is not always clear which references will finally be cited.

A paramount rule is that you should refer only to sources you have read in full text. Regrettably, this is not always the case, as can be seen by the many incorrect references that are widely disseminated by authors who uncritically copy reference lists made by other authors, without checking the originals. Whenever such chicanery is exposed, the author's credibility is tarnished and the work assessed accordingly. Surveys have shown that citing such erroneous references may be widespread, which creates extra work for those who strive to find original works cited.

5.7 IMPACT FACTOR

Is it possible to assess research quality objectively?

Starting in 1961, the ISI Science Citation Index provides a basis for the impact factor (IF), which is a measure of the citation frequency of the average article in a particular journal for a given year. The IF for a specific year is calculated as the ratio of citations over a specific year generated by articles published in that journal in the previous two years.

$$\text{Impact factor} = \frac{\text{Total number of citations}}{\text{Total number of citable articles}} \text{ in the specific journal}$$

The ISI IF was never intended to be a universal criterion for quality, even though it frequently is so regarded in practice. For example, scientists may consult IFs in deciding to which journals they should submit papers. Employers may use the IF as a criterion in hiring scientific staff. The IF is also involved in evaluating journal subscriptions and in allocating research funds. Quite naturally, allocating agencies seek assurance that funds are channelled to the scientific communities in which they will have the greatest benefit, and an employer is interested in hiring the best people. It is also natural that scientists wish to communicate their works as well as possible. But whether the ISI IF shall be a decisive criterion is uncertain (Garfield 2006).

The IF should be used with care, because:

- The IF does *not* reflect the *quality* of an individual article, but the average measure of the use of a journal as a whole for the two years that are included in the calculation. Usually, only a few articles generate half of the citations to the journal. Hence, articles not cited are also fully credited.
- It includes self-citations (graphs of self- and non-self-citations are now included for each journal).
- Preclinical journals are cited more often than clinical journals. Clinical researchers often cite preclinical articles. Understandably, the reverse is less frequent, as knowledge must be acquired before it is put to clinical use.
- The IF cannot be compared across disciplines and specialities owing to differing citation practices.
- Dynamic disciplines continuously producing new findings which supplant old knowledge (such as biochemistry and molecular biology) are favoured, owing to the time limit of the two previous years which is used in the IF calculation. Work of greater longevity is accordingly less favoured.

- Journals with supplements are at a disadvantage, because supplement series usually generate fewer citations and accordingly a lower citation frequency.
- Review articles are more often cited, which favours journals that publish many such articles.
- Journals with extensive correspondence sections have an advantage, because although letters to the editor are not taken into account as a basis of calculation (the denominator) for IF, citations to them are included.
- The IF is calculated for selected indexed journals, predominantly in English and published in the USA. Estimating coverage for various disciplines is difficult, but ISI claims more than 5700 titles in the subject areas of medicine and the natural sciences, of which some 3000 are medical journals. In any case, the selection of journals indexed in this single database is small in the overall picture of journals published. According to Ulrich's Periodicals Directory, today some 15 000 current journals are published in medicine and the biological sciences around the world.
- Might unfavourable mention be preferable to no mention at all? If, for instance, many research workers wish to comment on and reject a work exposed as research misconduct, this will result in many citations of that work, which thus increases the publishing journal's IF.

Quality assessment

The IF has too many weaknesses to justify the position of the hallmark that it has attained. As a quality assessment, it is preferable to assess studies according to how many – and who – cite the papers, rather than only considering the journals in which the papers are published. There is no easy way to assess quality.

> 'A true appreciation of the validity of quality judgements will, therefore, require a knowledge of the quality of the judges, for a committee of camels will never approve a horse' (Chargaff 1975–76, p. 331).

Access to ISI impact factors

The ISI Journal Citation Reports is available on subscription. Another approach can be to use a general search engine on the Internet and use 'impact factor' and the journal name as search terms. The home pages of journals often list their IFs.

Table 5.4 Queries

QUERY		STUDY DESIGN	SYSTEMATIC REVIEWS AVAILABLE?	DATABASE/SOURCE
How many …	Prevalence	Cross-sectional study	Very few/none	1. For systematic reviews and studies, Medline and/or other general databases[a] 2. National registers and statistics
How can we determine …	Diagnosis	Cross-sectional study (with a gold standard)	Some	1. Secondary journals (evidence-based journals) 2. Other Reviews and Technology Assessments (Cochrane Library) 3. Medline and/or other general databases[a]
How will it turn out …	Prognosis	Cohort study	Very few/none	1. Secondary journals (evidence-based journals) 2. Medline and/or other general databases[a]
				For queries on the side-effects of treatments (as with specific drugs)
Why …	Aetiology	1. Cohort study 2. Case–control study 3. Patient series	Some	1. Clinical Evidence 2. Secondary journals (evidence-based journals) 3. Other Reviews and Technology Assessments (Cochrane Library) 4. Medline and/or other general databases[a] Queries on other aetiologies: 1. Secondary journals (evidence-based journals) 2. Medline and/or other general databases[a]

(Continued)

Table 5.4 (*Continued*)

QUERY	STUDY DESIGN	SYSTEMATIC REVIEWS AVAILABLE?	DATABASE/SOURCE
What can be done …	Effect of intervention (treatment, prevention, rehabilitation)	Many	1. Clinical Evidence 2. Cochrane Reviews 3. Other Reviews and Technology Assessments (Cochrane Library) 4. Clinical Trials (Cochrane Library) 5. Medline and/or other general databases[a]
	1. Randomized controlled trial 2. Controlled trial without randomization 3. Cohort study 4. Case–control study		
How is it experienced …	Experience	Very few/none	1. Evidence-based nursing 2. CINAHL (good coverage in qualitative research) 3. Medline and/or other general databases[a]
	Qualitative studies		

[a]In this context, 'general databases' means those not limited to specific study designs.

5.8 PRINCIPAL BIBLIOGRAPHIC DATABASES

Choice of database

Several studies have spotlighted instances in which it is unwise to limit a search to a single database, such as Medline. A search in a single database may fail to find relevant information. It is wise, therefore, to search several databases to ensure the best possible perspective. Often, a database may be chosen according to the number of journals it indexes. Moreover, the characteristics, strengths and weaknesses of each database should be evaluated (Table 5.5).

Access

Most bibliographic databases require a subscription. The databases available to the staff within an organization usually depend on the contract the organization has with its database vendors. Contact your organization's library for details of the databases available and for assistance in accessing them.

Medline

Medline, maintained by the US National Library of Medicine (NLM), is the largest and most widely used bibliographic database in the medical and biological sciences. It contains more than 14 million references to articles published in more than 4900 journals and covers the fields of medicine, nursing, dentistry, veterinary medicine, pharmacy, healthcare and preclinical fields as well as associated disciplines. It is international, but most of the references (approximately 89%) are from journals published in English. Medline contains references from 1966 to the present day, and approximately 80% have abstracts in English. Medline is accessible via interfaces provided free by public vendors, such as PubMed, and through paid subscriptions via commercial vendors, such as Ovid and Silverplatter.

Medical Subject Headings: a useful search tool

MeSH is NLM's thesaurus, with approximately 24 000 descriptors, which are controlled subject headings in biomedicine. A thesaurus is a list of terminology for a particular field, with a set of terms or phrases connected with each other via general, specific and related terms. In a database, a thesaurus is organized in a tree structure (see Box 5.4) that is searchable at various levels, from the narrow (more specific) to the broader (more general), and the reverse.

MeSH terms are organized both by alphabet and by hierarchy. At the general level of the hierarchy are terms such as 'Anatomy', 'Nutrition disorders' and 'Enzymes and coenzymes'. At the more specific level are terms such as 'Ankle', 'Avitaminosis' and 'Calcineurin'. MeSH is updated once a year to reflect changes in medicine and medical terminology.

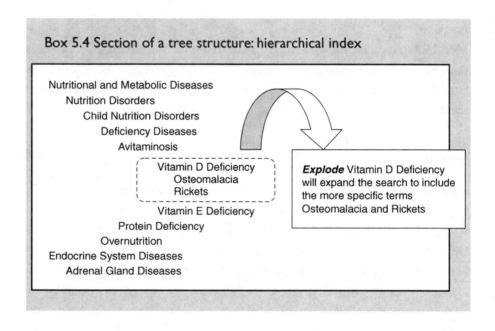

Box 5.4 Section of a tree structure: hierarchical index

Nutritional and Metabolic Diseases
  Nutrition Disorders
    Child Nutrition Disorders
    Deficiency Diseases
      Avitaminosis
        Vitamin D Deficiency
        Osteomalacia
        Rickets
        Vitamin E Deficiency
      Protein Deficiency
    Overnutrition
  Endocrine System Diseases
    Adrenal Gland Diseases

Explode Vitamin D Deficiency will expand the search to include the more specific terms Osteomalacia and Rickets

MeSH terms, or descriptors, are used to index the content of each article in Medline. There may be 10–20 descriptors associated with each article. Together, they provide useful ancillary information on the article, beyond that provided in its abstract. They also may afford an entry to other relevant search terms.

Search via MeSH involves synonym control and yields precise search results: one MeSH term spans many concepts, so a single term is sufficient for a search, regardless of which words or concepts the author(s) of the article may have used in the title or abstract or as a keyword. For example, a search using the MeSH term 'cerebrovascular accident' will cover all words and concepts associated with stroke. Applying the *explode* function will include all subordinate aspects of a general subject term. A *subheading* may be associated with a subject term in order to search for one or more aspects of a subject, such as:

- molecular biology/methods
- stem cell transplantation/ethics
- low back pain/radiography
- myocardial infarction/drug therapy.

PubMed

The Entrez text-based search and retrieval system developed by the National Center for Biotechnology Information (NCBI) at the NLM affords free access to several major databases, including PubMed, Nucleotide, Protein, Structure, Genome, Taxonomy and OMIM.

PubMed (www.pubmed.gov) provides access to the entire Medline from 1966 on, and in addition, includes OldMedline back to 1950. PubMed is updated daily with reference data supplied directly by publishers, often before a journal issue is released. These references often are available with abstracts, but they are not fully indexed in the Medline database and are not immediately assigned MeSH terms. Accordingly, PubMed is more comprehensive and up to date than Medline via other access means.

PubMed references are linked to full text articles whenever publishers have an agreement with NLM on electronic access via PubMed. Individual users may access the electronic journals to which their organization(s) subscribe.

Differences between PubMed and Medline are listed in Box 5.5.

OldMedline

The NLM also maintains OldMedline, a database containing references from biomedical research journals published from 1950 to 1965. References from Index Medicus are converted continuously and added to the database, although without abstracts. Limiting to the OldMedline subset enables access to the database.

> **Box 5.5 Differences between PubMed and Medline**
> - Medline includes biomedical articles published in some of the more general scientific journals. PubMed, however, includes all topics, including those outside the Medline frame, such as astrophysics.
> - PubMed includes in-process citations which provide a record for an article before it is indexed.
> - PubMed also includes links to biomedical journals published by PubMed Central in full text and are specifically assessed by NLM.

Cochrane Library

The Cochrane Library (www.thecochranelibrary.com) is the best source for finding systematic reviews and randomized control trials concerning the outcomes of healthcare initiatives. It is updated quarterly and comprises several databases, some of which are described below.

Cochrane Reviews

These are full-text systematic reviews compiled by the Cochrane Collaboration. Each review identifies and compiles the results of all relevant trials and studies within a particular topic. The criteria are that the studies are homogeneous in design, choice, handling of controls, etc. Cochrane reviews are updated whenever new scientific documentation is available. The database also contains protocols of reviews in the process of compilation.

Other Reviews

The database comprises published abstracts of high-quality systematic reviews from sources other than Cochrane. The abstracts are compiled by the NHS Centre for Reviews and Dissemination at the University of York, UK, the American College of Physicians' Journal Club and the journal Evidence-Based Medicine.

Clinical Trials

Clinical Trials is a database of controlled trials compiled by various Cochrane Review Groups. The references are gathered from several bibliographic databases worldwide, as well as through manual searches in journals. Clinical Trials is the best single source for controlled trials.

Economic Evaluations

The database comprises structured abstracts concerning economic evaluations of various measures within healthcare.

EMBASE (Excerpta Medica)

EMBASE is a European-orientated database with references from approximately 5000 scientific journals in pharmacology and medicine. More than half of the journals indexed are published in Europe. The topics covered concern drug research, toxicology, dependence and abuse, clinical and experimental medicine, healthcare, psychiatry, forensic medicine and biotechnology. EMBASE contains more than 11 million references to articles published from 1974 onwards. More than 80% of the newer references have abstracts and approximately 75% are from journals published in English. EMTREE is the EMBASE thesaurus, with around 52 000 subject terms, including MeSH terms, organized in a tree hierarchy. The overlap between EMBASE and Medline is 10–75% depending on the topic. (Publisher: Elsevier Science; further information at www.embase.com)

BIOSIS Previews

BIOSIS Previews comprises references to articles published in 5000 journals in the fields of biology and biomedicine. The base also includes references to books and US patents, as well as reports from meetings and conferences. BIOSIS Previews contains references from 1969 to the present day. Approximately 90% of the references have abstracts, and around 30% of the indexed journals are also indexed in Medline as well as in EMBASE. (Publisher: Thomson Scientific; further information at www.scientific.thomson.com/products/bp)

PsycINFO

PsycINFO comprises references to articles published in approximately 1800 scientific journals, as well as references to research reports, doctoral dissertations and conference reports. The database covers psychology and related topics including education, sociology, medicine, nursing, physiology, nutrition and psychiatry. PsycINFO contains references from 1887 to the present day, and in addition from books and chapters of books from 1987 onwards published in English. [Publisher: American Psychological Association (APA); further information at www.apa.org/psycinfo]

ISI Web of Science

The ISI Web of Science is an interdisciplinary database with references to papers published in approximately 8800 journals. It comprises three indexes: Arts and Humanities Citation Index (A&HCI), Science Citation Index Expanded (SCI-EXPANDED) and Social Sciences Citation Index (SSCI). SCI comprises references in most of the medical disciplines. SSCI coverage includes public health, psychology and psychiatry. The ISI Web of Science contains references from 1945 to the present day.

In addition to searching by topic, it is possible to carry out a citation search, i.e. to find to what degree a particular author or work has been cited (see also Google Scholar for citation search). (Publisher: Thomson Scientific; further information at www.isiwebofknowledge.com)

5.9 STAYING UP TO DATE

Updating: saving searches, alerts and Selective Dissemination of Information

Well-structured search strategies should be saved for future use. A search strategy that is restricted by a time limit, e.g. the last 30 days, before it is saved, can be used anew to rerun the search, e.g. once a month, in order to stay updated. This is a current awareness service, known as Selective Dissemination of Information (SDIs). Electronic journals often facilitate alert services such as subscribing to tables of contents delivered via e-mail. Remember, however, what was written earlier about using information from the most recent sources only.

Table 5.5 Overview of principal biomedical databases

	MEDLINE	EMBASE	ISI WEB OF SCIENCE	BIOSIS REVIEWS	PSYCINFO	CINAHL	COCHRANE LIBRARY
Time span	1966– (OldMedline 1950–)	1974–	1945–	1969–	1806–	1982–	Aims at comprehensiveness
Extent	>4900 journals from 70 countries. OldMedline: 1950–65; Pre-Medline: new articles not yet indexed	>5000 journals from 70 countries	>8700 journals and series, of which >3000 biomedical from 80 countries	about 5000 journals from 100 countries	about 2000 journals from 45 countries	<1800 journals	>4400 systematic reviews; >478 000 controlled trials
Scope	General medical database. Covers medicine, nursing, dentistry, veterinary medicine, public health, allied health	Biomedical and pharmacological database. Human medicine, clinical and experimental. Broad coverage of pharmacology, toxicology, drug therapy	Most medical disciplines, psychology, psychiatry, public health	Biology and biomedicine; pharmacology; 80% medically orientated material; 70% global coverage, 30% USA	Psychology, criminology, psychological aspects of education, sociology, healthcare, nursing, nutrition	Nursing, physiotherapy, occupational therapy, nutrition	Cochrane Reviews; other reviews (DARE); clinical trials; methods studies; technology assessments; economic evaluations
Focus	Biomedicine; clinical medicine	Medicine; pharmacology; drug research	Multidisciplinary research	Preclinical research in biology and biomedicine	Psychology; psychological aspects of disease	Nursing	Clinical medicine; EBM

(Continued)

Table 5.5 (*Continued*)

	MEDLINE	EMBASE	ISI WEB OF SCIENCE	BIOSIS REVIEWS	PSYCINFO	CINAHL	COCHRANE LIBRARY
Updating	Varies according to vendor: daily, weekly, quarterly	Weekly. New entries indexed earlier than in Medline	<1 week after publication	Weekly	Weekly	Weekly; CINAHL-Ovid monthly	Quarterly
Thesaurus	Medical Subject Headings (MeSH): 24 000 subject headings	EMTREE: >52 000 subject terms, of which >25 000 cover drugs and chemicals	Key Words Plus. No thesaurus, but lists author keywords and automatically generated keywords	BIOSIS Codes, MeSH, CAS Registry nos	Thesaurus of Psychological Index terms: >7900 subject terms	12 000 CINAHL subject headings organized like MeSH	Partly MeSH
Abstracts	>80%	>80%	Yes	About 90%	Yes	Yes	Yes
Benefits	Free access in PubMed interface. Pre-Medline: daily updates. General database for all healthcare workers	Best database for pharmacological topics; strong on European research; more subject terms assigned than in Medline	Cited reference searching; provides information on impact factor	In addition references to books, US patents, meeting reports and conference proceedings	Starting in 1987 also references to English books and book chapters; unique source for older material	Qualitative studies; includes references to books and book chapters, dissertations, videos and educational software	Updated systematic reviews in full text; best database for controlled trials

References	>14 million; +1.2 million OldMedline	>11 million	>36 million million	About 15.6 million	About 2.3 million	About 1	>521 000
Over-lapping	60% overlapping at journal level vs EMBASE; covers South America and Japan better than EMBASE	60% overlapping at journal level vs Medline; 10–75% overlapping depending on topic searched; covers Africa and Asia better than Medline	75% overlapping at journal level vs Medline	About 30% overlapping at journal level vs Medline; about 3000 unique journals vs Medline, EMBASE likewise		One-third unique material vs Medline	All controlled trials from Medline included, likewise from other sources; systematic reviews indexed in Medline

PubMed

PubMed search strategies may be saved in two ways, either by using MyNCBI or by setting up an RSS feed (RSS is the abbreviation for Really Simple Syndication, a web standard for the delivery of news and other frequently updated content provided by websites). An advantage with the MyNCBI service is that it may be accessed from any computer connected to the Internet. RSS feeds are available via an RSS reader (downloaded free from the Internet) or an e-mail client.

5.10 MEDICAL AND SCIENTIFIC INTERNET SEARCH ENGINES

Most bibliographic professional databases are available only upon subscription or on a 'pay-as-you-go' basis. Medline via the PubMed interface is an exception, as it is free. Although a great deal of information is available free via the Internet, in many situations its scope is insufficient, as you only get what you pay for. Restricting research to information freely available on the Internet is risky and undependable, as chance, not system, dictates the results.

Searching or surfing

After exhausting the search possibilities found in bibliographic databases, you may consider using Internet search engines to find supplementary information. Several dedicated search engines are available. However, as none of these searches the entire Internet, several should be used to make thorough searches. Three examples of search engines dedicated to scientific and clinical information are:

- Scirus (http://www.scirus.com): for scientific information only. A search engine for scientific information developed by FAST and Elsevier Science. It searches more than 300 million scientific websites. Hit lists are ranked according to relevance (word occurrence frequency and how many links there are to a resource).
- Google Scholar (beta version) (http://scholar.google.com): a search engine for academic literature with links to free material on the web. It is particularly strong in biomedicine, compared to other disciplines. It harvests PubMed

and a number of publishers and scientific associations, but its coverage is incomplete. Google Scholar is suited to rapid searches to gain a perspective over the material available, but does not replace subject searching in relevant databases. It can be a useful source for grey literature such as unpublished conference material. 'Cited by' is an interesting function that links to other theses and papers that have cited the original references.

- SumSearch (http://sumsearch.uthscsa.edu): SUMSearch simultaneously searches various medical Internet resources and databases and presents results in groups ranked according to form and relevance. It is principally for clinicians, developed for EBM.

Internet search, either by surfing or by using search engines, requires a high degree of awareness to assess the quality of sources. Critical vigilance is essential, as anyone can put anything whatsoever on the Internet. Some guidelines to keep in mind are listed in Box 5.6.

Box 5.6 Evaluating quality on the web
- Source description: who is the publisher?
- Extent: what is covered?
- Objective or just advertising?
- Updating?
- Is the content biologically plausible?

5.11 PERSONAL REFERENCE DATABASES

Efficient handling of references and manuscripts

Building a personal literature database is an essential part of a research effort. Bibliographic management software enables you to file, organize and retrieve your references. In addition, the software can be used as an effective support in preparing the results for publication. Reference Manager (RefMan) and EndNote are examples of flexible, often-used bibliographic management software. Their principal functions are described below; see the programme manuals for further details.

Reference Manager and EndNote

RefMan and EndNote are produced by ISI Researchsoft (www.isiresearchsoft.com) and have been specially developed to manage bibliographic references. They consist of an easy-to-follow and in part intuitive programme which handles some 30 reference types, e.g. references to articles, books, chapters in books and conference papers. Each reference is built up of a particular number of fields, according to its type. An EndNote library can include up to 32 000 references, while a RefMan database has unlimited capacity. Each single reference which is registered receives a unique ID number. Even though it is possible to set up an unlimited number of databases, it is recommended that all references are registered in one main database. If it is convenient for a specific purpose, a selection of references for a specific paper can be copied into a subdatabase. Placing all references in a single database eases the preparation of manuscripts, the generation of bibliographies and the moving of files.

Settings

Several options may be chosen upon setting up a reference database. You may specify how thoroughly references shall be compared when duplicates are to be identified, whether duplicates shall be included in the import of references, as from Medline, or whether they shall be filed in another database or rejected. You may also specify whether keywords are to be extracted from titles, abstracts or keyword lists, and so on.

Journal index

The installation package for RefMan and EndNote includes three text files with lists of journal titles in full and standard abbreviated form in medicine, chemistry and the humanities. It is advantageous to import a list into a newly established database before references are registered. The lists are useful aids in compiling bibliographies with journal names, full or abbreviated.

Entering references

There are several ways to register references in a database:

- adding references manually
- direct export of references from a bibliographic database or from an electronic journal
- import of text files with references downloaded from a search in a bibliographic database
- direct import from bibliographic databases by using RefMan or EndNote as a search engine.

Adding references manually should be the last resort. It is time consuming and prone to error. However, if the source is a book or journal article that is too old, or otherwise is not/to be registered in a bibliographic database, manual entry is the only way.

The easiest way is to export references directly from a bibliographic database or from an electronic journal. Some database vendors, such as Ovid and ISI Web of Science, offer a function that enables direct export from their webpages to a RefMan/EndNote database. Increasingly more electronic journals also offer this type of function for direct export of references. In this way the extra steps of downloading references into bibliographic software via text files are eliminated. Importing text files with references downloaded from a search in a bibliographic database is a good alternative to direct export. This method must be used if references are to be transferred from PubMed. It is important that the references are downloaded in a format that RefMan or EndNote can read. The programmes have numerous import filters that make it possible to read and import references in the formats found in the various databases.

It is also possible to import references directly from bibliographic databases using RefMan or EndNote as a search engine. Some of these databases, e.g. PubMed or Library of Congress, are available free of charge on the Internet. Otherwise, availability depends on subscription. To perform an Internet search in bibliographic databases through RefMan/EndNote is a simple and effective way of collecting references, but this method can be recommended only for very basic searches, e.g. in tracing the complete bibliographic details for one or more articles where the author, title, etc., are already known. Subject searches should be performed directly in the appropriate database (e.g. Medline/Ovid or PubMed), where it is possible to take advantage of the interface designed for that specific base, with its own special features. The results can then be transferred to RefMan/EndNote. Trials have shown that subject searches in Medline via RefMan/EndNote produce fewer results and incidental references than similar searches using the same strategies in the original database interface.

Cite While You Write

It is often propitious to cite references during the writing of a manuscript. It is therefore sensible to use bibliographic management software from the very beginning, as it saves time when the manuscript and its bibliography are being finalized for submission. Cite While You Write (CWYW) is a function that supports the easy insertion of citations in a manuscript, which is scanned thereafter for citations so that a bibliography can be compiled in a specified format. Citations are entered in the text, either via an application available in Microsoft Word or by marking references in a RefMan/EndNote database, so they may subsequently be inserted in the manuscript. The example in Box 5.7 illustrates the most typical construction of a citation in its temporary form while the manuscript is in progress: first the first author, then the year of publication and ID/sequence number, all enclosed in braces. Embedded in each citation, there is a link between the manuscript and the database from which the reference has been taken, which makes it possible for the manuscript to be formatted automatically in a specified bibliographic format.

Box 5.7 Temporary citation in a manuscript being written
Several assertions have now been put forward which must be documented with references. Here is the first assertion which must be supported by the following{Pitkin, 1999 11/id}{Seglen, 1997 4/id}. The article continues with a discussion and a statement of the approach to the problem and concludes with a reference to the following{Minozzi, 2000 1/id}.

Box 5.8 A manuscript formatted according to the Vancouver style

Several assertions have now been put forward which must be documented with references. Here is the first assertion which must be supported by the following (1–2). The article continues with a discussion and a statement of the approach to the problem and concludes with a reference to the following (3).

REFERENCE LIST

1. Pitkin RM, Branagan MA, Burmeister LF. Accuracy of data in abstracts of published research articles. JAMA 1999; 281(12): 1110–1111.
2. Seglen PO. Why the impact factor of journals should not be used for evaluating research. BMJ 1997; 314(7079): 498–502.
3. Minozzi S, Pisotti V, Forni M. Searching for rehabilitation articles on MEDLINE and EMBASE. An example with cross-over design. Arch Phys Med Rehabil 2000; 81(6): 720–722.

Bibliographic format: 'Output Styles'

'Output Styles' specify the presentation of citations and the corresponding literature list in a manuscript; for example, to what extent the citations in a text should consist of numbers which refer to a numbered literature list, or alternatively should consist of authors and years that refer to a literature list arranged in alphabetical order. 'Output Styles' also specify punctuation and the fields of a reference that are to be included. There are many predefined 'Output Styles' included in the programmes, specific styles for a range of medical journals, as well as bibliographic formats, such as Vancouver (see Box 5.8), Harvard and APA. If none of the existing styles is suitable then it is possible to define one's own, either by copying and editing an existing style or by devising a new one. RefMan has an aid to building up new styles, called 'Output Style Wizard'.

Transferring references between RefMan and EndNote databases

References may be transferred between the two reference tools using the RIS format employed both in RefMan and in EndNote. Accordingly, an EndNote library may be converted to a RefMan database, and vice versa.

REFERENCES

Chargaff E (1975–76) Triviality in science: a brief meditation on fashions. Perspectives in Biology and Medicine 19: 324–333.

Garfield E (2006) The history and meaning of the journal impact factor. JAMA 295: 90–93.

Pitkin RM et al. (1999) Accuracy of data in abstracts of published research articles. JAMA 281: 1110–1111.

FURTHER READING

McKibbon A (1999) PDQ Evidence-Based Principles and Practice. Decker, Hamilton, BC.

METHODS IN MOLECULAR BIOLOGY

Sigbjørn Fossum and Erik Dissen

'When man becomes capable of instructing his own cells, he must refrain from doing so until he has sufficient wisdom to use this knowledge for the benefit of mankind.'

(Nirenberg 1967)

6.1 INTRODUCTION

Living organisms are hierarchically constructed, with small molecules forming building blocks for macromolecules, which in turn are assembled into larger aggregates and ultimately into cells, tissues, organs and bodies. For each level the multitude of different possible organizational patterns and emergent properties necessitates distinct descriptive terminologies, investigational methods and manipulative techniques. Molecular biology can be defined as the scientific discipline covering the level between chemistry and biochemistry on the one side and cell biology, histology, anatomy and physiology on the other, intertwined with and overlapping the levels on each side. Until recently this level represented a largely unexplored territory.

By a series of remarkable advances in technical developments and novel discoveries during the past 50 years (Box 6.1) we are now gaining deep insight into this

Box 6.1 Landmarks in the development of DNA technology

1865	The Mendelian laws of genetics
1869	DNA discovered
1928	Transformation of bacteria
1944	DNA is the carrier of genetic information
1953	The DNA double helix: specific base pairing
1961	DNA renaturation and hybridization
1962	Restriction endonucleases
1965	The genetic code
1967	DNA ligase
1972	DNA cloning
1975	DNA sequencing
1976	Site-specific mutagenesis
1981	Transgenic animals
1983	The polymerase chain reaction
1985	Genetic fingerprinting
1988	Knockout animals
1990	First attempt at gene therapy in the human
1990	The human genome project launched
1995	Genomic sequences of first free-living organism (the prokaryote *Haemophilus influenzae*)
1997	Cloning of animals
1997	Genomic sequence of first eukaryotic organism (*Saccharomyces cerevisiae*)
1998	Genomic sequence of first multicellular organism (*Caenorhabditis elegans*)
2001	The human genome sequenced

organizational level. The progress has been astonishingly fast and the profundity of its consequences shattering. Until recently, the deficits in our understanding of how the molecules of life are organized and work made biology lag far behind other natural sciences in explanatory and predictive power. The current avalanche in insight into the basic levels of living systems represents a turning point in human knowledge on a par with the greatest discoveries in physics. It has opened the gates to a bioindustrial revolution and, of particular relevance to the medical researcher, is transforming medicine from a largely empirical art to a truly scientific discipline. In a wider perspective, for nearly four billion years the evolution of life forms has depended on the occurrence of unpredictable spontaneous mutations; now one of the descendant species has suddenly acquired tools to manipulate the genetic material at will.

6.2 RECOMBINANT DNA TECHNOLOGY

A cornerstone of molecular biology is a group of methods collectively termed recombinant DNA technology. The foundation was laid in the 1960s, with the discoveries of:

- restriction endonuclease (RE) enzymes, which made it possible to cut DNA at specific sites into shorter fragments amenable to analysis
- the techniques of DNA denaturation/renaturation and hybridization, by which specific pieces of DNA could be identified and pulled out from a large pool of cut fragments
- the enzyme DNA ligase, which allowed joining of fragments together into new combinations.

Combining the three techniques made it possible to amplify specific pieces of DNA in a cut-and-paste approach, inserting the fragment of interest into a DNA vector that can be transferred to a living cell in order to generate multiple copies. Isolation and multiplication of a specific DNA segment by this approach (DNA cloning) was made feasible in the early 1970s, and marked the birth of recombinant DNA technology.

Restriction endonucleases

Endonucleases are enzymes that cleave the phosphodiester bonds of the DNA or RNA backbone. Restriction endonucleases, often called restriction enzymes, are a subgroup of endonucleases that cleave double-stranded DNA only at sites with a specific nucleotide sequence. Several hundred different restriction enzymes have by now been identified, most of them with unique recognition sequences. The ones most frequently used in molecular biology typically recognize palindromic sequences of four, six or eight nucleotides, but enzymes with non-palindromic sequences are also used. They are named after the bacterial strain producing them, in the following manner: *EcoRI*, from *Escherichia* (genus) *coli* (species) strain *RY13*, I (order of isolation).

Some restriction endonucleases cut the two strands immediately opposite one another, generating blunt ends, e.g. *AluI*, which cuts the sequence AG↓CT in each strand (with the arrow indicating the site of incision). More common are staggered cuts, generating single-stranded overhangs referred to as sticky ends, e.g. A↓AGCTT for *Hind*III and GGTAC↓C for *Kpn*I. This is a useful property, as the ends will stick to (hybridize with) similar ends from other DNA molecules cut with the same enzyme, facilitating the generation of recombinant molecules.

Plasmid vectors

In medicine, a vector (from Latin, bearer) is a carrier (e.g. in malaria, a mosquito serves as the vector that transfers the *Plasmodium* parasite between the human hosts). In molecular biology, vectors are DNA molecules with the capacity to replicate when introduced into host cells and onto which we can load pieces of DNA to be amplified. The most commonly used vectors are derived from plasmids (Figure 6.1), which are extrachromosomal self-replicating DNA molecules. Plasmids occur naturally in most kinds of bacteria, and a wealth of different types exists. The plasmids normally used in gene transfer are circular and small, i.e. a few thousand base pairs (bp). By genetic manipulation of naturally occurring plasmids, artificial plasmids have been constructed with useful functional properties.

Three elements are shared by all: (1) an origin of replication, i.e. a sequence necessary for replication in the bacterial host; (2) a cloning site containing recognition sequences for one or more restriction enzymes, wherein the desired DNA fragment can be inserted. Usually, several different recognition sites have been put one after the other within a small region, referred to as the polylinker or multiple cloning site. This feature permits choices between different restriction enzymes (and makes possible unidirectional insertion of the DNA piece); (3) a

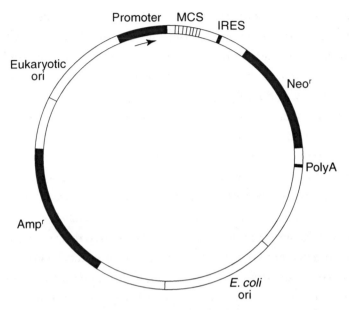

Figure 6.1 Plasmid vector designed for transient or stable expression in eukaryotic cells. A multiple cloning site (MCS) contains sites for several restriction enzymes. The promoter should be carefully chosen for efficiency in the cell or tissue of interest. Production of plasmid in *E. coli* requires a prokaryotic origin of replication (ori). Some vectors also contain a eukaryotic ori. Selection elements are included to allow growth of bacteria in Ampicillin (Amp^r) and of eukaryotic cells in G418 (Neo^r)

selection marker, i.e. a gene conferring a phenotype that can be selected for or against, usually a sequence conferring resistance to antibiotics. Optional elements added to most modern plasmid vectors are origins of replication allowing replication also in eukaryotic host cells (shuttle vectors) and promoter sequences allowing transcription into messenger RNA (mRNA) and hence protein synthesis (expression vectors). Through the years increasingly sophisticated plasmids have been generated, containing additional useful elements.

Viral vectors

Viruses are protein-coated infectious DNA or RNA particles with naturally evolved mechanisms for unassisted entry into host cells. Bacteriophages are viruses that infect and replicate in bacteria. These viruses can be used as vectors in molecular biology. The linear virus genomic DNA is purified, and a DNA fragment is inserted. Thereafter, the manipulated DNA is packed in empty cap-sid particles, and the resulting bacteriophage will infect and replicate in *E. coli* cells. Some laboratory strains of bacteriophage λ have genomes with added elements that allow in vivo excision of most of the phage-specific sequences and transformation of the resulting linear vector, plus insert DNA, into a shorter plasmid-like circular DNA called a phagemid.

Eukaryotic cells may in some cases be hard to transfect with plasmid DNA, and several different types of viruses are in common use that allow efficient delivery of DNA into most cell types. Viral vectors have clear advantages when we want to transfer genes into intact multicellular organisms such as mice or rats for research purposes; or humans in gene therapy. Viruses are specialized with respect to the type of host cells they can infect. This property, referred to as tropism, allows the researcher to generate viral vectors that will selectively infect certain cells or tissues, restricted to the desired species. Some viruses, such as retroviruses, stably integrate themselves into the host genome and replicate together with the host cell. One disadvantage is that with random sites of genomic integration the viral DNA may insert itself into, and functionally disrupt, host cell genes, ultimately leading to cancer. This is not a big problem with non-integrating viruses, such as aden-oviruses. However, in gene therapy, prolonged or lifelong gene expression is desir-able, requiring regularly repeated treatments if using non-integrating viral vectors. In gene therapy a variety of methods for delivering the desired genes is currently under intense investigation. Most of these are based on the use of viral vectors that have been modified in various ways, aimed at controlling factors such as tropism, nuclear entrance and sites of genomic integration.

Although viral vectors are useful, they generate the hazard of unwanted spread. Therefore, viral vectors are genetically modified so that after delivering their genetic material to a cell they are unable to replicate into new infectious virus par-ticles. Special safety conditions in the laboratory are nevertheless often required.

Transferring DNA into host cells

In a process called *conjugation*, plasmid DNA can be naturally transferred from one bacterium to another by direct physical contact between the donor and the recipient cell. Through their built-in mechanisms for entry into cells, viruses can also naturally transfer genetic information between cells, a process referred to as *transduction*. Bacteria can in addition spontaneously take up non-viral nucleic acids from the surroundings. To make this uptake more reliable and efficient for laboratory purposes, various methods have been developed. The most commonly used techniques for getting insert-loaded plasmids into bacteria are by generating transient pores in the cell membrane by brief electric pulses (electroporation), or by adding Ca^{2+} or other divalent cations to the medium. The genetic alteration of bacterial cells as a result of uptake of foreign DNA or RNA is referred to as *transformation*. For getting non-viral DNA into eukaryotic cells the currently most used techniques are electroporation, injection directly into the cell (microinjection) or complexing the DNA with molecules that facilitate the DNA transfer. Popular in use are cationic lipids, where the positive charge of the lipid heads interacts with the DNA and with the cell membranes. Non-viral transfer of DNA into eukaryotic cells is called *transfection*, in contrast to transformation, which refers to the changes occurring when eukaryotic cells become malignant, e.g. following infection of cultured cells with tumorigenic viruses.

In bacterial transformation, the new DNA is normally not integrated into the bacterial chromosome, but remains as an extrachromosomal plasmid. By using plasmids with selection markers, such as antibiotic resistance genes, and culturing the bacteria in selective media containing antibiotics, only bacteria containing plasmids will survive. In eukaryotic cells, transfection may be transient or stable. In *transient transfection*, many copies of the DNA are transferred into each cell, where they remain extrachromosomally and are expressed only for a short period, usually a few days. In *stable transfection*, the foreign DNA, which is either integrated into chromosomal DNA or stays extrachromosomally, is maintained by selection for resistance to cytotoxic drugs. Most common in use is a neomycin resistance element that allows eukaryotic cells to be grown in the presence of geneticin (also known as G418).

The prime example of a host cell: the laboratory workhorse *Escherichia coli*

The only host cell that is described briefly here is the intestinal bacterium *Escherichia coli*. This rod-shaped Gram-negative bacterium is the best studied bacterial species. Its full genome sequence (approx. 4.6 million bp) was reported in 1997. It is easy to work with, divides rapidly in simple media and is almost exclusively the bacterial species used in DNA technology. It is usually cultured in liquid media containing salts and amino acids together with vitamins and other metabolites. At optimal conditions, commonly used strains divide every 20–30 minutes, following an initial lag phase. The cells keep dividing exponentially (log phase) until the available nutrients are consumed along with the accumulation of waste products, leading to stagnation of growth (plateau phase).

A problem with cultivation in liquid media is that cells in which foreign DNA has been introduced may exhibit different cell division rates. Thus, within 24 hours a bacterium that doubles every 20 minutes will have overgrown one that uses 22 minutes a hundred-fold. To overcome this problem, single colonies (all offspring of a single parent cell) can be cultured in Petri dishes on media solidified by the addition of agar. Owing to limited access to nutrients in the medium immediately beneath the colony, the various colonies will eventually reach roughly similar sizes. Another advantage with this technique is that a single cell as a rule has taken up only one DNA vector molecule (which is subsequently replicated within the cell). Each separate colony therefore contains a unique DNA clone, so that by picking solitary colonies, single amplified DNA clones can be isolated.

The *E. coli* strains in laboratory use are genetically crippled to impede their spread. A variety of different strains has been developed for molecular biology purposes. For instance, to counteract the inherent ability of *E. coli* to identify and destroy foreign DNA, genes that encode proteins involved in this process have been inactivated by mutations in laboratory strains. Popular laboratory strains have been modified to be easily transformed, produce high copy numbers of DNA and low amounts of carbohydrates, and lack recombination enzymes, yielding high-quality DNA that is easy to purify.

6.3 DNA AND RNA: ISOLATION, IDENTIFICATION, SYNTHESIS AND ANALYSIS

The development of the techniques delineated above for cutting, pasting and amplifying DNA set the stage for a variety of applications that allowed researchers to probe deep into the organizational principles of living organisms. Before going into these applications, some other relevant techniques for working with DNA and RNA are briefly described.

Isolation of DNA and RNA

DNA and RNA have distinct physical properties permitting their separation from other macromolecules. Of particular importance are the phosphate groups of the DNA/RNA backbone, which will release H^+ ions at neutral pH. DNA and RNA are consequently negatively charged. In contrast to lipids and most proteins, DNA and RNA are therefore not soluble in organic solvents (such as phenol and chloroform), but highly soluble in water, and this property can be used to separate nucleic acids from lipids and proteins. Although the phenol/chloroform isolation method has traditionally been popular, it is time consuming, involves flammable and toxic liquids, and has largely been replaced by chromatographic methods, based in particular on anion-exchange resins.

Anion-exchange resins consist of positively charged solid particles, to which the highly negatively charged DNA and RNA molecules bind particularly strongly. Sugars and proteins, which bind with moderate affinity, can be washed off the resin. Highly purified nucleic acids are then eluted using buffers with the appropriate pH and salt concentration. Another traditional method for DNA/RNA purification is based on differences in densities between RNA, supercoiled DNA, linear DNA, proteins and lipid (with densities decreasing in the order listed). By ultracentrifugation in caesium chloride density gradients, these macromolecules can be separated into discrete bands.

When working with nucleic acids it is important to avoid degradation due to contamination with DNases and RNases, i.e. enzymes that degrade DNA or RNA. In particular, RNases are ubiquitously present and furthermore resist autoclaving. Preventing RNase activity therefore represents a primary concern in the handling of RNA, requiring strict measures and working routines.

Visualization: Southern and northern blots

DNA and RNA molecules can be separated for analytical and preparative purposes by electrophoresis in agarose or polyacrylamide gels. The DNA or RNA samples are loaded into wells lying side by side in the gel. An electric field is applied to the gel, and the negatively charged nucleotides migrate towards the cathode. The gel matrix serves as a sieve that sorts the molecules according to size, with smaller molecules migrating more quickly than the larger. Agarose gels are most widely used. Polyacrylamide gels are more time demanding, but give better separation, in particular of shorter molecules, and are therefore used in sequencing, mutation detection and separation of short fragments (< 100 bp). Very long DNA molecules ($> 20\,000$ bp) tend to get stuck in the gel. Their release from the gel matrix is facilitated by altering the directions of the electric field [pulsed field gel electrophoresis (PFGE)], causing the long thread-like molecules to wriggle through the gel in zigzag movements. Fragments up to 10 million bp in length can thereby be separated, so that, for example, individual yeast chromosomes can be visualized and isolated.

DNA and RNA are visualized in the gel using fluorescent dyes such as ethidium bromide. In each lane there may be molecules of different lengths. If the number is limited, each will be seen as a distinct band along the lane. If the number is very large (such as with restriction enzyme-digested genomic DNA) the bands merge into a diffuse smear.

Distinct fragments can be identified based on their nucleic acid sequence with a DNA probe. Two strands matching each other with respect to base pairing are complementary. The two strands can be made to separate into single strands, or *denature* by heating, immersion in denaturing buffers, or both. By cooling or changing buffers, the two strands can be made to rejoin, or *renature*. Making use of the specificity by which the strands will rejoin only to complementary sequences, an exogenous single-stranded DNA (a probe) with sequence matching one of the fragments in the smear can be added to single-stranded target DNA. This probe will form a *hybrid* together with the target DNA fragment that has a complementary sequence. To allow visualization of the probe, it can be labelled with radioactive nucleotides (^{32}P, ^{33}P or ^{35}S) that can be detected by autoradiography or haptens that can be detected by staining with antibodies.

To preserve the electrophoretic separation of the molecules during denaturation and hybridization, the DNA (or RNA) must be immobilized by blotting onto a membrane with high binding capacity for nucleic acids. The transfer of DNA from gel to membrane was first described by E. M. Southern, and the method for detection of specific, electrophoretically separated DNA fragments is consequently called *Southern blot*. For RNA, the similar method was subsequently referred to as *northern blot* (and for proteins, *western blot*). DNA or RNA may also be applied to membrane for detection by hybridization with probes without prior electrophoretic separation. If pipetted directly onto the membrane the method is referred to as *dot blot*; if transferred from virus colonies it is called a *plaque lift*, and from bacterial colonies a *colony lift*.

The polymerase chain reaction

The polymerase chain reaction (PCR) technique is applied to amplify only specific segments of DNA. Enzymes catalysing DNA replication (DNA polymerases) are used to generate copies of DNA in a cyclic fashion, so that the number of DNA copies is doubled for every cycle. Because DNA polymerases require a primer at the 5′ side, the region amplified can be controlled by adding synthetic DNA primers (single-stranded oligonucleotides of typically 15–30 nucleotides) with the appropriate sequence. Because DNA replication has to occur on both strands, a pair of primers is used in PCR, one complementary to each strand. This primer pair defines the flanks of the region that is amplified.

In contrast to DNA cloning described above, PCR takes place entirely outside living cells. Only template DNA, primers, the four deoxynucleotides (in triphosphate form dTTP, dCTP, dATP and dGTP), DNA polymerase and a suitable buffer are needed. The reaction is repeated in a cycle consisting of three steps: (1) DNA denaturation, where the reaction mix is heated to 94°C, causing separation of the two DNA template strands; (2) annealing, where the temperature is lowered so that the primers can hybridize with the template DNA. Because the primers are present in huge surplus, they will hybridize with the template before renaturation of the template DNA; and (3) elongation, which is performed at the optimal working temperature of the DNA polymerase, usually around 72°C. The key to the method working without having to add fresh enzyme for each new cycle is using thermostable DNA polymerases isolated from prokaryotic cells that can tolerate very high temperatures. The first exploited for this purpose, *Taq*, isolated from *Thermus aquaticus*, which lives in hot springs and hydrothermal vents, is still the most common in use. *Taq* is efficient and robust, but with a high error rate as it lacks 3′ to 5′ exonuclease activity. Proofreading thermostable

polymerases are available, but the price to pay for higher fidelity is lower efficiency than with *Taq*.

The PCR is run in machines that shift rapidly between temperatures, and
where the temperatures, lengths of each step and number of cycles can be pre-
programmed. Each cycle only takes a few minutes. Except for the first two cycles,
the targeted DNA is almost doubled with each cycle, giving an exponential
increase in the number of molecules, so that after 30 cycles the number of copies
has increased around a billion-fold. This simple, rapid and relatively inexpensive
technique is used for a variety of purposes:

- *In diagnostics*, to identify the presence of a particular sequence (e.g. belong-
  ing to viruses or bacterial strains) in a sample.
- *In cloning* of cDNA (see below) or of fragments of genomic DNA. The pieces
  of DNA are first amplified by PCR, and can then be ligated into a plasmid
  vector. Pieces of DNA encoding parts of different proteins may be ligated
  together into the plasmid, for generation of chimaeric proteins. The purpose
  may also be construction of plasmids with novel properties.
- *For quantification of mRNA*: see reverse transcriptase (RT)–PCR, below.

Paramount to successful PCR is proper design of primers. Computer programs are used to minimize problems such as primer dimerization or hairpin
formation due to complementarities between parts of the primers. A recurrent
problem in PCR is false priming, i.e. the primers bind to other sites than the
intended. This can be corrected by raising the annealing temperature to a point
where the primers only bind to completely complementary sites, or by designing
longer primers. To minimize false priming further, the so-called hot-start
and/or touchdown principles can be applied (explained in available laboratory
manuals).

DNA sequencing

The currently leading methods for analysing the sequence of a DNA molecule
are reminiscent of PCR. A reaction mixture is prepared as described for PCR, but
with a single primer, so that only one of the two strands is amplified. The key

element, developed by Sanger and co-workers, is the use of dideoxynucleotides that when incorporated into the newly synthesized DNA strand terminate further elongation. This leads to random termination sites, and the generation of newly synthesized single-stranded DNA fragments in a discrete spectrum of different lengths. Thus, for example, a fragment of 231 nucleotides containing a dideoxycytidine is indicative that the base in position 231 is a C. By electrophoresis the order of chain lengths can be established, and thereby the order of bases in the template DNA. Each of the four dideoxynucleotides is labelled with a different fluorescent marker; the full spectrum of different lengths is separated by electrophoresis in narrow tubes, and the presence of different fluorochromes corresponding to different fragment lengths is identified with a laser source and detectors for each colour. This continuous wave-like information is fed directly into a computer, that will translate the wave patterns into DNA sequence strings. The DNA polymerase step in the sequencing reaction is performed in a cyclic fashion, reducing the amount of template DNA required. Modern DNA sequencing machines can reliably distinguish between fragments with lengths up to about 1000 bp and have up to 384 parallel capillaries that can be run simultaneously. As they can perform up to 12 runs every 24 hours, the daily sequencing output per machine can be millions of base pairs.

Reverse transcription and cDNA libraries

Reverse transcriptases catalyse the synthesis of DNA using RNA as a template (thereby working in the reverse direction of transcription of RNA from DNA). In the laboratory, reverse transcriptases are mainly used to make DNA copies of mRNA. Most mature mRNAs have a long poly(A)-tail attached to the 3'-end. This property can be exploited to isolate relatively pure mRNA by hybridization to dT-oligonucleotides bound to a solid phase. If the goal is to make a collection of DNA copies of the whole set of mRNAs, reverse transcription can be performed on the mRNA while still bound to the oligo(dT), which will function as primers. If the aim is to amplify a particular sequence, specific primers are used. Synthesis of the first strand results in the formation of DNA/RNA hybrids. The RNA is removed and second strand synthesis performed using a DNA-dependent DNA polymerase. The result is double-stranded DNA, referred to as cDNA, for complementary DNA (to mRNA). The cDNA may be ligated into plasmids. A collection of cDNA synthesized from the whole set of isolated mRNAs is referred to as a cDNA library. Different cells, tissues and organs express different genes, and thus produce different mRNAs. cDNA libraries are therefore cell, tissue and organ specific.

Artificial chromosomes and genomic libraries

Libraries can also be made from genomic DNA. The genomic DNA is digested with restriction enzymes and the fragments are ligated into vectors. With plasmids and viral vectors, the foreign DNA that can be inserted is limited in size (<20 000 bp for most plasmids; <50 000 bp for bacteriophage λ). For some purposes, such as sequencing of whole genomes, it is advantageous to use even larger pieces of DNA. To this end artificial chromosomes have been developed. Yeast artificial chromosomes (YACs) permit cloning of DNA fragments exceeding a million base pairs. YACs are technically difficult to work with, and a serious disadvantage is their tendency to recombine. Phage- and bacteria-derived artificial chromosomes (PACs and BACs) are easier to work with, and BAC vectors are largely chosen in current genome sequence projects, incorporating fragments of 150 000 bp on average. Note the differences between cDNA libraries and genomic libraries: the former are made from mRNA and will therefore vary according to cell or tissue source, whereas the latter are made from genomic DNA, which with a few exceptions is identical in all cell types in a given organism.

6.4 PRACTICAL APPLICATIONS OF DNA/RNA TECHNOLOGY

The described techniques are used alone or in combination for a large variety of practical purposes. Space restricts this presentation to brief introductions to the following examples: whole genome sequencing; identification of novel genes; analyses of gene expression profiles and programs; functional analysis of genes, including introduction of mutations, generation of chimaeric constructs and transgenic animals; studies of genetic variation within populations and identification of genes associated with disease; and DNA fingerprinting.

Whole genome sequencing

Whole genome sequences from both prokaryotes and eukaryotes are currently churned out at a rapid pace from large dedicated high-throughput laboratories where most of the steps have been fully automated. The sequences are, as a rule, immediately made publicly available via databanks (see Section 6.6). Before sequencing the genomes are cut into smaller fragments, which are cloned. In order to see which pieces are neighbours, we need to generate fragments that overlap in a tiling pattern. Accordingly, the genome is chopped up in such a way that each part of the genome is represented on several different clones. To eliminate sequencing errors, genome sequences are usually reported with a six to 10 times coverage, with the implication for large eukaryotic genomes that many hundred million fragments need to be cloned and sequenced.

It may then seem surprising that it is not the physical handling and sequencing of the large numbers of clones that represent the greatest challenge, but the final assembly of the sequenced fragments. The reason is that the eukaryotic genomes are littered with interspersed repeated sequences, in the human making up almost half of the genome. Although residing on different sites in the genome, two recently duplicated repeats may be almost identical. The question then arises whether slight differences between two sequences stem from amplification or sequencing artefacts, whether they reflect different genomic sites (loci) or whether they represent alleles (i.e. are from the same site on homologous chromosomes).

The assembly problems led to two different main strategies for genome sequencing. The method preferred by the Human Genome Project is the *hierarchical shotgun sequencing* two-step method. In this approach, genomic DNA is first cut into large pieces (on average approx. 150 000 bp), which are inserted into BAC vectors used to transform *E. coli*, where they are replicated and stored. The map position of each BAC insert is then determined. Because these fragments are generated in a random fashion, they will tend to overlap in a tiling pattern. In the second step each BAC is cut into smaller pieces, which are ligated into plasmids, cloned and sequenced. At this stage there is no attempt to identify the order of pieces before sequencing, so the clones to be sequenced are selected at random (shotgun sequencing). The sequences are aligned by data programs so that identical sequences overlap. Once sequenced sufficiently many times to produce high-quality data, the contiguous pieces are assembled into finished sequence. In the alternative single-step method, *whole genome shotgun sequencing* (applied by the private company Celera), the BAC step is skipped. This direct approach was developed on prokaryotic genomes which are smaller in size and contain less repetitive DNA. Here genomic DNA is directly randomly sheared into small pieces which are cloned into plasmids and sequenced on both strands. The sequences are fed into computers, aligned and assembled into finished sequence.

The genome projects continuously report their data to large web-based databases with user-friendly interphases, where several aspects of information are integrated for each gene (further outlined below).

Identification of novel genes

In molecular biology a gene can be defined as a region of DNA encoding a functional RNA molecule, where most encode mRNAs instructing protein synthesis. Although the human genome has been (almost fully) sequenced, there still remain genes to be discovered. In species where the full genome sequence is available on databanks, novel genes are primarily identified by computer analysis (*in silico* cloning). Two main approaches are used. The first predicts the existence of genes *ab initio*, i.e. without previous knowledge other than general rules for how genes are organized, and applies primarily to protein encoding genes. The second is based on sequence similarities (homology) with genes already known. The input sequence can be used to search for the corresponding (orthologous) gene in a different species or for a related but different (homologous) gene in the same species. Duplicated genes in the same species are called paralogs. Computer programs compare nucleotide or protein sequences by alignment to sequences in the databases and calculate the statistical significance of matches. Most commonly used are variants of the Basic Local Alignment Search Tool (BLAST) program, designed to detect regions of local similarity between sequences. Alternatively, the input sequence may be from larger genetic regions or even whole genomes, in which case all sequences with functional significance are targeted, with the presumption that these will be more resistant to mutations than non-functional sequences and hence better conserved during evolution. As an example, sequencing of the pufferfish (*Takifugu rubripes*) genome was performed mainly for the purpose of cross-species comparisons. The lineages leading to pufferfish and humans split approximately 450 million years ago. The pufferfish was chosen because it has the smallest vertebrate genome known, less than one-seventh the size of the human genome, but with an estimated equal number of genes to the human. By mapping regions showing high sequence similarities between the two species, more than 1000 novel human genes were detected. In addition, the finding of large clusters of syntenic genes, i.e. genes held together in the same chromosomal region during this vast amount of time, shed light upon the evolutionary processes forming the two genomes.

Homology screening is also performed in vitro, based on hybridization to DNA probes and with cDNA libraries usually ligated into bacteriophage λ-derived vectors. These are packed into active viruses, which are spread thinly over a dense layer of *E. coli* grown on agar plates. Virus particles infecting a bacterium will multiply, lyse the host and spread to neighbouring bacteria. The lytic spots or plaques created in the *E. coli* layer are rich in λ-DNA with cDNA inserts. The DNA is immobilized by blotting onto a membrane (plaque lift), permitting denatured and hybridized with a probe. If the library contains sequences of sufficient similarity (homology) to the probe, it will bind to the membrane. The positions where the probe binds are used to identify the corresponding plaques on the agar plate, for amplification of the particular clone, which can then be sequenced and further analysed. Plasmid cDNA libraries are screened in a related fashion (colony lifts).

Three other widely used methods to identify novel genes are expression cloning, subtraction cloning and positional cloning. In expression cloning, cDNA or genomic DNA ligated into expression vectors (see Plasmid vectors) is isolated based on their biological activity in vivo. Subtractive cloning is a technique for pulling out genes expressed in one cell population but not in another. In positional cloning the chromosomal location of a functional trait, such as inheritable disposition to a disease, is first mapped by linkage analysis. The map position on the chromosome is then used to clone the gene.

Analysis of gene expression

Essentially what make cells different is the different selections of genes they transcribe and express. The first key to the function of a new gene often comes by studying in which cells it is expressed.

Investigations of gene expression are based mainly on measurements of transcriptional activity, i.e. on levels of gene-specific RNA produced in cells and tissues. A source of error when comparing amounts of specific RNA is variations in the amount of total RNA isolated from the samples. The amounts of each sample are therefore adjusted with respect to levels of RNA from housekeeping genes, which ideally should be transcribed at roughly equal levels in the cell types studied, unaffected by external influences. Northern blot, described above, is the traditional method for quantitative measurements of RNA. The disadvantages with this method are that it is time consuming, requires relatively large amounts of RNA and can give spurious results owing to cross-hybridization, in particular when dealing with related members of a multigene family. The latter problem is obviated in a method called RNase protection assay, which requires full identity between sample and probe. Based on simplicity and improved sensitivity, these techniques have now largely been replaced by methods based on RT-PCR.

Semiquantitative and quantitative reverse transcriptase-polymerase chain reaction

Reverse transcription as part of generating cDNA libraries was described above. A first strand cDNA copy is first made, using a reverse transcriptase together with an oligo-dT or a gene-specific primer. Thereafter, PCR is performed with a gene-specific primer pair, the products are separated by electrophoresis on agarose gels and stained with ethidium bromide, and the relative band intensities are estimated by visual comparison or measured by densitometry. Ideally, the band densities in the different lanes should perfectly reflect the amount of mRNA in the different samples. There are many pitfalls with this approach. The near-exponential characteristics of the amplification make it vulnerable, and small differences in reaction conditions between tubes can have large effects. In addition, the number of cycles must be limited to avoid exhausting the reagents, but still sufficient for reliable detection of PCR product. With the use of internal controls to normalize for sample to sample variations in total RNA amounts and reaction efficiency, it is possible to reach reasonable, semiquantitative estimates of specific RNA in the samples. The method is widely used as it does not require sophisticated apparatus, and is rapid, sensitive and, depending on the appropriate choice of primers, highly specific. Because only the endpoint of the PCR reaction is measured, the quantitative accuracy of the method is limited. It is increasingly being replaced by real-time or quantitative PCR (qPCR). These are based on the detection of fluorescent reporter molecules (dyes that bind double-stranded DNA or sequence-specific probes) that accumulate in step with the PCR product with each cycle of amplification. The fluorescent intensity is continuously measured during the reaction, allowing the gradual accumulation of PCR product to be monitored (hence the term real-time). The data are fed into computers and the starting concentration of substrate is estimated from the fluorescent curves.

Microarrays

The electrophoretic separation of mRNA in the northern blot technique is time-consuming and limits the number of samples that can be processed simultaneously. Advantages are that it yields information about the lengths of the mRNA to which the probe hybridizes, which can be useful as a safety test for specificity, and that it reveals the presence of splice variants or incompletely spliced pre-mRNA. If specificity, incomplete splicing and splice variants can be checked out as problems or minimized by design of the probes, the electrophoretic step can be omitted and the mRNA attached directly to a solid surface (dot blot). This permits the simultaneous testing of RNA from many different cell types or tissues attached as orderly collected spots.

It is also possible to invert the system, with the probes rather than the RNA attached to the surface, and to miniaturize it, so that a small piece of plastic or glass or a silicon chip contains thousands of spots with different probes, each spot corresponding to an individual gene. This systematic deposition of probes is called a microarray (Figure 6.2). The probes have been selected to bind specifically to only one mRNA, and consist of synthetic oligonucleotides or short PCR product fragments. In an inverse approach to northern blot, cDNA synthesized from mRNA is labelled and hybridized to the microarrays. Hybridization to each spot (actually, pairs of spots with the same gene probe in duplicates) is detected, and a signal indicates that the respective gene is expressed and represented among the mRNAs in the cell type or tissue investigated. The power of the microarray technique is the ability to analyse expression of thousands of genes in a single mRNA sample.

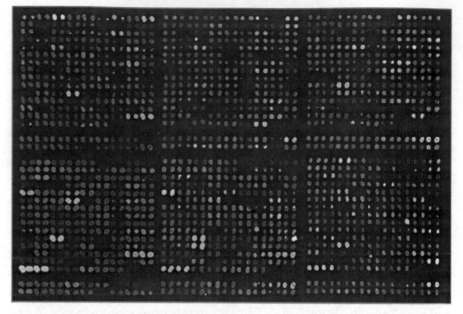

Figure 6.2 A microarray hybridized with cDNA from a sarcoma (green fluorescent label) and corresponding normal tissue (red fluorescent label) (Picture generously provided by Professor Ola Myklebost, The Norwegian Radium Hospital)

In two-channel microarrays cDNA made from two different samples, e.g. a tumour and the normal tissue from which it originates, are labelled with two different fluorochromes, mixed and hybridized together to a single microarray. The array is scanned with lasers coupled to photodiodes detecting emitted light and the information fed into computers. The system allows detection of upregulated or downregulated genes in the test sample versus the control, but does not give information about absolute expression levels. For this purpose single-channel microarrays are used. Microarrays covering complete genomes or selected groups of genes are commercially available from private companies or produced by dedicated core facilities. Note the difference between the northern blot (or its simplified dot-blot equivalent) and the microarrays. The former monitors the expression profile of a single gene in different tissues, the latter the expression profile of thousands of genes in a single tissue (or compares two tissues). DNA microarrays are an example of high-throughput technology with great possibilities in biomedicine. It is still in its infancy, however, where not least the huge amount of information per sample presents novel problems. There are technical problems with probe design, and limited dynamic range (normal, functional expression levels differ markedly between genes). Another challenge is balancing the likelihood of false-positive results, i.e. the chance occurrence of deviant or interesting results, against false-negative results, i.e. avoiding overlooking real deviant results. The many actors in this rapidly growing market also confer problems with standardization and exchange of data.

Functional analysis of the encoded products

In the human, the majority of genes (DNA regions encoding functional RNAs) have by now been identified, of which the largest group encodes mRNAs/proteins. Identifying a gene or a protein is just a starting point. For each gene or protein, we need to know its function, how it interacts with other genes in forming a body and making it work, its involvement in disease and whether it can be targeted as part of treatment or prevention of disease. Full elucidation of only a single gene represents a formidable task. With an estimated number of 25 000 protein-encoding genes, this will keep an army of researchers busy for decades to come. Techniques dealing directly with proteins are described in Section 6.5. Here we shall point to some techniques involving manipulations at the level of DNA or RNA.

Vectors for functional studies: expression vectors

One way of studying gene function is to induce its expression in cells in a controlled manner. Genes can thus be turned on or selectively inactivated in vitro in cell cultures or delivered in vivo to cells in intact organisms. Introduction of genes into cells with plasmid or viral vectors was described above. In most cases the object of study is a protein-encoding gene, and the gene is usually inserted in the form of a cDNA. The vector must contain suitable promoters (expression vectors), ensuring efficient transcription in the particular cell type under study. If the purpose with the transfection is to produce copious amounts of a protein for biochemical studies or immunization, it may be advantageous to make gene constructs that allow the protein to be secreted (in case it is a cytoplasmic or nuclear protein not normally secreted) and that contain added sequences (tags) allowing easy purification by affinity chromatography. Plasmids that contain the desired elements (promoters, selection markers, tag-encoding sequences, secretion signals, etc.) are either commercially available from a large number of suppliers or can be made in the laboratory by modification of available plasmids.

Modifying the inserted genes can be informative

To understand the function of a protein, manipulating the natural amino acid sequence can be very informative. Deletion of regions or replacement of single amino acid residues can be carried out by modifications of the inserted sequence. Often we want to produce only part of the protein. For example, if the gene encodes a transmembrane protein and the purpose is to produce secreted versions of the membrane external part, it is necessary to delete the parts encoding the transmembrane region and the cytoplasmic stalk. Moreover, it can be useful to construct chimaeric proteins, where parts of two different proteins are joined together in a single polypeptide. Rearrangements of larger cDNA fragments can be done in a cut-and-paste fashion, using restriction enzymes. A convenient approach is to use PCR to amplify the regions of interest. By adding restriction enzyme sites to the ends of the primers, the amplified product can be ligated into the multiple cloning site of the vector. The purpose can be illustrated with an example. Let us say we are dealing with a transmembrane receptor, where we have indications that it is involved in transmitting signals to the cell upon ligand binding, but the ligand is unknown. We can then make hybrid or chimaeric constructs that consist of the signal-transducing region of our receptor joined to the ligand-binding part of a different receptor with known, available ligand. The construct is ligated into vectors for transfection into suitable host cells where the signalling properties can be studied. If we want to identify the unknown ligand(s), we can make constructs the other way round, gluing the ligand-binding parts together with signalling parts taken from other proteins. The system is set up so that readily observable cellular signals report hits when the transfected cells are exposed to candidate ligands. Alterations of one or a few residues generally are not done by restriction enzymes, but instead by site-directed mutagenesis. Today this is commonly done by a PCR-like method using primers that cover the mutation site, but have a different (desired) nucleotide sequence.

Gene knockdown by targeting specific mRNAs

A different approach to elucidating the function of a protein is to look at what happens when it is no longer present, or its expression is significantly reduced. This is called a knockdown experiment, and is performed by specifically targeting the corresponding mRNAs for degradation, stopping them from being translated. Previously this was done by exposing cells to *antisense oligoribonucleotides*. These are oligonucleotides that are complementary to the mRNA we want to inhibit. The method has gradually been replaced by two other, more efficient methods with a similar aim. The first exploits *ribozymes*, which have a catalytic RNA domain which cuts and degrades other RNA molecules. The catalytic domain can be directed against specific mRNAs by adding flanking sequences complementary to the target. Cells can then be transiently or stably transfected with the construct. The second method is referred to as *siRNA* (small interference RNAs), exploiting an evolutionary conserved eukaryotic defence system against viral infections. The siRNAs used for research purposes are synthetic 19–22 bp long double-stranded RNA, made complementary to the mRNA targeted for inactivation. The siRNAs bind to a RNA-induced silencing complex (RISC), which then is guided to and degrades the target mRNA. With properly designed siRNAs this method can be very efficient, and is to some degree even used in vivo in multicellular organisms such as the nematode *Caenorhabditis elegans* and the fly *Drosophila melanogaster*.

Transgenic and knockout animals

The establishment of techniques for genetic manipulation of experimental animals has during recent decades emerged as an invaluable source for functional in vivo studies, with the two main methods popularly referred to as *transgenic* and *knockout* animals (see Box 6.1). In the former technique, a gene is transferred into the genome by the use of retroviral vectors or by microinjection at the zygote (fertilized egg) stage. The *transgene* (transplanted gene) is stably integrated into the genome and will be present in all the cells of the body, including the germ cells. The transgene is therefore heritable. A problem is that the gene is inserted at more or less random sites in the chromosomes. As the majority of loci are transcriptionally silent in mature cells, the chromosomal position effects are usually manifested as unwanted transcriptional silencing. It is possible to shield against these effects by flanking the transgene with insulator sequences. By incorporating suitable promoters, transgene expression can be guided to specific cell types. Genes may be transplanted from other species, but usually the transgene is copied from a gene endogenous to the animal, in which case the experimental model gives information on what happens when the gene is overexpressed.

Knockout animals are generated by experimental inactivation of one or more chromosomal genes. Inactivation is performed on embryonic stem (ES) cells, which are totipotent stem cell lines cultured in vitro. In brief, a defect copy of the targeted gene is transfected into the ES cells. By homologous recombination (a process normally called upon in cells for repair of double-stranded chromosomal breaks) the copy invades and replaces the normal gene. The gene construct must contain a positive selection marker, allowing selection of cells with integrated copies of the replacement construct. Most commonly used is the neomycin resistance gene. Successful site-specific replacements are relatively rare events, whereas unwanted insertions into non-homologous sites are frequent. Incorrect insertions can be screened for by PCR, or alternatively, by adding a negative selection marker to the replacement vector. The negative marker is placed outside the region of sequence similarity between the vector and the targeted locus. The rarity of correct insertions necessitates transfection of many cells, explaining the requirement for ES cell lines, which limits the technology to species where such cell lines are available. The knockout technique is technically more difficult, but offers analytical possibilities beyond the simple transgenic method.

Identification of genes associated with disease: single nucleotide polymorphism typing

The previous section started with a gene and looked at methods to study its function, including its phenotypic effects in vivo. Here we shall look at the reverse problem: starting with phenotypic differences between individuals, we want to identify the genetic contributions to this variability. In the perspective of general biology, genetically induced differences in phenotypes constitute the basis for selection and evolution of life forms. In medicine, identifying disease genes is of utmost value for diagnostic, therapeutic and preventive purposes. Whereas numerous mutations that cause single-gene diseases have already been identified, it is more difficult to identify disease genes and mutations for traits manifested through the interaction by multiple genes, where each locus involved contributes quantitatively [referred to as quantitative trait loci (QTL)].

Genetic loci associated with human disease have in some cases been discovered by the demonstration of chromosomal translocations or deletions of small parts of a chromosome (microdeletions). More commonly, genetic loci associated with diseases are identified by studying whether polymorphic DNA markers co-segregate with the trait. Studies of co-segregation are performed on samples of patients compared with normal controls, with the aim of identifying polymorphic sites in the genome (loci for which different alleles exist) where one allele goes together with the disease (more precisely, there is a statistically significant skewed distribution of alleles between the patient and the control groups). Previously the markers used were variations in length of tandem repeated DNA segments [variable number of tandem repeats (VNTR)] and restriction fragment length polymorphisms (RFLPs). With novel technology facilitating the detection of single nucleotide polymorphisms (SNPs), SNP typing is the preferred method. SNPs, which are alterations of single nucleotides in the genome, constitute the main source of genetic variation between two individuals of the same species. In the human genome, SNPs occur every 100–300 bases, mainly in the form of substitution of one base with another. Most SNPs are diallelic, i.e. have only two alleles. Most SNPs have no effect on cell function, but some may predispose to disease.

There are various methods for typing SNPs, one of which is primer extension. The principle is shown in Figure 6.3. The genomes of the total human population harbour more than 10 million SNPs. When large numbers of SNPs are investigated simultaneously, the chances are great that some will show statistically significant spurious associations, unless P-values are set sufficiently low. This creates the converse problem, of overlooking potentially interesting observations. To obtain reasonable test strength sufficiently many persons must be investigated. Owing to limitations in cost and capacity, testing every SNP in every person is precluded, necessitating the selection of a subset of SNPs for typing.

To aid researchers with this problem, the International HapMap Project was launched. The aim of this project is to identify SNPs inherited together. The background is that the crossing over that occurs between homologous chromosomes during meiosis tends to occur at particular points rather than at random sites. As a result, blocks of sequence can be inherited through a large number of generations

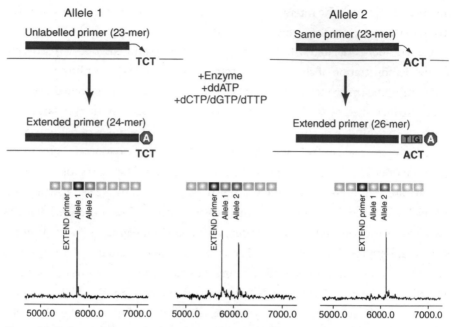

Figure 6.3 Primer extension. At the polymorphic site (SNP) there is either an A or a T. A primer (here 23 nucleotides long) is positioned close to the site and PCR run as described above, except that the added nucleotides are dCTP, dGTP, dTTP and ddATP (i.e. in contrast to a sequencing mix, no deoxyATP, only dideoxyATP). In the example shown, the template with a T immediately after the primer will just incorporate the ddA-nucleotide, and further extension is stopped. The template with an A at the polymorphic site will incorporate a dT-nucleotide, permitting further extension until a template T is encountered. In the case shown, the two PCR products will be a 24mer and a 26mer, respectively. The differences in size are detected by mass spectrometry (bottom panel: AA homozygote to the left, AT heterozygote middle and TT homozygote to the right). By designating primers with different lengths, several SNPs can be analysed simultaneously (multiplexing). The system can be fully automated, with PCR run in, for example, 96-well trays, and the PCR products fed directly into the spectrometer. In this way thousands of SNPs can be analysed per day by a single machine

without being broken up. The blocks vary in size, but most are thought to be 10 000 bases or more in length. The DNA sequence of a block constitutes a *haplotype*.

DNA fingerprinting

Eukaryotic genomes contain many hundred thousands more or less randomly distributed tandemly repeated sequences. The repeat unit can vary in length from just one to more than 100 bp. When the repeated units are $\leqslant$10 bp the sequences are referred to as short or simple tandem repeats (STR), simple sequence repeats (SSR) or microsatellites (the latter designations by some authors restricted to units $\leqslant$5 bp). The different alleles are easily demonstrated by PCR, using primers directed against the normal, non-repetitive DNA sequences flanking the repeat sequence, and with the fragment lengths identified by gel or capillary electrophoresis. Most STRs are polymorphic with respect to number of times the unit is repeated, and multiple allelic forms of the STRs often exist in the population. STRs are therefore highly useful as DNA markers in family linkage studies as well as in forensic medicine for determining paternity, clarifying alleged kinships in cases of illegal immigration and for DNA fingerprinting in criminal cases.

6.5 PROTEIN ANALYSES

Nucleic acids are physically relatively homogeneous. Specific genes or mRNAs are easily detected by hybridization with complementary probes and their locations are limited to the nucleus, cytosol and mitochondrial matrix of the cells where they are produced. Proteins, in contrast, appear as an unmanageable jumble, necessitating a wide repertoire of investigational methods. The main challenges are:

- the wide range of physical properties, with water solubility as a major problem
- the wide range in protein concentrations, which for cellular proteins may vary with a factor of 10^6 (e.g. actin versus transcription factors) and for serum proteins with a factor of 10^9 (albumin versus cytokines)
- the problem of making a comprehensive set of reporter ligands or probes for quantification by protein arrays or mapping of cell and tissue localization
- determinations of three-dimensional (3D) structures, which currently cannot be predicted, but must be determined by resource-demanding techniques, mainly X-ray crystallography and nuclear magnetic resonance
- the complexity of interactive networks between proteins.

Protein purification

Methods for protein purification exploit (main method in parenthesis):

- Differences in surface properties, such as distribution and accessibility of charged, polar and hydrophobic groups, which results in differences in solubility in solvents where variables are salt concentration/ionic strength, pH and content of detergents (precipitation)
- Differences in size and shape (gel filtration, SDS gel electrophoresis)
- Net charge (ion-exchange chromatography, isoelectric focusing)
- Biological properties, mainly affinity for other molecules (affinity chromatography with natural ligand or specific antibody bound to solid phase).

Some of these methods, as well as several other, classical biochemical techniques are also used for protein *characterization*. Most of these will not be dealt with here, where we shall limit the presentation to brief descriptions of a few selected techniques for protein expression analysis, for identification of interacting proteins, as well as some methods included in what is loosely referred to as proteomics.

Protein expression and recombinant proteins

In this section, protein expression refers to the directed synthesis of a large amount of a specific protein. For this purpose a DNA sequence, usually cDNA, encoding the protein is ligated into an expression vector. The vector must contain a promoter that will be triggered in the host cell, and the insert must be correctly orientated relative to the promoter. Three main types of host cells are used: prokaryotic cells, mainly *E. coli*; insect cells; and mammalian cell lines. Prokaryotic cells and mammalian cell lines will be briefly discussed here.

The main advantage of *E. coli* is its simplicity in use: it is easily transformed, divides rapidly, thrives in cheap culture media and can produce large amounts of the desired protein. The disadvantage is that the protein may not be correctly folded or modified. Many eukaryotic proteins need assistance from molecular chaperones for proper folding or to inhibit premature physical contact between hydrophobic parts of incompletely folded neighbouring proteins, which may lead to their forming insoluble aggregates, referred to as inclusion bodies. Most extracellular eukaryotic proteins, whether secreted or constituting the external part of transmembrane proteins, are normally glycosylated, but not in the transformed prokaryotic host cells. In some circumstances a lack of glycosylation is beneficial, in particular when growing protein crystals for use in X-ray crystallography.

Expression in mammalian cell lines can be transient (usually lasting only for a couple of days), stable or inducible. To ensure proper folding and post-translational modifications the host cell should ideally be as closely related to the cell type normally producing the protein as possible. In practice the choice of host cell is limited, as it needs to be established as a cell line that continues to grow and divide in *in vitro* cultures. Besides, many eukaryotic cell types are notoriously difficult to transfect. Two commonly used mammalian cell lines for protein expression are COS-7 and 293T. These lines produce constitutively large amounts of the viral protein SV40 T-antigen (simian vacuolating virus 40 large tumour antigen), involved in viral genome replication, leading to prolonged high expression of plasmids containing the SV40ori. Adding signal sequences for secretion directs the protein to be secreted into the culture medium. By adding epitope tags (see *Vectors for functional studies* in Section 6.4) recognized by antibodies, the secreted protein can be enriched to high purity by affinity chromatography. Alternatively, the Fc-tail of IgG can be added to the C-terminus of the recombinant protein, permitting purification with protein G. This is particularly useful if the protein is to be used for immunizing mice for production of monoclonal antibodies, as the mouse Fc-tail will not be immunogenic in mice.

Protein analyses by electrophoresis, western blotting and immunoprecipitation

Proteins, like nucleic acids, can be separated by electrophoresis. Unlike nucleic acids, which are uniformly negatively charged, proteins differ greatly in their net charge at a given pH. This is because the side-chains of amino acids have very different properties; some have acidic groups, some basic, while some are neutral and hydrophobic. To enable separation based on size, as well as to solubilize largely hydrophobic proteins, the negatively charged detergent sodium dodecyl sulphate

(SDS) is added to the protein sample. Because SDS binds relatively uniformly to most proteins, the net negative charge will be proportional to size. SDS-bound proteins are thus easily separated according to size by electrophoresis in poly-acrylamide gels, in a process known as SDS–polyacrylamide gel electrophoresis (PAGE). Detection of protein can be done unspecifically in the gel by staining methods (Coomassie staining, silver staining or fluorescent dyes). Visualization of a specific protein is commonly done by transferring the proteins in the gel to a polyvinylidene fluoride (PVDF) membrane, a process known as western blot-ting. The membrane can then be exposed to specific antibodies or other probes that specifically bind to a particular protein of interest. One popular use of west-ern blot is the identification of protein interactions. In immunoprecipitation, an antibody is added to a protein sample (typically a cell lysate). The antibody binds to the specific protein of interest, and is then coupled to a solid phase that allows washing away unbound proteins. If the binding forces between two interacting proteins are sufficiently strong, new interacting proteins can be identified from the purified or immunoprecipitated protein sample by subsequent SDS-PAGE and western blot analysis.

Two-dimensional electrophoresis

In addition to separating proteins according to size (or, more correctly, molecular mass), they can be separated based on their net charge. Every protein has an iso-electric point (pI), that is, a pH at which its net charge is zero. In isoelectric focus-ing (IEF), a protein sample is applied to a horizontal medium that carries an immobilized linear pH gradient. An electric field is applied. A protein with net positive charge will then migrate towards the anode while the pH in the medium increases, until the net charge of the protein is zero, and the movement stops. Conversely, a negatively charged protein will migrate towards the cathode until the concentration of H^+ ions renders it uncharged, and the electrophoretic move-ment is arrested. Isoelectric focusing can be useful alone, but has become very useful to improve the power of protein separation. In two-dimensional protein electrophoresis, the isoelectric focusing is first performed on a strip, which is then immersed in SDS, inserted at the end of a polyacrylamide gel and subjected to SDS-PAGE with the electric field at a 90° angle. Because most proteins will have their unique combination of pI and molecular mass, gel separation in two dimensions allows (provided the separation technology is optimal) separation of every single protein into a unique spot. By standardizing the conditions, and analysing stained two-dimensional (2D) gels by computer, proteins that are differentially expressed between two different cell extracts can be identified. Single spots of interest can then be directly cut out from the gel, purified and sequenced.

Protein–protein interactions

When studying the function of a particular protein, it is often desirable to iden-tify other proteins that it binds to and interacts with. One way of identifying such interacting proteins is immunoprecipitation, as outlined above, followed by western blot or protein sequencing. Because this approach has its limitations, an alternative method, *two-hybrid assay*, is sometimes preferred. This can be per-formed in yeast or in mammalian cells. In brief, a gene construct is made that encodes the protein of interest (protein X) coupled to a different protein, e.g. the DNA-binding domain of a transcription factor. A cell line is stably transfected with this hybrid gene. Then, a hybrid cDNA library is made, where the transcription-activating part of the transcription factor is coupled to all the various proteins encoded in the cDNA library (Y proteins). The cell line that already expresses protein X is transfected with the hybrid cDNA library. If one of these library hybrid proteins (Y) binds to our protein X, the joining of two necessary func-tional subunits of the transcription factors in close proximity will lead to tran-scription of a marker gene that can be easily detected, such as green fluorescent protein (GFP) or the enzyme luciferase. Several other similar methods exist, which are more useful in studying the normal function of two presently known partner proteins.

Proteomics

The term proteome refers to the entire collection of proteins that are expressed at a given time in an organism, a tissue or a cell. Proteomics is the simultaneous studies of large sets of proteins, aiming at identifying all proteins that constitute a proteome and understanding their functions and interactions. In the human, the Human Proteome Organisation (HUPO) was launched in 2001 to coordinate international collaboration to achieve these ambitious goals. An important role of HUPO has been to standardize protocols and information output formats. The PRoteomics IDEntifications (PRIDE) database, is a public data repository for proteomics data, complying with standards set by HUPO.

With its basis in high-throughput technologies, proteomics deals with all aspects of protein biochemistry and biology, including separation, identification, quantification, analyses of sequences, modifications and interactions, determination of structure and mapping of cell and tissue distributions. Central tools for protein identification are 2D electrophoresis followed by mass spectrometry; for determination of 3D structure X-ray crystallography and nuclear magnetic resonance; for interaction studies affinity chromatography, two-hybrid assays, phage display and fluorescence energy transfer; and for cell and tissue mapping labelling methods such as immunostaining or incorporation of radioactive or fluorescent markers combined with imaging methods like light and electron microscopy, X-ray or nuclear magnetic resonance (NMR) tomography, positron emission tomography (PET) scanning and confocal microscopy. Some of these techniques have been dealt with above. In the following the presentation will focus on expression analyses, mass spectrometry and in situ mapping.

Expression proteomics and protein expression arrays

In expression proteomics the goal is to catalogue the complete set of proteins in a given cell type or tissue, including how it is changed consequent to disease or medication. For several reasons this is at present an unattainable goal, in particular for multicellular eukaryotes. First, even in the human there still remain novel proteins/protein encoding genes to be identified. Secondly, it is estimated that approximately 60% of our protein-encoding genes represent complex transcription units, in the sense that they can give rise to different proteins dependent on alternative initiation, splicing, termination or editing events. Thirdly, post-translational modifications are widespread. Therefore, at present we have no account of how many different protein variants eukaryotic cells can produce, with estimates varying from 50 000–60 000 to many hundred thousands. Finally, for the majority of proteins there is a lack of specific tools (e.g. reporter ligands) to identify them.

Each cell type expresses more than 10 000 different proteins, with different cell types expressing dissimilar, overlapping subsets. Measurements are complicated owing to wide dynamic ranges in physical properties and concentrations, as well as a lack of specific reporter ligands for the majority of the proteins. Methods used should exhibit

- high resolving power, including the ability to distinguish between modified variants of the same protein
- high sensitivity
- suitability for high-throughput analysis
- the ability correctly to reflect relative protein concentrations.

In principle, expression patterns of a large number of proteins can be measured by microarrays. In contrast to DNA arrays, however, probes cannot be generated based on sequence complementarity. Instead, the reporter ligands must be able specifically to recognize 3D features specific for each protein. The reporter ligands most commonly used are *monoclonal antibodies* (mAbs). There is considerable work behind the generation of mAbs, and although many mAbs are now commercially available, they are expensive and sometimes difficult to produce owing to the low immunogenicity of the proteins.

Alternatives to antibodies are reporter ligands generated by *phage display* and *nucleic acid aptamers*. In the phage display technique random DNA expression libraries are made so that the peptides produced are fused with the coat protein of the bacteriophage vector and therefore will be displayed on the surface of the viral particle. Protein X, against which reporter ligands are to be produced, is used to coat the surface of a plastic dish. The phage-display library is then added to the dish. After a while the dish is washed, removing non-adherent phages. DNA isolated from attached phages contains sequences that encode peptides mediating binding to protein X. The DNA is then isolated and used for mass production of the peptides.

The nucleic acid aptamer technique exploits the folding properties of single-stranded polynucleotides (RNA or single-stranded DNA). Just like chains of amino acids they can form complicated 3D structures with surfaces that can fit into complementary surfaces of other macromolecules, such as proteins, so that the molecules are hooked together by summation of multiple weak chemical bonds. The advantage of using polynucleotides is that they can replicate themselves. By repeated rounds of in vitro affinity chromatography to protein X followed by amplification of the bound aptamers, with gradual increases in binding stringency conditions for each round, high-affinity aptamers can be 'evolutionarily' selected. There are other methods for generating reporter ligands for protein detection, but currently with very limited use. The field is under intense investigation, but it will take many years before we have comprehensive expression arrays for proteins similar to DNA microarrays.

Mass spectrometry

Previously, proteins separated by 2D electrophoresis were identified mainly by immunostaining or sequencing by Edman degradation. Now the main analytical tool is mass spectrometry. The protein spots of interest are cut out of the 2D gel and digested with an endopeptidase, usually trypsin, which cleaves the C-terminal to the basic amino acids lysine and arginine, generating peptides of suitable lengths for further analysis. The collection of resulting peptides can then be analysed by peptide mass fingerprinting. Here, the absolute masses of the peptides are measured by mass spectrometry, which separates ionized molecules according to mass to charge (m/z) ratios. The results are compared with deduced masses of theoretical peptide fragments generated by in silico translating all known protein encoding genes (from the same species) and 'digesting' these with the same enzyme. The comparisons are statistically analysed and the best matches presented.

The peptides can also be further degraded, permitting sequencing of the proteins. This is done by tandem mass spectrometry, involving multiple rounds of mass spectrometry. In the first step the peptides generated by trypsin digestion are isolated one by one. In the second they are stabilized while being fragmented, e.g. by collision with a gas. In the third step the masses of the new fragments produced are analysed. The great accuracy of the mass measurements permits precise identification of the amino acid composition in the peptides, from which the sequence of the original peptide can be deduced based on the series of fragments present in the mixture. The technology of mass spectrometry, with respect to instruments, analytical methods and applications, is developing at a rapid pace. Besides extreme accuracy other advantages with mass spectrometry are exquisite sensitivity and high throughput.

In situ mapping of proteins

The precise cell or tissue localization of proteins is of considerable interest when elucidating their functional roles or mapping potential pathogenetic deviations. The main reporter ligands for in situ mapping of protein are mAbs (see also *Expression proteomics*, above). For detection the mAbs need to be conjugated to reporter groups, such as fluorescent dyes, enzymes or gold particles. Current technology permits the concurrent detection of only a few different reporter groups, limiting the number of proteins that can be mapped at a time. For modern demands of fast access to large amounts of information this is unsatisfactory. Novel types of reporter groups are therefore currently under development, in particular nanoparticles and quantum dots. Nanoparticles down to the size of a few nanometres (nm), which is the size of most proteins, can be made, with antibodies and reporter groups attached to their surface. Quantum dots are tiny semiconductors that can replace organic fluorescent dyes as reporter tags. By adjusting quantum dot sizes, fluorescent groups varying in emitted colours as well as in intensities can be made. Several different quantum dots can be attached to the nanoparticles, making it possible to allocate particles coated with a particular antibody a unique mixture of tags out of many thousand possible combinations. There are great expectations for this technology, which in theory allows the simultaneous in situ mapping of large numbers of proteins.

6.6 BIOINFORMATICS

'In the 1980s, trying to track down the cystic fibrosis gene took us about 10 years of very hard work. There were probably 100 researchers involved and millions of dollars were spent. With the database that's now available an average post-doc working in a good lab would be able to accomplish that in a couple of weeks.'

(Collins 2001)

The annual informational output in biomedicine increases almost exponentially. In particular, the large genome projects have generated enormous amounts of information. In 2006, GenBank, the National Center for Biotechnology Information (NCBI) and its collaborating databases, the European Molecular Biology Laboratory (EMBL) and the DNA Data Bank of Japan (DDBJ), announced reaching the milestone of 100 billion sequenced bases (from over 165 000 organisms). The greater part of this is unprocessed data, representing veritable goldmines laid bare for scientists. The success of the genome projects has depended heavily on the concomitant rapid progress in information technology, as one of the great challenges in these projects has been how to store, sort and make the huge amounts of data readily accessible for further refinement and analysis by the scientific community. In the wake of the biotechnology/information technology booms, bioinformatics has developed as a central research discipline. The official human genome project information website defines bioinformatics as 'The science of managing and analysing biological data using advanced computing techniques', and emphasizes its pivotal role in analysing genomic research data. The recent availability of large numbers of sequenced eukaryotic and prokaryotic genomes has spawned the subfield of comparative genomics, depending on techniques for global and local sequence alignments. Other applications of bioinformatics are virtual evolution, metabolic networks, molecular 3D predictions and morphometrics. An increasingly larger part of biomedical research is now performed *in silico*, according to one authority, 'freeing the biomedical researcher from the tyranny of pipetting'.

Space does not allow description of the many analytical tools available in bioinformatics. Databases and computer programs covering most needs are freely obtainable via the web, such as the homepages of NCBI, EMBL and DDBJ. Via the NCBI homepage, databanks can readily be reached containing information on DNA, protein or whole genome sequences, protein 3D structures, human and mouse SNPs and human genetic diseases. These homepages also offer the highly useful sequence alignment search tool BLAST. Ensembl, a joint project between the European Bioinformatics Institute and the Sanger Institute, produces and maintains annotation on the main eukaryotic genomes of interest to most biomedical researchers. Their webpages are also highly recommended to novices as they offer genome information in a user-friendly way. UniProt (the Universal Protein Resource) is a databank dedicated to information and analysis of proteins. A variety of analytic tools is accessible from web servers such as the EMBL homepage. These range from simple, user-friendly interactive web-based programs to highly sophisticated programs that can be downloaded to run on local UNIX-based computers. One example of the latter is the integrated work package EMBOSS. While many of the UNIX-based programs are at the time of writing difficult to use for the non-specialist, they offer great power and flexibility in use. Alternative, simpler, more user-friendly integrated work packages are commercially available to run in Windows or Macintosh environments. With bioinformatics becoming an increasingly important tool in many biomedical research fields, courses offering more in-depth information as well as practical training are recommended.

REFERENCES

Collins F (2001) Speech of 26 June 2000 announcing the human genome draft sequence, quoted with permission.

Nirenberg M (1967) Will society be prepared? Science 157: 633.

CHAPTER 7

STRATEGIES AND METHODS OF BASIC MEDICAL RESEARCH

Bjørn Reino Olsen and Haakon Breien Benestad

7.1 INTRODUCTION

What is basic medical research? Does it differ from basic research in general? Can it be distinguished from translational and clinical medical research? An introduction to the strategies and methods of basic medical research needs answers to these questions. Yet, no generally accepted answers are available, in part because advances in biomedical research over the past 50 years have led to eroding traditional boundaries between disciplines and research fields, clouding the distinction between basic medical and basic biological research. Moreover, new and ever more powerful molecular biological, genetic, biochemical and morphological methods have enabled the exploration of the fundamentals of cellular and biochemical mechanisms using patient-derived biological materials, making the distinction between clinical and basic research less distinct. For these reasons, the definition of basic medical research in this chapter is pragmatic: it is research that usually is conducted in medical, dental or veterinary schools or institutes, with the aim of eliciting deeper understanding of fundamental normal or pathological processes at the molecular, cellular, organic and organismic levels.

This definition emphasizes understanding and insights at many levels. Consequently, the strategies and methods used to reach such understanding must be effective at all levels of organization. The methods obviously must be able to generate data, but since the goal is to obtain intellectual insight, the data must contribute to the understanding of the biological and pathological 'logic'. Usually this means that the results permit testing of one or more structured ideas or hypotheses. In this view, gathering biological facts alone is seen to be of lesser worth and accordingly is given low priority. Also, because the aims of basic medical research projects are to gain insights into biological processes, the methods used should be the most amenable to providing answers to the questions asked. Successful medical research projects are characterized by relevant questions and clearly stated problems that prompt the use of adequate, effective methodology, and not the other way round (Boxes 7.1 and 7.2). Letting thorough knowledge of and experience with a method trigger research questions, arguably seldom, leads to outstanding research.

Box 7.1 Basic requirements of a research problem

- A clearly stated problem should permit clear description of overall goals and specific aims.
- All hypotheses (structured ideas) should be testable.
- The problem should be amenable to data collection, and the methods selected should be those most efficient in testing the hypotheses.

Box 7.2 The importance of important problems

'The purpose of scientific enquiry is not to compile an inventory of factual information, nor to build up a totalitarian world picture of natural Laws in which every event that is not compulsory is forbidden. We should think of it rather as a logically articulated structure of justifiable beliefs about nature. It begins as a story about a Possible World – a story which we invent and criticise and modify as we go along, so that it ends by being, as nearly as we can make it, a story about real life …

The scientific method is a potentiation of common sense, exercised with a specially firm determination not to persist in error if any exertion of hand or mind can deliver us from it. Like other exploratory processes, it can be resolved into a dialogue between fact and fancy, the actual and the possible; between what could be true and what is in fact the case.

It can be said with complete confidence that *any scientist of any age who wants to make important discoveries must study important problems. Dull* or piffling problems yield dull or piffling answers. It is not enough that a problem should be "interesting" – almost any problem is interesting if it is studied in sufficient depth. A problem must be such that it *matters* what the answer is – whether to science generally or to mankind.'

(Medawar 1979)

7.2 LONG-TERM GOALS AND SPECIFIC AIMS

Significant research projects entail sustained effort of long duration; science is a marathon, not a sprint. Hence, clearly stated long-term goals should be defined at the outset (Box 7.3).

Box 7.3 Goals, objectives and aims

Goals reflect a vision of what the research seeks to accomplish in the long run.

Example: 'The long-term goals of these studies are to identify the molecular mechanisms responsible for the occurrence and characteristic features of infantile haemangiomas in humans.'

Objectives are the more immediate outcomes of studies.

Example: 'The objective of the proposed studies is to obtain insight into the molecular mechanisms responsible for the initial rapid growth and subsequent slow involution of haemangiomas.'

> *Specific aims* serve as practical guideposts for daily, weekly and monthly research efforts.
>
> *Example*: 'Aim 1: Screen for genetic loci in familial haemangiomas by linkage mapping and loss-of-heterozygosity (LOH) analyses; Aim 2: Screen for candidate gene mutations in sporadic haemangiomas.'

The long-term goals serve as remote reminders of where daily scientific work is headed. However, whether their attainment is three to five years in the future (for a doctoral dissertation) or longer (5 to 10 years for a junior faculty member to establish an independent laboratory) or 10 to 20 years (for a professor hoping to significantly impact a chosen field of research), the goals (no matter how well defined) are too distant to serve as practical guideposts for the daily, weekly or monthly efforts. So, specific aims should be generated to organize shorter term efforts.

The aims serve as short-term goals, chosen so that when successfully manoeuvred, they enable the investigator to reach specific objectives on the way towards the long-term goals. They should be structured so that they are interrelated but not sequentially interdependent. Successful completion of one aim should not be a prerequisite for starting work on the next aim. In fact, it should be possible to work simultaneously on two or three specific aims. Also, aims should be defined so that the specific outcomes of work on one aim do not make work on another aim intractable.

For example, in a project on the pathogenetic mechanisms of infantile cutaneous haemangiomas, benign tumours of capillaries that grow rapidly for a few weeks and months after birth and then slowly regress over the next few years (Box 7.4), it would be wrong to plan the generation of a gene knockout model of haemangiomas in mice, following studies to identify gene mutations that are responsible for haemangioma formation (Box 7.3). If gene mutations are responsible for haemangiomas (analogous to genetic alterations responsible for other tumours), generating a mouse knockout model would make sense only if haemangiomas are caused by loss-of-function mutations.

The aims implicitly prescribe effective experimental methods for their fulfilment. Methods and detailed procedures should logically result from the specific

Box 7.4 Rapid growth and slow involution of infantile haemangiomas

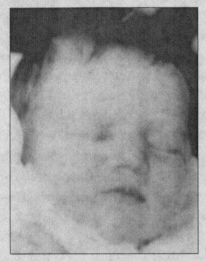

At birth

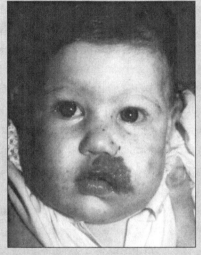

5 months

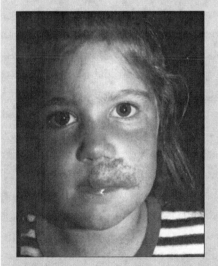

6 years

16 years

From Vikkula et al. (1998)

aims, and not be chosen only because the investigator knows how to perform them. For these reasons, aims are best described in statements that both imply hypotheses to be tested as well as the techniques to test them, without being overly focused and method orientated. For example, for the research on the molecular mechanisms responsible for the initial rapid growth and subsequent slow involution of infantile haemangiomas, the statement 'Screen for genetic loci in familial haemangiomas by linkage mapping and loss-of-heterozygosity analyses' is a better description of an aim than 'Sequence genes x, y, z with DNA from tissues of individuals with haemangiomas' (Box 7.3). Although the work to be done ultimately involves sequencing DNA, the first statement implies the hypotheses that inherited germline mutations engender haemangiomas in some families (and therefore can be mapped), and that second-hit somatic mutations (detectable as loss-of-heterozygosity) may be present in haemangioma lesions. The statement also prescribes the technical approach for addressing the hypotheses, namely using linkage mapping and loss-of-heterozygosity analysis.

The description of a specific aim around a testable hypothesis works well whenever hypotheses are tested. In more descriptive studies, it may be better to state the specific aims in series of questions. For example, if the research were focused on the natural history of infantile haemangiomas rather than the molecular mechanisms responsible for their formation, one may ask: 'What is the phenotypic variability among individuals with sporadic and familial haemangiomas?' A hypothesis-driven aim is not necessarily superior to, just different from a descriptive aim. But it helps to be clear concerning the category of aims pursued, as different aims lead to different goals.

Hypothesis-driven aims can lead to mechanistic insights; descriptive aims can lead to new hypotheses for subsequent testing. Hypothesis-driven research frequently is seen as the gold standard of biomedical research, and many editors and editorial boards will not consider anything else for publication in their journals. Nonetheless, descriptive studies undoubtedly will continue to be important. For example, the natural history of disease is essential to judging the effects of therapeutic interventions, and descriptive analyses of gene expression profiles during development and growth are essential to the understanding of the tissue-specific effects of mutations and of new drugs that are designed to interfere with gene action.

What is a reasonable time period that a specific aim might cover? This depends on the overall scope of the project. For three-year research projects funded by the National Institutes of Health in the USA or foundations in many countries, it is not uncommon to have three aims, each covering a year of work; for projects that are funded by one-year grants, it may be advantageous that each embodies shorter milestones (three to four months).

7.3 BACKGROUND AND SIGNIFICANCE

In a good research project, each long-term goal and each specific aim should address a significant need for better understanding of an area of investigation. Identification of such a need should be based on a comprehensive, in-depth, critical analysis of the contemporary understanding as reflected in the literature. This preparatory phase of a research project is often glossed over and postponed until the experimental phase is completed and a manuscript is about to be written.

All too often, studies are triggered by reading of recent papers in a 'hot' area. If the papers are exciting and lead to some obvious next questions to be addressed, someone with befitting expertise may plunge into the work. This approach to selecting a research project, of letting the work of others define what is to be done next, generally does not generate sustained research at a high level of quality and originality. One reason is that if an exciting paper opens up obvious and important questions, it is likely that the authors of the paper have already extended the work to those questions by the time the paper appears in press. Therefore, any study that is initiated by others upon reading the paper is probably 'too little, too late'. Another reason is that papers that upon first reading seem to break new ground often turn out to be less important in the light of subsequent work and more careful analyses.

Research projects based solely on the possibilities for their funding should also be avoided. Ample funding is essential for sustained research efforts, but it is wrongheaded to let the research area, the long-term goals and the specific aims be shaped by requests for grant applications in specific areas. At best, funding areas or themes are decided by committees of scientists, or at worst by governmental bureaucrats, based on considerations that may be unrelated to scientific needs.

In advance of starting a research project, think about its significant facets: what is scholarly intriguing, what areas are less well developed and therefore challenging, what questions have noteworthy biological and medical ramifications, and so on. A rewarding research project starts by pondering its *background and significance* and, indeed, a statement with just that title usually is required in applications for grants to funding agencies. Funding aside, a description of the background and significance of a project is a useful guide, even if the work is already funded and no grant application is needed (Box 7.5).

Box 7.5 Points to keep in mind during the literature review for a research project

- Do not limit the review to papers that are available only electronically.
- If necessary, go back 100 years – significant discoveries in anatomy, histology and physiology were made that long ago.
- Put your ideas and thoughts concerning the rationale for doing the research and the arguments for selecting specific methods in writing, even if you are not preparing a grant application.
- Think broadly – avoid defining a project so that it principally is a follow-up of specific results published in a paper by scientists elsewhere.

In reading the background literature, be strategic. Read with the purpose of extracting the elements that will contribute to the logical foundation, the rationale, for the research planned. This requires a study of primary publications. A quick reading of recent reviews is insufficient. Another pitfall to be avoided is the commonplace practice of restricting the literature review to papers accessible in electronic databases. Remember that the greater part of the body of medical research of the past 100 years comprises work done before desktop computers became ubiquitous and information technology tools such as Medline and PubMed became available. Including older works in the background reading helps to prevent the awkward rediscovering of what others have reported 50 years or more ago.

When the overall rationale for a research project has been resolved, its significance is apparent. So then it is time to define specific aims, which are the experimental steps that most readily will advance understanding in the chosen area of research in the context of the available resources, the experience of collaborators, the past contributions by the investigator and the state of the art in the field.

7.4 EXPERIMENTAL STRATEGIES AND METHODS

The experimental strategies for addressing the hypotheses and questions of a research project are crucial to its ultimate success. Inadequate methods can easily spoil a project with an unsurpassed hypothesis; innovative methods can enhance a relatively insignificant project and lead it to serendipitous findings. Many investigators are most comfortable with methods they know well and tend to use them even though they may be inadequate. To avoid this cul-de-sac, think about experimental methods as broadly as possible, seek expert advice on unfamiliar methods, learn new techniques as necessary or collaborate with other experts as needed. Above all, be open to using new and technically challenging approaches, and remember that with enough time and effort any technique can be learned and effectively applied.

The experimental strategies should include alternatives to the planned procedures that may be used should the results come out as expected as well as when they differ from expectations. Any significant finding should be corroborated in different ways by different techniques. Likewise, the failure of an experimental technique should be examined by comparison with other techniques.

7.5 PILOT STUDIES

Scientific research might be said to be the systematized pursuit of unknowns. This is particularly true in the medical and biological sciences, in which means and ends usually are not known exactly at the outset. Consequently, a research study may have many unknowns due to its size, its nature or both. So, often a study is preceded by a pilot study, which is a miniature version of the full-scale study that aims to assess its methods and warn of potential drawbacks.

For example, in the project aimed at identifying the molecular mechanisms responsible for the formation of sporadic infantile cutaneous haemangiomas (see Box 7.4), it was clear that the lesions consist of capillary tubes with rapidly proliferating endothelial cells plus pericytes and other stromal cells (Box 7.6). But the lesions may be ascribed to a primary defect in endothelial cells or may be a response to an abnormality in pericytes/stromal cells or in keratinocytes in the overlying skin. As these two underlying causes each would dictate a course of study,

Box 7.6 The cells of proliferating haemangioma lesions

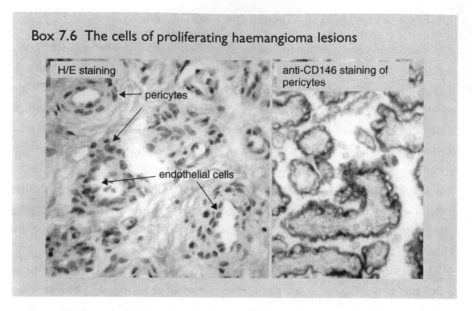

a pilot study was conducted to determine the primary driver of the abnormal angiogenesis (Boye et al. 2001).

It aimed to find whether the primary lesion in haemangiomas lies within the endothelial cells, through testing the hypothesis that haemangiomas are caused by somatic mutations in genes that regulate endothelial cell behaviour, including their proliferation. If correct, the hypothesis predicts that all endothelial cells within lesions are clonally derived from a single abnormal cell. Therefore, in haemangioma lesions of female patients, all endothelial cells should carry the same inactivated X-chromosome (maternal or paternal). To determine whether this is the case, cells were isolated from biopsies of haemangioma lesions and endothelial cells were separated from other cells with a lectin-magnetic bead procedure (Box 7.7).

Using a polymorphism (variable length of a CAG repeat sequence) within the androgen receptor gene on the X-chromosome to distinguish between the maternal and paternal X-chromosome and cleavage with a methylation-sensitive restriction enzyme to distinguish between active and inactive X-chromosomes, it was demonstrated that endothelial cells in sporadic haemangiomas are indeed clonal. In contrast, pericytes/stromal cells from within the lesions were not clonal. That provided strong evidence for a primary defect in endothelial cells and defined a specific strategic direction for the project and a rationale for defining

Box 7.7 The primary defect in haemangiomas

An abnormality intrinsic to endothelial cells?

Caused by somatic mutations?

If answers are Yes and Yes, the prediction is that hemangiomas are clonal expansions of abnormal endothelial cells

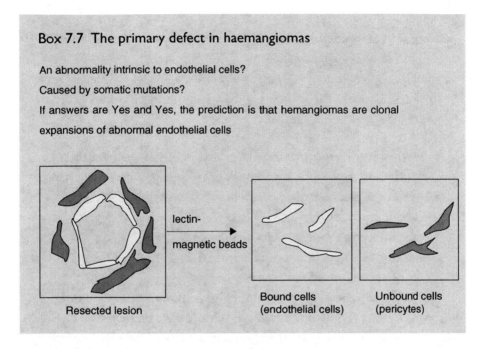

specific aims with the objective of identifying the nature of the endothelial cell abnormality.

As this example illustrates, pilot studies must be conducted as rigorously as full-scale studies. Indeed, a pilot study provides the foundation on which a full-scale study rests and consequently should be just as complete and reliable, with proper controls, positive and negative.

7.6 RULES FOR BASIC MEDICAL RESEARCH PROJECTS

At the beginning of this chapter a pragmatic definition of basic medical research was given as research 'with the aim of eliciting deeper understanding of fundamental normal or pathological processes at the molecular cellular, organic and organismic levels'. This definition can accommodate research on a variety of problems using a wide range of techniques (Box 7.8).

Specific advice on research across this broad spectrum of science is beyond the scope of this chapter. However, at the start of a research project, a few rules (Box 7.9) may be kept in mind.

Box 7.8 A seventeenth century view of the experimental
sciences — still valid?

The *History of the Royal Society*, published in 1667 by Thomas Sprat
(1635–1715), one of the founders of the Society, included a durable definition of the experimental sciences:

> But they are to know, that in so large, and so various an *art* as this of
> *experiments* there are many degrees of usefulness: some may serve
> for real and plain *benefit*,
> without much *delight*,
> some for teaching without apparent *profit*;
> some for light now, and for use hereafter;
> some only for *ornament* and *curiosity*.
> If they will persist in condemning all *experiments*, except those
> which bring with them immediate *gain* and a present *harvest*;
> they may as well cavil at the Providence of God, that He has not
> made all seasons of the year,
> to be times of *mowing, reaping,* and *vintage*.

The first and second of the five rules in Box 7.9 emphasize the need to focus
on 'real' problems of biological systems, that is problems related to in vivo
processes as compared to problems arising from in vitro model systems. For
example, details of gene regulation in cultured cell lines may not reflect gene regulation in vivo and should therefore be studied only if direct in vivo relevance
can be demonstrated. The third rule emphasizes the need for bold choices of
problems to work on. The fourth rule is to divide challenging projects into
smaller steps without loosing sight of the final goals. The fifth and final rule is to
adopt an open, collaborative attitude in dealing with ideas and research projects.
Secrecy does not produce great basic medical science; openness, intellectual gen-
erosity and free exchange of ideas, information and reagents are the triggers of
exciting, innovative science.

Box 7.9 Points to keep in mind when choosing a research theme

- Select biological problems that apply to cells, tissues, organs or organisms in vivo.
- Stay away from questions that have emerged primarily from in vitro studies, unless they clearly apply to in vivo situations as well.
- Do not be afraid of choosing big, challenging problems – why waste time and efforts on minor questions?
- Think strategically – break big problems into a series of smaller steps. Stay focused on the big, distant goals, and at the same time, keep your eyes and ears open, as serendipitous observation may take the research in a direction that is more productive than the one planned.
- Do not be so afraid of having ideas 'stolen' as to prevent productive exchange of ideas and collaborations – the best ideas that result in the most significant research projects usually result from discussions among groups of individuals.

REFERENCES

Boye E et al. (2001) Clonality and altered behavior of endothelial cells from haemangiomas. Journal of Clinical Investigation 107: 745–752.

Medawar PB (1979) Advice to a Young Scientist. Human Action Wisely Undertaken. Harper & Row, New York.

Vikkula M et al. (1998) Molecular basis of vascular anomalies. Trends in Cardiovascular Medicine 8: 281–292.

FURTHER READING

Comroe JH (1977) Retrospectroscope. Insights into Medical Discovery. Von Gehr Press, Menlo Park.

McCabe LL, McCabe ERB (2000) How to Succeed in Academics. Academic Press, San Diego.

Medawar PB (1969) Induction and Intuition in Scientific Thought. Jayne Lectures for 1968. American Philosophical Society, Philadelphia.

Schatz G (2006) Jeff's view on science and scientists. Essays from FEBS Letters. Elsevier BV, Amsterdam.

CHAPTER 8

CLINICAL RESEARCH

Eva Skovlund and Morten H. Vatn

Medical studies are of two types, *observational* and *experimental*. Clinical trials study the effects of interventions, so they are experimental studies on humans.

8.1 CONTROLLED CLINICAL TRIALS

The designation *clinical trial* usually applies to any type of planned medical experiment involving humans. The intent of a clinical trial is to determine which treatment is best suited for patients with a specific diagnosis. Results from a sample of patients are used to draw conclusions on a population of present and future patients.

A greater part of all clinical trials conducted concerns the effects of medicinal treatments and are often initiated by the pharmaceutical industry. Clinical trials may also be used to study other modes of treatment, such as surgery, radiation therapy or lifestyle counselling.

A long series of clinical trials must be conducted to document the effect of a new treatment. The clinical part of the development of a new drug usually occurs in four phases (Box 8.1). Phase I: as a rule, a trial is conducted on a small group of healthy, volunteer test subjects. Its purpose is to study toxicity and side-effects. In Phase II, small groups of patients are used to study the effects of

> **Box 8.1** Phases in the development of a drug
> - Phase I   Toxicity and side-effects
> - Phase II   Dose–response
> - Phase III   Comparison with established treatment or placebo
> - Phase IV   Postmarketing

various doses of a drug. When a drug is believed to have shown promising effects in these early phases of its development, it is desirable to test the new treatment against the standard treatment. Phase III comprises larger clinical trials that compare the effects of two or more different treatments, either active drugs or placebo. These trials are used to document the effect of a new drug before the manufacturer applies for marketing authorization. Some trials have combined Phases II and III, called Phase II–III or Phase IIb, to investigate dosage in relation to assessment of effect in a randomized, double-blind study. Phase IV, postmarketing, covers both long-term studies of safety and pure marketing studies.

Phase III trials are those that most often come to mind when clinical trials are mentioned, so in the remainder of this chapter, 'clinical trial' is used synonymously with 'Phase III study'.

Protocol

A detailed protocol must be written before a clinical trial starts. Its purpose is to state the aim of the study, document its trial plan, its inclusion and exclusion criteria, the treatments and clinical examinations, the evaluation of response to treatment, randomization, blinding, number of patients, plan for statistical analyses and reporting of adverse events. Moreover, it includes relevant practical and administrative procedures. A protocol may be required to obtain official approval to conduct a clinical trial as well as to apply for funding of a trial, both vital aspects for the researcher.

The principal parts of the protocol are described below.

Trial plan

The protocol shall describe the background for the study and state why it should be done and how it shall be conducted. The first phase of planning should set three main aspects:

- Which treatments will be compared?
- Which patients will be included?
- How will adverse events be recorded?

Moreover, the study design is defined and the required number of patients estimated (Box 8.2).

Box 8.2 Principal points of a protocol
- Purpose of the study
- Trial plan
- Inclusion and exclusion criteria
- Number of patients, power
- Description of treatment
- Patient follow-up
- Randomization, blinding
- Outcome variables
- Adverse events
- Statistical analysis
- Administrative matters
- Ethics

It is important to describe how the data are to be processed (statistical analysis), how dropouts or withdrawals will be treated and how deviations from the protocol will be addressed. In addition, patient information must be compiled to provide the basis for an informed consent statement to be signed by patients before they take part in the study.

For studies that go beyond routine clinical treatment, approval must be sought from the relevant ethics committee and, if necessary, from the data inspectorate and the national medicines agency.

Patient characteristics and the patient's actual condition at the outset are pertinent to the study and accordingly are recorded as the study baseline. Some studies include a run-in period to identify fluctuations in advance of treatment or to assess patient compliance with instructions on medication intake. Such a period provides better rationale for deciding whether a patient is eligible for inclusion in the study. That said, the drawback of such a strategy is that excluding patients with certain characteristics reduces the generalizability of the study results to the general population of patients. Occasionally, an initial pilot study is conducted before a clinical trial starts. A pilot study may have various purposes, such as testing procedures in practice or estimating the variability of an outcome variable to calculate the number of patients to be included.

For most trials, the number of patients included is decided in advance. That leads to a fixed sample design. If the intended number of patients is large, data may advantageously be analysed one or more times in the course of the trial. If conclusions on the effect of a treatment can be drawn early on, new patients need not be included. Such analyses, conducted before the intended number of patients is included, are called interim analyses and require the use of special statistical methods. If the number of patients intended is not set in advance and data are continuously analysed, such as each time a response for a patient is recorded, the trial is termed sequential.

The effect of a new treatment is assessed through comparing it with the effect of the standard treatment. There are two principal types of comparative study. In a parallel-group study, one group of patients receives a new treatment, and another group of patients receives a standard treatment. The effects of the treatments are compared between the two independent samples of patients. In a cross-over study, all patients receive both treatments, but in different periods of time during the study. Only the order of the treatments varies among the patients. The patients then are their own controls, which reduces random variation. Accordingly, cross-over studies require fewer patients than do parallel-group studies. Nonetheless, cross-over studies are less frequently used, in part because their applicability is limited to studies of treatments of illnesses that are chronic and stable. Moreover, there may be a carry-over effect, if the effect of treatment with one medication extends into the next treatment period. In both parallel-group studies and cross-over studies, more than two treatments may be compared.

In a parallel-group study, the comparison is known as 'between individuals', while in a cross-over study the comparison is 'within individual'. The within-individual comparison has the advantage that the variation for an individual patient over time usually is less than that between different individuals. Cross-over studies are less well suited to long-duration treatment, as with increasing time, there is increasing risk of patients withdrawing from the study. As cross-over studies usually involve few patients, the relative loss when a patient leaves the study is greater than in a parallel-group study involving a larger number of patients.

In addition to being used for comparing treatments with different drugs, randomized clinical trials are used to compare a medicinal treatment with other forms of treatment and to compare other types of intervention, such as different surgical treatments. These types of clinical trial bring new complications of standardizing interventions and organizing studies, and often it is difficult or impossible to conduct a blinded study. Other examples include various forms of care (hospital-based compared with home-based) as well as prevention (e.g. vaccination) and screening trials.

Historical controls

It is not unusual to try to conduct a study without a control group. The comparison then is between patients receiving a new treatment and patients who previously received a standard treatment. The patients previously treated are termed historical controls. Using historical controls is tempting, but entails difficulties, the chief of which is ensuring fair comparison of treatments. If the two groups compared differ in other ways than the treatments administered, it is difficult to say whether a seemingly better outcome in the group receiving the new treatment is due only to that treatment. Sources of bias may be differences in the patient selection or systematic changes in surroundings over time, and may tend to exaggerate the effect of the new treatment.

As a rule, a historical control group is not subjected to inclusion and exclusion criteria, whereas the inclusion of patients in a group receiving a new treatment is far more restrictive. It is also possible that the current patients are of a different type than those treated a while ago. Another problem is that the quality of recording of the treatment effect for the historical controls may be substantially lower, as they were originally not part of a trial. Retrospective collection of information seldom can overcome the problem. Moreover, as a rule, the criteria for evaluating response may change with time.

Accordingly, historical controls should be used as little as possible. Historical controls might provide a basis for the generation of hypotheses, but should not be used in reaching conclusions concerning an effect.

Randomization

Unbiased comparison of treatments is ensured by random allocation of patients to treatments (or to sequence of treatments in cross-over studies). The purpose of randomization is to prevent systematic differences between the treatment groups. Simple randomization is equivalent to tossing a coin, but usually random number tables are generated on a computer in order to allocate patients randomly to the groups of a trial. Selection bias may be avoided by obtaining patient consent to take part in the trial before randomization begins.

The most common randomization is one-to-one, that is, approximately equal numbers of patients in both groups. However, in comparisons of new and established treatments, occasionally more patients may be included in the new treatment group, as the new treatment is the less well known. Likewise, in studies including a placebo group, the randomized groups may be numerically unequal because one wants fewer patients to receive placebo.

Simple randomization does not necessarily result in the groups being equally distributed in age, gender and other characteristics that may be significant, but any differences will be due to chance and not to systematic selection of patients to different groups. Stratified randomization may be used when more closely similar distributions are required.

Stratification

Stratified randomization keeps patient groups balanced for prognostic factors. It is most often done in randomized blocks. First, a few important prognostic factors are identified. For example, in a breast cancer study, the characteristics can be oestrogen receptor (positive or negative) and tumour size (T1 or T2). Stratification based on these two factors, each of which has two levels, results in four (2×2) strata, or subgroups. Within each stratum, patients are randomized in blocks.

Let the two treatments be called A and B. A customary block size is four, for which there are six permutations of the sequence of patient enrolment: AABB, ABAB, ABBA, BBAA, BABA, BAAB. This ensures that after every fourth patient is enrolled, two will have received treatment A and two will have received treatment B. Hence, balance is attained between treatments A and B. An example of block randomization in four strata in a breast cancer study is shown in Table 8.1.

Table 8.1 Randomized blocks within strata

OESTROGEN RECEPTOR $+T_1$	OESTROGEN RECEPTOR $+T_2$	OESTROGEN RECEPTOR $-T_1$	OESTROGEN RECEPTOR $-T_2$
A	A	A	B
B	A	A	A
B	B	B	A
A	B	B	B
B	B	A	B
B	B	B	A
A	A	B	B
A	A	A	A
A	A	B	A
B	B	B	B
A	B	A	A
B	A	A	B
.	.	.	.
.	.	.	.

Unfortunately, when the experimenter knows the block size in a trial that is not blinded (discussed below), it may be possible to anticipate the treatments given to some patients. In any event, that is obvious for the last patient enrolled in each block. Hence, the experimenter may know which treatment a new patient will receive. Such knowledge may consciously or unconsciously affect the decision of whether a particular patient should be included in the trial. The result may be selection bias. Larger blocks lessen that problem, but there is also a greater risk of imbalance in a stratum in which few patients are enrolled.

Stratification with a large number of prognostic characteristics should be avoided. The number of strata increase rapidly with an increasing number of factors. For example, including a third prognostic factor in the breast cancer study example above, such as number of positive lymph nodes, grouped in three levels (0, 1–3 and 4+), will increase the number of strata from four to 12 ($2 \times 2 \times 3$). A difficulty with a large number of strata is that some strata may have very few patients (or none at all), which breaks down the entire concept of balance between treatments using randomized blocks.

Stratified randomization can be resource demanding. As a rule, stratification is considered unnecessary in larger studies, because full randomization is expected to provide good balance with a large number of patients. Moreover, stratification should be avoided if the prognostic significance of a particular factor is uncertain.

Blinding

Randomization of patients into treatment groups does not ensure unbiased comparison of treatments. Systematic bias may arise whenever a physician knows which patient has received which treatment. Particularly when response is subjectively evaluated, knowing the details of the treatment a patient receives may lead to the physician's (or the patient's) confidence in the treatment consciously or unconsciously affecting the evaluation of its effect.

Ideally, all randomized trials should be double-blind, so that neither the patient nor the treating physician knows the details of the treatment given. If a standard treatment is non-existent or largely ineffective, as often is the case for cancer treatments, both the patient and the physician may be overly optimistic about the effects of a new treatment. Consequently, knowledge of the treatment given can influence a patient's response. If there is no standard treatment, the patients in one group may be given a placebo, which is an inactive substance having the same appearance and preferably the same smell and taste as the active drug.

Blinding clinical trials is sometimes difficult, such as in clinical cancer research, in which few blinded studies are actually run. This may be because a treatment is so toxic that it is essential to know when it is given to a patient, or it may be because dosing regimens differ considerably. Assume, for instance, that treatment A is two tablets taken twice daily and that treatment B is one capsule three times a day. This difficulty may be overcome by using a double dummy technique in which one patient will be given the active treatment A tablets and placebos for B, while another will be given active treatment B capsules and placebos for A. Each patient takes six tablets and three capsules daily, but only the tablets or the capsules contain the active drug, depending on whether the patient is randomized to treatment A or B. This procedure is straightforward for oral treatment with few tablets or capsules, but can be difficult or possibly ethically unacceptable if the medication has to be injected.

Sometimes a treatment has specific side-effects that clearly reveal its use, even though the research plan is double blind. Trials that are not blinded or in which the randomizing code is broken can be subject to bias, both in influencing patient expectations and in evaluation of the effect of a treatment. Bias in evaluation can be avoided by having one investigator administer treatment and another evaluate patient responses.

Which patients should be included in the statistical analysis?

It is desirable to record as much relevant information as possible on all patients enrolled in a clinical trial, but in practice, much information may be lacking at the end of a trial. Some patients may have dropped out of the trial, some may have taken excessive or insufficient doses of the medication studied, or some may have completely refrained from taking the medication they were randomized to receive. In some cases, it may be necessary to stop treating a patient, such as when adverse events occur.

It is vital to take a decision on the patients to be included in the statistical analysis before viewing the effects of a treatment. There is no set criterion concerning which patients should be included in an analysis, but plans for handling insufficient data should be made in advance and preferably be described in the protocol, lest systematic error may arise due to the lack of a plan. For instance, the statistical analysis may overestimate treatment effect if it does not include patients who withdraw from treatment owing to lack of efficacy.

Strategies for deciding which patients shall be included in statistical analyses divide roughly in two categories. A *per-protocol* analysis includes only compliant patients who take a predefined percentage of the planned dose (such as 80–120%) and who fulfil the inclusion and exclusion criteria of the protocol. Such an analysis may be used to estimate the 'true' effect of a treatment. In real life, many patients do not take prescribed medications, so an *intention-to-treat* analysis includes all randomized patients, regardless of whether or not they have taken the treatment prescribed. This can be viewed as a pragmatic approach that underestimates rather than overestimates differences in treatment effect. Conclusions regarding the effect of a treatment often are drawn from a form of intention-to-treat analysis.

Choice of endpoints

One or just a few efficacy variables, or endpoints, should be chosen to evaluate a treatment effect. The results of a study with many efficacy variables may be difficult to interpret. Nonetheless, data on many relevant variables are often collected, and efficacy could be assessed in different ways with different variables. In such cases, it is wise at the planning stage to distinguish between primary and secondary efficacy variables. Typically, a study will have one primary and two or more secondary endpoints.

The recording of endpoints assumes standardized, reliable, valid clinical methods, including laboratory tests, clinical recording of symptoms and indications, or the filling in of questionnaires. In any event, reproducibility and accuracy must be acceptable. The standardization of a method also presupposes familiarity with conditions that can affect its stability and area of application.

Multiple endpoints

In most clinical trials, effects are measured in various ways. For example, the outcome of a cancer treatment may be expressed as overall survival, cancer-specific survival, progression-free survival, tumour response according to specific criteria, toxicity or quality of life.

Analyses of a large number of endpoints can be problematic, because separate analyses at the 5% level will increase the probability of false-positive findings. With a significance level of 5%, one out of 20 tests will be expected to show a statistically significant difference, even when the treatments are equally good (or equally poor). The more tests performed, the greater the probability that one or more of them will by chance have a low P-value. The Bonferroni correction is one way of reducing the probability of false-positive findings. Each P-value is multiplied by the number of tests to be performed, and the adjusted P-values are compared at the 5% level. If, for instance, four tests are planned and a total level of 5% is sought, each calculated P-value is multiplied by four. (Instead, one may divide the total significance level by the number of tests and use the quotient for each test, that is, for four tests, compare each P-value with the 1.25% level.)

A disadvantage of the Bonferroni correction is that it is conservative. It overcorrects slightly, particularly when the various endpoints are correlated. A better strategy may be to reduce the number of endpoints or predefine one or two primary endpoints. It can be tempting to select endpoints retrospectively for which the statistical analyses have given low P-values, but that arguably is cheating, as it results in overestimation of the difference in treatment effects.

Instead of identifying a single primary endpoint, several endpoints may be combined in a composite endpoint or joint variable, such as in assessments of the quality of life. That said, it may be problematic to define a clinically meaningful composite or joint variable, so care must be exercised in interpreting the analytical results.

Subgroup analyses

It is natural to ask whether an observed difference between two treatments depends on particular patient characteristics, such as gender, age or another prognostic factor. A commonplace, though unsuitable, way of addressing that question is to divide the patients into subgroups that are analysed separately.

One of the drawbacks of subgroup analysis is that many subgroups may be defined and accordingly many P-values calculated, which increases the risk of false-positive findings. Conversely, as subgroups are often small, relatively large real effect differences may go undetected because the power (probability of revealing differences) is too low (see Section 8.3). Presenting results for various subgroups may be useful, but the presentation should be ancillary to comparisons based on all the patients, and the results accordingly interpreted cautiously. In any event, subgroup analyses should be planned in advance.

Separate significance tests do not answer the question as to whether a given prognostic factor affects the treatment effect. Instead, one should perform a test for interaction, that is, a significance test that assesses whether there is a difference between treatment effects in different subgroups. This can be done by including an interaction term in a statistical model, such as multiple regression analysis. A typical result of including such a term may be that treatment A has a better effect than treatment B and that the difference of effects is greater for women than for men.

Interim analyses

Many clinical trials are conducted over several years. Including enough patients may take a long time, and in some cases, patients must be followed for several years before the effect of a treatment can be evaluated. An example is clinical cancer research, which is primarily concerned with studying the effects of treatments on survival. In such cases, it could be valuable to conduct one or more interim analyses on the effects of a treatment before the planned number of patients have been included in the study. If a difference between the effects of treatments is identified early on, it is desirable from an ethics point of view as well as financially sensible to stop the trial. Thus, fewer patients will be included in the group having the poorest treatment effect, and an earlier start may be made in making the best treatment available to all.

Interim analyses should be made only when planned in advance. To see why this is so, consider a study planned with 100 patients, 50 in each of two groups, with analysis scheduled until responses have been recorded from all 100 patients. Then assume that it takes longer to include patients than initially assumed; after data from 50 patients have been acquired, it is tempting to analyse the results to date. If a significant difference then can be shown between the groups, it is tempting to conclude that one treatment is superior to the other. If, however, no significant difference is shown, the study will continue to enrol patients. Hence, there may be several opportunities to identify differences between the treatments. If each test is performed at the 5% level, the total probability of a false-positive finding (rejecting the null hypothesis even if it is true) will be considerably greater than the 5% sought. The probability of false-positive findings increases with repeated testing of accumulated data, as shown in Table 8.2. In the example of 100 patients, a hypothesis test is conducted after data are available on the first 20, say, and the test is stopped if a significant difference is found at the 5% level but continued with the enrolment of 20 new patients if no significant difference is found. That scenario may be repeated until data have been acquired from 100 patients, so there may be up to five tests. With that strategy, the total probability of detecting a difference even though none exists will be 14%, not 5% as initially planned.

Table 8.2 Probability of false-positive finding

MAXIMUM NUMBER OF TESTS	TOTAL SIGNIFICANCE LEVEL (%)
1	5
2	8
3	11
5	14
10	19

The computed total significance levels in Table 8.2 are based on responses that follow a normal distribution with known variance, but other distributions produce similar results.

The drawback of increasing probability of false-positive findings may be circumvented by reducing the nominal level for each interim analysis. The maximum number of analyses is set before the trial begins. Interim analyses are usually performed at set intervals, either in time or in numbers of patients enrolled. Interim analyses are often called group sequential tests. A simple group sequential method that ensures that the total level is maintained has been proposed by Pocock (1983). Here, the level for each analysis is reduced, as shown in Table 8.3. If a maximum of five analyses is to be performed, each individual interim analysis must have a significance at the 1.5% level in order that the trial may be stopped with a declared

Table 8.3 Nominal level in interim analysis with a total significance level of 5%

MAXIMUM NUMBER OF TESTS	ANALYSIS NUMBER	NOMINAL SIGNIFICANCE LEVEL	
		POCOCK	O'BRIEN & FLEMING
1	1	0.05	0.05
2	1	0.029	0.0051
	2	0.029	0.0475
3	1	0.022	0.0006
	2	0.022	0.0151
	3	0.022	0.0472
4	1	0.018	0.00004
	2	0.018	0.0039
	3	0.018	0.0184
	4	0.018	0.0411
5	1	0.016	0.000005
	2	0.016	0.0013
	3	0.016	0.0085
	4	0.016	0.0228
	5	0.016	0.0417

difference of treatments having a total significance level of 5%. There are several group sequential and pure sequential methods (in which an analysis is performed for each recorded patient response). The pure sequential method differs from most group sequential methods in that the trial may be stopped early not only if a significant difference is identified, but also if it may be concluded that there hardly is any difference in the effects of the treatments.

The disadvantage of the Pocock method is that it somewhat reduces power. In practice, other stopping rules (O'Brien & Fleming 1979) are more often used. They differ from the Pocock method in that the nominal significance level varies between the interim analyses. The level for the first analysis is chosen to be very low, while that for the last analysis is nearly 5%. The advantage of the method is that power is not lost and that there is no need to include more patients than would have been the case had the trial been conducted with a fixed sample size, that is, without planned interim analyses. Examples of the O'Brien & Fleming stop rules are also listed in Table 8.3.

8.2 PUBLICATION BIAS

Statistical significance is often regarded as conclusive proof that two treatments have different effects. Regrettably, that line of thinking often leads to erroneous conclusions. In published papers, there is a trend for authors to include significant results but not mention insignificant ones. Such omissions may be deliberate, but most often occur because the author is not aware of the inherent bias. Moreover, trials showing statistically significant differences are more frequently published than trials with a 'non-significant' outcome. That worsens the situation. In addition, journal editors and readers tend to give more weight to significant results.

Were all trials published, one might easily suspect that rare divergent positive findings were due to chance. In the real world, for every published paper reporting significant differences, there might be 10 other trials that showed no difference between treatments which remain unpublished. This leads to the assumption that the differences reported are real and, in turn, to drawing erroneous conclusions on efficacy. It is likely that a worrisomely large portion of the differences reported in the medical literature actually comprises false-positive findings.

8.3 ESTIMATING SAMPLE SIZE

Power

Before an experimental or observational study starts, it is essential to estimate how many patients or people need be enrolled to reach valid conclusions on an effect. The first question to be answered is how large the difference in the effects of two treatments must be to have clinical relevance.

As illustrated in Table 8.4, one of four results may occur in a hypothesis test. Two events, A and D, are desirable, while two others, B and C, result in erroneous conclusions. The errors that may occur are rejecting the null hypothesis even though it is true (B) or accepting it when it is false (C), often called type I error

Table 8.4 Possible results of a hypothesis test

	H_0 IS TRUE	H_0 IS FALSE
H_0 accepted	A	C
H_0 rejected	B	D

and type II error, respectively. The aim should be to keep the probabilities of these errors small. The upper limit to the probability that B will occur, $P(B)$, is the significance level of the trial, which as a rule is set at $\alpha = 5\%$. Likewise, β often designates $P(C)$. The power is the probability of detecting a certain treatment effect; it is designated $P(D) = 1 - \beta$ and is the probability of rejecting the null hypothesis when it in fact is false. This probability should be high, and in planning a trial it is usually set to 80%, 90% or 95%.

The first step in planning a trial is to select the significance level and power relevant to the least observable difference considered clinically relevant. Then the number of patients to be enrolled is determined, and the trial is conducted. Finally, statistical analyses are performed on the data.

The smaller the difference sought, the greater the number of patients to be enrolled in the trial. Unless a trial is large, small differences have low power, and even clinically interesting effects run considerable risk of going unseen. The conclusion that two treatments have equal efficacy cannot automatically be drawn, even though the null hypothesis is not rejected, as perhaps there were too few patients to detect a difference. Only when the power is high can the claim be made that not rejecting the null hypothesis corresponds to there being no difference in the treatment effects.

Binomial response

Assume that of the patients receiving a standard treatment, 40% ($p_s = 0.4$) survive for five years, and a trial is set up to determine whether a new treatment can increase the survival rate to 50% ($p_n = 0.5$). See Box 8.3 to calculate sample size.

Box 8.3 Sample size, binomial response

A simple and useful equation for calculating sample size is

$$n = \frac{p_n \cdot (1 - p_n) + p_s \cdot (1 - p_s)}{(p_n - p_s)^2} \cdot c$$

in each group. The constant c depends on the choice of significance level and power (probability of detecting a certain difference between two treatments).

Relevant values of c for various significance levels and power are given in Table 8.5.

Table 8.5   Constant c

SIGNIFICANCE LEVEL	POWER		
(TWO SIDED)	0.80	0.90	0.95
0.10	6.2	8.6	10.8
0.05	7.9	10.5	13.0
0.01	11.7	14.9	17.8

Assume that in the example above we choose a significance level of 5% and a power of 80%. Accordingly, in each group it is necessary to enrol

$$n = \frac{0.5 \cdot 0.5 + 0.4 \cdot 0.6}{0.1^2} \cdot 7.9 = 387 \text{ patients}$$

Continuous response
Pairs of observations

Assume that a cross-over study is conducted in which all patients receive two types of treatment in randomized order. Let Δ be the difference in treatment effect sought and σ_d the standard deviation of the difference between the two treatment effects. It may be difficult to make a valid assumption on variability before the trial is done, but earlier publications or a pilot study may provide an estimate. Sample size is calculated as shown in Box 8.4.

Box 8.4 Sample size, continuous response: pairs of observations

The number of patients to be enrolled is

$$n = \left(\frac{\sigma_d}{\Delta}\right)^2 \cdot c$$

where values of c are listed in Table 8.5.

Assume that a cross-over study of two asthma drugs is planned. The peak expiratory flow rate (PEFR) in litres per minute is the primary efficacy variable. A clinically relevant difference in effect is set at $\Delta = 30$ l/min, and we assume that $\sigma_d = 50$ l/min. The significance level is set at 5% and the power at 80%. This gives $c = 7.9$. So the study must enrol

$$n = \left(\frac{50}{30}\right)^2 \cdot 7.9 = 22 \text{ patients}$$

Two independent samples (parallel groups)

> **Box 8.5 Sample size, continuous response: parallel groups**
> When two independent samples are to be compared, the number of patients n in each group is:
>
> $$n = 2 \cdot \left(\frac{\sigma}{\Delta}\right)^2 \cdot c$$
>
> where, as before, Δ is the clinically relevant difference sought, c is a constant that depends on the significance level and power, and σ is the standard deviation of the observations, assumed equal for the two groups.

As a rule, considerably more patients are needed in a parallel-group study than in a cross-over study. Assume that a parallel study of two asthma drugs is planned, similar to the cross-over study described above. A clinically relevant difference in PEFR is assumed to be 30 l/min, while the standard deviation is assumed to be $\sigma = 60$ l/min. The standard deviation is larger than in the cross-over example, because the parallel study takes individual data from different patients in the two groups, whereas the cross-over study used within-patient differences. As it is reasonable to assume that observations on a single patient are positively correlated, σ_d is usually less than σ. The estimated number of patients to be enrolled then is

$$n = 2 \cdot \left(\frac{60}{30} \right)^2 \cdot 7.9 = 63$$

patients in each group, that is, 126 in all.

Unequal group size

Most often, randomization is 1:1, so the two groups are of the same size. However, in some cases, unequal randomization is desirable, so the two groups are of different sizes. One such case may be when a trial includes a placebo group that is kept small to minimize the number of patients given inactive treatment. Another case is when more extensive experience is sought with a new drug, so a larger number of patients will receive it.

For a set total number of patients, 1:1 randomization always achieves the highest power. As shown in Table 8.6, if instead, randomization is 2:1, the loss of power is small. For greater differences in the numbers of patients in the two groups, it may be wise to increase slightly the total number of patients enrolled.

Table 8.6 Loss of power with unequal sample size

RATIO F	
NEW TREATMENT:CONTROL	POWER
1:1	0.95
2:1	0.92
3:1	0.88
4:1	0.84

First, n is calculated as if the groups were of equal size. Then, a modified sample size n' is calculated. Assume that f is the ratio between the numbers of patients in each of the two groups. The total number of patients to be enrolled then is

$$n' = \frac{n \cdot (1 + f)^2}{4f}$$

and the sample sizes in each of the two groups are

$$\frac{n'}{(1+f)} \text{ and } \frac{f \cdot n'}{(1+f)}$$

For instance, randomizing in a ratio of 3:1 leads to a modified number n' of

$$n' = \frac{4}{3} \cdot n$$

Computing sample size based on precision of estimates

Instead of basing computation of the sample size on the power sought, the computational problem may be formulated in terms of the precision in an estimate. The question then is, for example, how wide shall the 95% confidence interval of an estimate be accepted to be, or, in other words, how imprecise an estimate will we accept?

Binomial response

> **Box 8.6 Precision of the estimate of a binomial p**
> The 95% confidence interval of a binomial p is
>
> $$\hat{p} \pm 1.96 \cdot SE(\hat{p})$$
>
> where $\hat{p}$ is the estimated proportion, n the total number of observations, and $SE(\hat{p}) = \sqrt{(\hat{p}(1-\hat{p}))/n}$. Let the interval length be $2d$ (that is, the 95% confidence interval is $\hat{p} \pm d$). The number of observations n then is
>
> $$n = \left(\frac{1.96}{d}\right)^2 \cdot p \cdot (1-p)$$

The magnitude sought, p, is unknown. To estimate the number of observations necessary, we must either reckon a reasonable value of p or set it $p = 0.5$, which is conservative and results in the largest possible value of n for a given interval length, because the product $p(1 - p) = 0.25$ is maximum. The product is lower for all other choices of p. If p cannot be estimated in advance, it is reasonable to use $p = 0.5$ in estimating n. If, however, p is presumed to be relatively small (such as <10%), $p = 0.1$ is more suitable in calculating n.

This type of calculation is the basis of assessing the number of people to be questioned in a political opinion poll and is also often used for Phase II trials.

Example: Assume that we wish to know the proportion of general practitioners whose first choice in treating sinusitis is to prescribe penicillin. We intend to send out a questionnaire and accordingly need an estimate of how many general practitioners should be queried. First, we must stipulate a precision of the estimate, in other words, how wide the 95% confidence interval of the proportion is accepted to be. Assume an interval length of 0.1 (that is, $d = 0.05$) and set $p = 0.5$ because the magnitude of the proportion is unknown. We then estimate that

$$n = \left(\frac{1.96}{0.05}\right)^2 \cdot 0.5 \cdot 0.5 = 385$$

general practitioners should be included in the study.

Continuous response

The 95% confidence interval for the true mean is

$$\bar{x} \pm t_{0.05,n-1} \cdot \frac{s}{\sqrt{n}}$$

where $\bar{x}$ is the estimated mean, s the standard deviation, n the number of observations, and t the relevant fractile in the t-distribution. For practical purposes, $t_{0.05}$ is often replaced with $z_{0.05} = 1.96$, and this approximation holds when $n \geqslant 30$.

Box 8.7 Precision of an estimate of the true mean

The number of observations necessary to attain the required precision of an estimate of the true mean is

$$n = \left(\frac{1.96 \cdot \sigma}{d} \right)^2$$

where, as before, d is the half length of the confidence interval and σ is the standard deviation.

Regrettably, in practice σ is unknown and must be estimated, for instance, from earlier studies, if such exist.

8.4 'NON-INFERIORITY' STUDIES

The above text discussed clinical trials that seek to document the superiority of one particular treatment compared to another. That is often the case, but an increasingly greater part of clinical trials aims to show that a new treatment is not inferior to a standard treatment.

This aim is a consequence of the regulations on approval of new drugs. According to European legislation on medicinal products, it is sufficient to document that a new product has an effect not worse than the product already approved and that its benefit:risk ratio is acceptable.

In documenting that a new drug is not inferior to one already on the market, it is not sufficient to demonstrate that the P-value for a comparison of the efficacy of two treatments is not statistically significant, i.e. $p > 5\%$. A high P-value may be due to the two treatments being (approximately) equally good; but it also may be due to a type II error, which corresponds to not detecting an actual treatment difference even if it exists. The risk of type II error increases with the diminishing number of patients included in a trial.

In a trial aiming to show that the effect of one treatment is superior to that of another, the null hypothesis would be that the effects of the two treatments do not differ. The alternative hypothesis would be that the effects differ, and tests are almost always two sided. In general, the null hypothesis is the one that the trial seeks to reject (Box 8.8). When the aim is to demonstrate the equivalence of two treatments, the null hypothesis is that their effects are unequal. That can be illustrated by the following example.

Box 8.8 Hypotheses

Null hypothesis

- the hypothesis that one seeks to refute

Superiority study

H_0: Treatments A and B are equally effective.

H_A: Treatments A and B are not equally effective.

Here, the alternative hypothesis H_A is two sided: A could be superior to B, or B could be superior to A.

Non-inferiority study

H_0: Treatment A is inferior to treatment B.

H_A: Treatment A is as good as or better than treatment B.

Here, the alternative hypothesis H_A is one sided. All that we need to know is that treatment A is not inferior to the standard treatment B.

But what does 'not inferior to' really mean? Two different drugs or treatment strategies rarely have identical efficacy, and even if they had, so documenting would be impossible without a huge study of tens of thousands of patients in each group. Therefore, in practice one defines how large the efficacy difference may be for the two treatments to be regarded as equally effective, and thereafter a trial is conducted to show that the difference in treatment effects is too small to be clinically relevant. This maximum difference usually is termed Δ.

Agreeing on an acceptable value of Δ is not easy. A fixed limit cannot be set because the choice will depend on the relationship between efficacy and safety (and arguably also the costs of the two treatments). However, all agree that two drugs can be said to have similar efficacy when the difference between their effects is considerably less than the difference between the effects of a standard treatment and placebo.

Even though a non-inferiority trial actually tests against a one-sided alternative hypothesis, in practice decisions are usually made based on a two-sided 95% confidence interval but deal only with the lower limit of the interval. This is equivalent to performing a one-sided test at the 2.5% level. Various results possible in non-inferiority studies are illustrated in Figure 8.1.

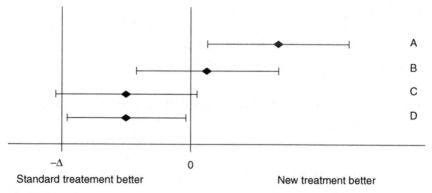

Figure 8.1 Four examples of assessment of non-inferiority using a 95% confidence interval for the difference

Study A shows a significant difference in the effects of two treatments. The lower end of the 95% confidence interval of the effect difference is greater than 0. Study B shows non-inferiority. The lower end of the confidence interval is less than 0, so there is no significant difference between the two treatments. Moreover, the lower limit of the confidence interval is greater than $-\Delta$, so the greatest plausible difference between the two treatments is less than the difference defined as the minimum difference of clinical relevance. Study C shows no significant difference between the two treatments, but nonetheless does not demonstrate non-inferiority, because the confidence interval overlaps $-\Delta$. The difference between the two treatments may then be larger than that accepted for equivalence. Moreover, the point estimate is on the negative side, i.e. the result implies that the new treatment has efficacy inferior to that of the standard treatment. Study D shows the same point estimate of the effect difference as does study C, but nonetheless has a paradoxical result. The new treatment is documented as non-inferior because the confidence interval does not include $-\Delta$, but at the same time, the new treatment is statistically significantly inferior to the standard treatment (upper limit of the confidence interval <0). Here, the estimated effect difference must be judged against the disadvantages of the two treatments, such as safety problems, in addition to the consequence that a patient may receive inferior treatment. For life-threatening diseases, it is obviously more difficult to accept a treatment that is 'almost as good as' a standard treatment, even though it may be less expensive. The difference between studies C and D is that study D had probably enrolled more patients and therefore gave a more precise estimate of the effect difference.

8.5 GENERALIZATION

The great advantage of controlled clinical trials over epidemiological studies is that randomization and blinding avoid an association between treatment and external factors that may influence the results of a study (confounding). Therefore, it is possible to draw conclusions on direct causal relationships between an intervention and an effect, and this is why controlled clinical trials are often said to have high internal validity.

However, a drawback of controlled clinical trials is that they have low external validity, which implies that the results of a study cannot necessarily be generalized. Clinical trials often employ strict criteria for including patients, with regard to age, other illness than that studied, and intake of other drugs. Hence, the trial group comprises a selected part of the population that will receive the treatment if authorized, and it is not obvious that a result based on a sample of included patients is representative for the entire population concerned. Various levels of patient selection are illustrated in Figure 8.2.

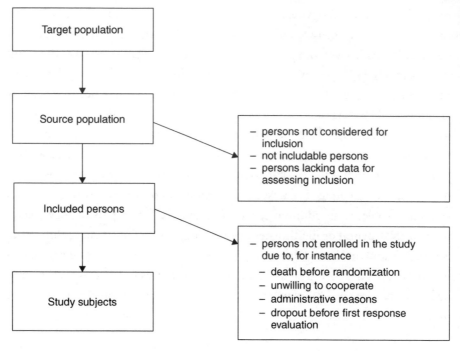

Figure 8.2   Selection levels

Whenever only a small percentage of the relevant patients is included, or if a large percentage of the patients who fulfil the inclusion or exclusion criteria is not enrolled in the study, there may be cause to question the generalizability of the results of a study. Some patients do not want to be randomized, and in other cases, the treating physician may decide that a patient be given a specific treatment. All such exclusions can lead to selection bias and consequently to weakening the generalization of the results of a controlled clinical trial.

REFERENCES

O'Brien PC, Fleming TR (1979) A multiple testing procedure for clinical trials. Biometrics 35: 549–556.
Pocock SJ (1983) Clinical Trials. A Practical Approach. John Wiley, New York.

GUIDELINES

Note for guidance on good clinical trial practice (CPMP/ICH/135/35): http://www.emea.europa.eu/pdfs/human/ich/013595en.pdf

Note for guidance on general considerations for clinical trials (CPMP/ICH/291/95): http://www.emea.europa.eu/pdfs/human/ich/029196en.pdf

Note for guidance on statistical principles for clinical trials (CPMP/ICH/363/96): http://www.emea.europa.eu/pdfs/human/ich/036396en.pdf

CHAPTER 9

EPIDEMIOLOGY: CONCEPTS AND METHODS

Dag S. Thelle and Petter Laake

9.1 INTRODUCTION

In the 1990s, the mortality rates of cardiovascular disease in men declined markedly in the Nordic countries, as shown in Figure 9.1. Nonetheless, the mortality rate for men in Finland remains higher than the rates in Denmark, Norway and Sweden. For women, the rate in Denmark rose over the decade, while the rates in Finland, Norway and Sweden fell. Why do such differences occur? Why do Finnish men differ from men in the other Nordic countries? Why is cardiovascular disease increasing among Danish women? Are these differences related to treatment? Are there differences in the registration and coding of mortality among the four countries? Do differences in living conditions, external environment or even genes give rise to these inequalities? These and similar questions are the concern of epidemiologists.

When it comes to the issue of why diseases occur there are two principal questions that one may ask. For an individual patient, one may ask why just this individual, and not a neighbour or a friend, became ill at this particular time. For population groups that differ in risk of disease, one may ask why they differ. Why is there more disease in one group than in another? The answers to these two questions, aimed at either an individual or at groups, need not be the same. The single individual became ill at a certain point in time because of a number of conditions

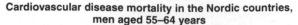

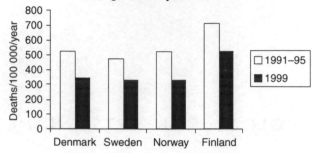

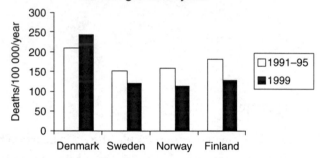

Figure 9.1 Cardiovascular disease mortality in the Nordic countries, men and women aged 55–64 years

related to genetic disposition, lifestyle and external environments, whereas the differences between populations more often are determined by the external conditions under which they live, such as social and economic inequalities. Genetic differences may, of course, also play a role when it comes to diversities between populations. However, the principal causes of the differences are in living conditions and other external factors.

9.2 DEFINITIONS

There are many definitions of epidemiology. Among the more common are:

- Epidemiology is a scientific discipline concerning the occurrence of disease in the population.

- Epidemiology is the scientific discipline concerning the occurrence and variation of disease in the population regarding factors determining this variation, and how health problems can be treated and controlled.

The origin of the word epidemiology is related to that of the word epidemic. An epidemic is defined as an increase in occurrence of disease compared to its normal rate of occurrence in a particular population. Traditionally, epidemic has applied to contagious diseases, but today any condition that exhibits elevated occurrence may be called an epidemic. There is no set rule as to how many cases must occur before being declared an epidemic, but any increase relative to a comparable population may be designated epidemic. The word is a composite of *epi* and *demos*, which mean 'on or upon' or 'over and above' and 'population or people'. Hence, epidemiology may be interpreted as something that takes people totally unaware (casts a spell on people).

9.3 THE ROLE OF EPIDEMIOLOGY

Epidemiology describes variations in morbidity and mortality in time and space and attempts to identify the variables that may explain the variations, including inborn and environmental factors. Epidemiologists are also involved in risk assessment, such as estimating the risks of leukaemia that may accompany living near a nuclear power plant or how radon in basements and building materials may affect the risk of cancer. The field consists of a descriptive part as well as an analytical part and is aimed at aetiological research, but it is also closely associated with basic biological science, clinical specialities, statistics and the social sciences. So, epidemiology is a quantitative medical discipline that bases its methods on logic and statistics. The concepts of epidemiology are useful in addressing a broad array of problems, and the methods applied by epidemiologists are fundamental in assessing the knowledge essential to clinical medicine and public health. Epidemiologists are less concerned with pathogenesis; their main interest is in the aetiology of disease and health problems, variables or factors that trigger disease events and the conditions predisposing to such events. Aetiological research implies that one can determine both the direction for and the strength of the association between a certain variable and a disease.

Epidemiological studies may be regarded as having three aims, as listed in Box 9.1.

> ## Box 9.1 The three aims of epidemiological studies
> - Describing disease occurrence and distribution of disease in population groups and development over time (trend).
> - Identifying cause of disease (aetiology).
> - Undertaking experiments to assess the effects of treatments or preventive efforts.

Study designs

The choice of study design depends on the problem to be studied, such as how prevalent a health issue may be in a population and what resources are available for the study of it. The most common study types and their designations are listed in Table 9.1 and described further in Section 9.7.

Table 9.1 Study designs

STUDY DESIGN	OBSERVATION UNITS
Observational studies	
Ecological studies ('correlation studies')	Groups
Cross-sectional studies	Individuals
Case–control studies	Individuals
Cohort studies, longitudinal studies	Individuals
Experimental studies	
Randomized, controlled trials	Individuals
Intervention trials	Individuals
Interventions in populations or communities	Groups

Epidemiological studies can be classified as observational or experimental, and the observations made may be of either individuals or groups of individuals. Observational studies can have an administrative aim, such as describing the occurrence of disease in a population as precisely as possible, or be aimed at identifying aetiological associations.

9.4 POPULATION AND SAMPLE

For epidemiological purposes, a population may be considered as groups of observation units, all with measurable or quantifiable characteristics. The observation units may be human beings, groups of individuals, families or any other unit that we define and describe. The population may be seen as an immensity as well as a dynamic, and a study is always based on a sample of it. The sample affords the opportunity to examine a factor that we seek to describe. Statistics provide us with the procedures to draw conclusions and inferences of the population based on the study sample. The characteristics to be measured or registered in a sample are called variables. The types of variables to be measured determine the inferences that may be drawn. A summary of the observations from a sample is a descriptive statistic, that is, the mean value that says something about the true and unknown mean of the population. It is intuitively obvious that the larger the sample, the more accurately a sample mean reflects a true value for a population. The magnitude of the sample and the accuracy of the measurement methods determine the confidence with which we can say something about the phenomenon under study. The principles for assessing sample size are discussed in Section 8.3 of Chapter 8. These principles are the same for epidemiological as for clinical studies.

Samples from the same population provide a number of mean values that are likely to differ from sample to sample. It is obvious that all these means represent the true mean for the population, but no one of them is fully correct. The variation of the means of all possible samples is called the standard error of the mean (SEM), which is equal to the sample standard deviation (s) divided by the square root of the number of observations (n):

$$SEM = \frac{s}{\sqrt{n}}$$

The confidence interval is an estimate of the confidence that we have in the mean value of the sample. With a certain probability, it shows the highest and lowest values that the true value of the population may have. The estimated 95% confidence interval around an unknown mean value is $\bar{X} - 1.96$ SEM, $\bar{X} + 1.96$ SEM, assuming that the measurements follow a normal distribution (see Section 11.6 of Chapter 11).

9.5 MEASURES OF DISEASE OCCURRENCE, ASSOCIATION, RISK AND IMPLICATIONS

The concept of risk is often used in describing the association between a certain factor and the probability for later disease or harm. To express a certain factor in terms of a risk factor, the population studied is divided into two or more groups, such as those who are exposed and those who are not exposed, or by grading of the actual exposure. The importance of the factor may be assessed when we know how many of the exposed and how many of the non-exposed actually are ill or became ill during a certain period of time. This enables us to estimate the risk associated with the exposure for the factor. The occurrence of a disease and the exposure to a potential risk factor may be summarized in a table, such as Table 9.2. This sort of table is called a contingency table, because it displays contingencies between the data entered in it.

Table 9.2 Association between exposure and disease

DISEASE	EXPOSURE		TOTAL
	NON-EXPOSED	EXPOSED	
Healthy	a	b	$a + b$
Affected	c	d	$c + d$
Total	$a + c$	$b + d$	n

Table 9.2 is a 2×2 contingency table because it displays two rows and two columns of data. This table is similar to Table 11.4 of Chapter 11 and enables estimation of the epidemiological measures of interest, provided that a suitable and valid study design has been used (see also Chapter 4). The numbers of affected subjects among the exposed (d) and the non-exposed (c) provide information about the occurrence of disease according to exposure. These numbers provide the basis for assessing the disease risk. The ratio of disease risk among

the exposed to the non-exposed is a measure of the association. The direction of the association, that is whether the factor increases or reduces the risk of disease, is indicated by the ratio being greater or less than one. Epidemiological measurements can be classified in three categories:

- disease occurrence
- association or effect measurements
- measures of importance or implication.

Measures of disease occurrence are used in describing causal relationships and in descriptive analyses of the evolution of disease occurrence or mortality over time.

Effect measures or associations yield the strength of the association between the variables, such as between a disease and the possible cause of it. Such measures are based on comparisons between disease occurrences in groups with various exposures to a certain factor. The measure is expressed as the risk or the probability for an event during a certain time period. An example could be a study in which we are interested in the association between breast cancer and the use of oral contraceptives. The observations permit estimating the risks for breast cancer among those who are exposed and those who are not exposed to oral contraceptives; the ratio of risks expresses the strength of the association between disease and exposure.

Measures of the importance express the impact of a certain disorder or exposure on a population. This is a measure of public health interest, that is, in assessing the possible effects of preventive efforts. For example, for breast cancer, we might ask how many cases can actually be attributed to the use of oral contraceptives.

The epidemiological measures are listed in Box 9.2 and discussed further in Section 9.8.

Box 9.2 Examples of different epidemiological measures

Disease occurrence	Association	Importance
Incidence	Relative risk	Excess risk
Prevalence	Odds ratio	Attributable risk
	Incidence rate ratio	
	Regression coefficient	

Epidemiology is about denominators, numerators and time

Risk assessment is based on quotients. A central concept is the assessment of proportions or prevalence and rates; all are quotients. Prevalence or a proportion is the quotient of the number of affected subjects in the numerator to the number of individuals who are at risk in the denominator. A proportion is dimensionless and ranges from near zero to one, and consequently often is expressed as a percentage up to 100%.

A rate expresses the change of a quantity relative to the change of another quantity, often time. An example is the speed of a car at an instant in time expressed as the distance covered divided by time elapsed (km/h). A rate always has a derived dimension and thereby may assume a wide range of values. For the present purposes, constant or average rates are more practical and will be used in the rest of this chapter. The rate is the ratio of the number of new cases in the numerator and the total length of follow-up time (observation time) in the denominator. A denominator may, for example, be the number of person-years estimated over a certain period. The follow-up time may be estimated for a group or for the population in the study. For instance, annual mortality is the rate for a time period of one year based on the number of people who die during that year. The follow-up time is estimated by the sum of the time lived through that year for all the individuals in the population. For each individual who dies, time is from the beginning of the year until the date of death. The total follow-up time is the denominator. The average population or mid-year population is often used in estimating the number of person-years. The error induced is minimal. Valid results from comparisons of populations depend on using the same criteria for disease endpoints or cause of death and on the populations included being similar in age and gender. The numerators of the rates are based on the definitions of disease events, harmful effects or deaths (endpoints). The same diagnostic criteria must be applied to all the subjects in order to include them in the numerator of the quotient.

The ratio between risk for disease among those exposed and those not exposed to a factor provides a measure of the association between a possible causal factor for disease and the disease. It can be estimated as the ratio between proportions or rates, for example, the risk of dying from a coronary heart disease during a 10-year period for smokers compared to non-smokers. The ratio between the risks in the two groups is a measure of relative risk for disease.

Therefore, studies often assess the gravity of a particular exposure. However, the particulars of the diagnoses or endpoints must be independent of knowledge of the exposure. This independence between exposure assessment and endpoint ascertainment is prerequisite to drawing conclusions concerning possible associations between causal factors and disease or injury.

Measures of disease occurrence (incidence, prevalence and duration)

There are several measures for disease and their occurrence in the population, of which three estimates are of principal interest:

- number of ill or injured people
- the size of the population
- the duration of the condition.

The *incidence rates* are used to explore why diseases and illnesses vary between groups in the population. The incidence is defined as the number of new affected individuals in a certain population during a defined period, and the incidence rate is the number of new events divided by the total observation time during the period.

At any one time, the proportion of a population ill with a defined disorder gives the prevalence of it. For example, in Norway each year, about 100 people are infected with HIV. The incidence rate for HIV infection can be expressed as the number of:

- cases per year
- cases per person-year
- cases per person-day.

The numerator of the HIV incidence rate is the total number of new cases. The total length of follow-up time for the whole population is the denominator. So the dimension is cases per year or per person per day. We assume that the population remains constant over the period, save for the few with the disease under study.

The size of the examined population will, for all practical purposes, be reasonably stable, and the average population can be used as a denominator.

Incidence rates for diseases or deaths can be estimated from cohort studies. Follow-up time may be difficult to estimate, because the time interval to the end-point for an individual of the group – be it death, emigration or another reason for ending observation – often is not known. The difficulty is overcome by making the time interval a censored variable, that is, a variable that is observed only under the specific condition of the endpoint being attained.

The subjects included in the survey from start to end have the whole study period as their follow-up time. The sum of these individual follow-up times is the total observation time.

Incidence assessment assumes that we know how many people are taken ill in a population, but this sort of information is often lacking, and we may only have access to the number of deaths or to the number of people who die from a particular disorder. This implies that we cannot estimate the incidence rate but must rely on mortality rates. Mortality rates can be looked upon as incidence rates for death. The total for deaths from all causes is usually expressed as the number of deaths per 1000 people per year. Mortality can be expressed as cause specific or broken down by age, groups, gender and occupation depending on the associations studied.

The duration of a disorder will, together with the incidence rate, determine how large a proportion of the population is affected by the disorder at any time, that is, the prevalence. Short duration implies decreasing prevalence. Increasing prevalence is seen when the duration of a disorder lengthens even when incidence rates are constant. Disease prevalence declines for two obvious reasons: patients may die or they are cured. It does not matter which of these factors operate, as both affect the duration of the disease and thereby its prevalence. There are two measure of prevalence: point prevalence, which is the probability for an individual to be ill in a population at a time (t), and period prevalence, which is an expression of a probability that an individual will ever have been ill during a defined period. This is often used as a substitute for risk when the time for start of the disease is not known. Any measure reducing the fatality of the disorder but not regaining complete health will imply that the prevalence of the disorder is increasing, in which case increased pressure on the health service is unavoidable. Correspondingly, if the fatality increases, the prevalence of disorder will decrease.

9.6 VITAL STATISTICS

The importance of vital statistics in an epidemiological setting

Vital statistics comprise systematic registration of information regarding single individuals in a population with regard to birth, death, causes of death, marriage, divorce, area of residency and moving residence. These data were originally collected for administrative purposes, seldom with epidemiological aspects in mind. Today, it is also reasonable to include other types of register in the vital statistics, such as census data and disease and accident registers. The main uses of vital statistics in epidemiology are listed in Box 9.3.

Box 9.3 Vital statistics are used for three main rationales
- Description of the population mortality trends.
- Comparison of age- and gender-specific death rates during a time period.
- Description and comparison of cause-specific, age- and gender-specific death rates for different groups in the population.

Vital statistical data may be combined with other information about a population to provide the bases for ecological studies. Ecological studies involve comparisons of aggregated data and are not based on information about each single individual in the population. This implies that we can estimate the occurrence of disease in a group and assess it against aggregated data of known exposure variables, which may provide insights and enable the development of new hypothesis. Ecological studies have also been designed as observation studies, but the term is also applied to other types of studies. Ecological studies are often based on vital statistics that are readily and inexpensively available. An ecological study is the first study conducted before starting specific studies for testing hypothesis. Epidemiologists often ask whether more than one factor will change an individual's risk for disease or injury. Vital statistics contain information on the causes of death and consequently partly on disease, so they may be used as a register for endpoints in an epidemiological study. Linking vital statistics data to a data file containing information on individual lifestyles, habits, environments and social

251

conditions, and possibly also genetic properties, forms the basis for a follow-up of a cohort study. This may be the most important use of vital statistics in present-day epidemiological research.

Mortality

Mortality rates are analogous to incidence rates, but the event here is death instead of new cases of disease. Three categories of mortality statistics are:

- A: death due to disease X
- B: death of other reasons but with disease X
- C: death of other reasons without X.

Total death or all-cause death in a population during a certain period equals the sum of A, B and C. Mortality can be described in two ways; either as a death rate or as a death risk (the probability of dying during a certain period). The way it is expressed depends on what we are actually studying, mortality under A, B or C. The mortality rate associated with disease X can refer both to those who die from disorder A and to those with disorder B. It often is impossible to distinguish between A and B in a sample. Another measure of interest is disease lethality or fatality rate. This is the mortality rate in a group of patients that designates how many die from a certain disorder among those who have that disorder during a certain period.

Other measures of status of health in a country are infant mortality and perinatal mortality, which are often used as indicators for living conditions, quality of the healthcare system, and mother and child care. Infant mortality is the number of deaths among children under the age of one year per 1000 live births during the same period. Perinatal mortality is the number of stillborn and deaths during the first week after birth per 1000 born, including stillborn in the same period. Stillbirths are the number of dead babies per 1000 babies born, including the dead during the same period. The first year of life is usually divided into different periods, most often:

- under 24 hours
- under one week

- under four weeks (the period with neonatal deaths)
- four weeks to less than one year (postneonatal deaths).

Cause of death statistics

The bases for the statistics on the causes of death are data entered in death certifi-cates. For example, the Norwegian death certificate has two parts for the cause of death: part I for the main cause or the underlying cause and part II for other con-ditions contributing towards death but not the direct cause of it. In the part for the main cause of death, all conditions leading to death are meant to be entered, as the intention is to identify the cause that started the disease process. Death cer-tificates are handled in different ways in different countries. However, the classifi-cation of diagnoses has been internationally standardized. The origin of the present-day system dates from the 1850s, and the first edition – the international list of causes of death – was adopted by the International Statistical Institute in 1893. The current International Classification of Diseases (ICD) was adopted by the World Health Organization (WHO) in 1948. The classification system has been revised a number of times and the current system is edition 10.

The number of revisions reflects the advances in medicine and consequently in the classification of diseases and causes of death with regard to aetiology, pathogenesis and morphology. How much of the risk of dying from coronary heart disease should be attributed to the fact that a person had diabetes for many decades? It may be reasonable to put diabetes as the condition leading to myocardial infarction, taking into account the considerable risk induced by dia-betes. However, as the mechanism for increased risk is not fully clear, diabetes will be recorded in part II of the death certificate. For a more detailed descrip-tion of the problems with regard to registration of cause of death, coding and classification, see the current edition of the International Classification of Diseases (ICD 10) (WHO 1994).

Morbidity register

A register of diseases comprises more than reports of cases. It includes permanent documents for all submitted individuals, all of which are followed up in statistical tabulations of occurrence, frequency and survival. Various disease registers are maintained around the world. The most common are inpatient registers, cancer registers and birth registers, and registers of infectious diseases, some of which aim at chronic disorders such as tuberculosis. A full description of the registers is beyond the scope of this book, but is available on the WHO website at www.who.int/classifications/icd/en.

Morbidity registers or mortality registers?

Mortality is a good indicator of the occurrence of disease provided fatality is high. When the majority of those with a disorder die in a short time, mortality reflects the incidence of the disease. This is not always the case, and fatality may change over time and may well vary between groups. Morbidity registers are therefore the obvious choice if the development of different disorders in the population is to be followed. However, permanent morbidity registers are large and costly. The value of such registers usually becomes apparent with time when trends have become evident and may be analysed. The standard of the registration and the validity of the information collected must be consistent with time and consequently under permanent control. This is one of the reasons why there are relatively few permanent disease registers operating on a global basis. The establishment of a register of chronic disorders only to support studies of their incidence may not be a sufficient rationale for these kinds of register. One can, however, see the establishment of such registers as useful for particular population segments as part of the follow-up after surveys and screenings.

9.7 STUDY DESIGNS OF EPIDEMIOLOGICAL STUDIES

The possibility of there being an association between a certain factor and a disease or health damage usually comes from clinical observations or case studies where a number of patients with the same disorder seem to have been exposed to the same factors. Case histories are detailed descriptions of the patients' disease histories. They usually comprise small groups of patients and often concern rare disorders. They permit rare diseases to be surveyed and thereby offer insights into their occurrence, prognosis and treatment options. Case histories, however, give limited insight with regard to aetiology. There are certain situations where one may draw conclusions regarding aetiology as they are so obvious, such as work accidents or other similar events. For other situations, the association of the causal relationship is far less clear. A healthy group of control subjects is needed to assess whether an affected group of people differs from healthy people with regard to more than the disease in the study. In principle, there are three main study designs that aim at describing associations. Box 9.4 combines Table 9.1 and Box 9.2 and gives a simplified overview of different types of studies with their epidemiological methods. Note that different studies give different epidemiological measures. One way to describe these studies is to place the observer in the present time studying events that may have occurred in the past (case–control, retrospective studies), orientated towards the present (cross-sectional studies) or looking forward into the future (cohort, longitudinal or prospective studies). This has given rise to a number of conflicting descriptions and classifications of study design. Hence, there may be cohort studies that are retrospective in the sense that their data have been collected in the past, or there may be case–control studies in which the data of both patients and controls have been collected with a future perspective in mind.

Box 9.4 Principal differences between case–control studies, cross-sectional studies and cohort studies

	Studies		
	Case–control	*Cross-sectional*	*Cohort*
Population	Start with affected subjects	All subjects are included	Start with healthy subjects
Measures			
Incidence	No	No	Yes
Prevalence	No	Yes	No
Association	Odds ratio	Relative risk, odds ratio	Relative risk, odds ratio, incidence rate ratio

Case–control studies

Design and structure

The classic approach to identifying disease causes has been to compare patients with a corresponding healthy control group in a case–control study. These studies compare a number of conditions in the two groups to determine how they differ in regard to a number of possible aetiological factors. The prerequisites for such conditions to give meaningful results are that the control subjects are sampled among healthy subjects who would have been included as cases if they had become ill. Therefore, the ideal requirement for a control group is that its subjects should belong to the same population as the patients. Case–control studies are retrospective, looking back in time. They attempt to elicit aspects of the patients' past, before they became clinically ill.

Advantages

The main advantage of case–control studies is that they are relatively simple to carry out. Moreover, they are inexpensive, reasonably rapid and suited to infrequent conditions. They permit examining large numbers of possible risk factors. They also may be organized as multicentre studies, whereby a number of research groups can participate with patients and controls.

Disadvantages

The major disadvantage of the case–control study is the inverse direction of the data on exposure, which is collected after the patient has become a case. The fact that someone has become sick may affect both the responses to questionnaires as well as biological valuables (biomarkers), which may be examined among both patients and control subjects. The observer – the examiner or the researcher – will be well aware of the problem or the research question and thereby also know who among study subjects belong to the cases and who belong to the control group. Thus, the examiner may also be more thorough and assess the patients differently from the control subjects, or those affected may give answers that are coloured by their being affected.

There are myriad possible sources for systematic error in case–control studies that may affect study outcome. Systematic error or errors that affect results in a certain direction are called bias. Thus, the inverse time aspect is a major problem in interpreting the results of retrospective or case–control studies.

Classical examples include the relationship between stress and myocardial infarction or the use of drugs during pregnancy and the risk of congenital malformations. A person who has just survived myocardial infarction and who is being asked about stressful events immediately before the infarction may give completely different answers from a person in the control group chosen for another reason, such as having had a skiing accident. The affected persons will search for causes of their affliction and may give responses coloured by these situations. Mothers of babies with congenital disorders will scrutinize their past with regard to the use of drugs during pregnancy, whereas mothers with healthy babies will be less prone to do so. Hence systematic error, or recall bias, influences results. To avoid these types of systematic errors, one must establish examination procedures that reduce the possibility of clinical disease affecting the answers. One of the practical problems in case–control studies is selection of control subjects. The following three options involve direct contact (by mail or telephone):

- Neighbours in the same residential area as the patient are contacted, inviting them to take part in the control group. Patients themselves contact friends or relatives in the same age group.
- A random sample of the population based on census registers is invited, including background factors that also are available for the patients drawn from the same cohort.
- Patients with other diagnoses from the same hospital as the original patients are invited to take part as control subjects.

To choose a patient from one's own institution is a relatively easy way to solve the problem. It may, however, induce insecurity with regard to whether the control subjects are representative of the population from which the patients are recruited, and that the risk factors one wishes to examine may be related to the conditions affecting the control subjects. Assume that you wish examine whether cigarette smoking causes an increased risk of bladder cancer, and the control subjects are selected among patients in a lung department. This would imply that the smoking habits for the control subjects would be likely to deviate from the general population, and the results of the study would be invalid.

Matched case–control studies

One of the concerns when planning a study is to avoid the influence of factors that may contribute to the increased risk of disease but otherwise are of no particular interest in the study. One approach is matching, in which the control subjects and the patients are selected to be as similar as possible with regard to factors such as age, gender, residential area and social class. However, the drawback of matching is that one may control for factors related to the variables that one wishes to assess, thereby reducing the possibility of demonstrating associations. One example is that of a study of the importance of cigarette smoking for myocardial infarction. Assume that you match patient and control subjects with regard to social class. This may reduce the possibility of showing the association between smoking and myocardial infarction, as smoking is associated with social class.

Cross-sectional studies

Design and structure
A cross-sectional study collects all data at one point in time. This permits the prevalence of a condition or variables of interest to be assessed.

Advantages
A cross-sectional study is intuitively clear and permits a large number of variables to be examined. The methods can be standardized by the researcher, and clearcut definitions may be applied to the exposure and endpoints.

Disadvantages
By definition, a cross-sectional study has no dimension of time, so it cannot support conclusions on the risk of disease or on the inference on causality. One way to circumvent this drawback is to formulate questions that assess an individual's past, such as enquiring about previous lifestyle, occupation or other exposures. This enables the researcher to categorize individuals with regard to previous exposure

even if the collection of the information occurred at one point in time. The interpretation of cross-sectional studies may, to a large extent, be similar to that of case–control studies. The lack of the dimension of time implies that the cross-sectional or prevalence studies are poorly suited to examining conditions of shorter duration.

Cross-sectional studies may also exhibit recall bias, because disease or assessment of disease may influence the response pattern and even the biological variables. A cross-sectional study in Norway, found that those who were most overweight in the population were also those who reported consumption of low-fat milk. It is unreasonable to imply that low-fat milk will increase your body weight, but the finding reflects attempts at slimming. Here, the findings of cross-sectional studies should be interpreted carefully. An increasing problem in cross-sectional studies as well as cohort studies is the low response rate or attendance rate, an issue that may invalidate the results of the studies with regard to prevalence assessment.

Cohort studies

Design and structure

A cohort study is one that follows a defined group (cohort) over a given period. The usual approach is to start with healthy or unaffected subjects. The main purpose is usually to assess the possible effects of various external or internal factors on the risk of disease. Information on exposure about each individual in the population is collected at baseline. The individuals or examinees are followed over time, and the final analysis contains a comparison between those who have remained healthy and those who have become ill. These studies imply variables at baseline that differ between the examinees. A homogeneous cohort, with regard to a variable, makes it impossible to assess the impact of the variable as a disease risk-associated factor. For instance, there is no way to associate disease with smoking if everyone in the cohort has been exposed to it (at the same dosage) since the age of 15 years up to the end of the study.

Advantages

The principal advantage of the cohort study is that it includes the dimension of time, which permits drawing conclusions with regard to causal associations. The studies are intuitively easy to understand, numerous variables can be assessed at the same time and one can apply standardized methods with clear definitions with regard to both exposure and endpoints. The ascertainment of exposure and that endpoints are independent as all data on exposure were collected while the subjects are still healthy. Prospective cohort studies are assumed to give a more valid and less biased result than the other study designs.

Both incidence rate and incidence rate ratio can be estimated from cohort studies. Large cohort studies permit more than one type of endpoint to be assessed, whereby multiple disease problems as different exposure variables might be examined.

Disadvantages

Several practical and inferential problems are associated with cohort studies. Diseases with a low incidence rate are not amenable to drawing causal conclusions, as too much time is needed to accrue a sufficient number of events upon which conclusions may be drawn. Cohort studies are, however, suitable for addressing risk factors that are stable over time and for assessing diseases that are relatively frequent or of certain duration.

Cohort studies are relatively costly and usually require long follow-up, as well as an apparatus for follow-up and updating of the database. Another weakness is that the registration of exposure variables can be inflexible. Data on exposure are collected at baseline and then used to construct the different exposure categories. Changes that the participants in the study may undergo during the follow-up period may not be taken into account, such as change of residential area, change of occupation or work (if occupational exposure is of interest). These changes are likely to be unknown to the researcher and lead to underestimation of the risk assigned to that particular exposure. Assume that we are following a group of subjects where some are cigarette smokers and others have never smoked. During the follow-up period, some of the smokers stop smoking. This implies that there will be a number of quitters in the smoking category. The analysis of the associations between disease and smoking will be based on the information that was collected at baseline as we have no information on who actually stopped smoking and when. The result may be a reduced risk of disease, as incidence is lower than that actually associated with smoking. This misclassification usually gives an effect measure that underestimates the true association between smoking and disease. Repeated assessments of the exposure values during the follow-up are one way to try to avoid this kind of systematic error.

The ascertainment of exposure including the degree or dose of exposure is crucial to establishing the baseline cohort. Cohort studies will, as all other epidemiological studies, contain three principal elements:

- choice of target population containing the study population
- methods for exposure ascertainment
- registration of endpoints or events.

Exposure ascertainment and registration may comprise data collected directly from individuals via questionnaires, biological tests and measurements, or may be data already collected via other registers, such as census data and income data. These data can be collected in standardized ways to prevent systematic errors. The work involved, however, may be substantial. One example of an area where it is difficult to obtain good valid data is registration of nutrition and other lifestyle habits. Extensive multipage questionnaires covering myriad topics have been developed. They may be cumbersome to use if a cohort group comprises thousands of subjects. Data collection should therefore be simplified or organized at a level that is acceptable to the participants. Registration of disease events and endpoints is time and resource demanding in a cohort study, as in practice during the follow-up period, all individuals must be monitored to ascertain whether they have had any of the disorders of interest or whether they have died or emigrated before the end of the study (censored).

9.8 EFFECT MEASURES IN EPIDEMIOLOGICAL STUDIES

This section will discuss three measures of association:

- relative risk or risk ratio (RR)
- odds ratio (OR)
- incidence rate ratio (IRR)

and three derived measures:

- excess risk
- attributable risk (AR)
- population attributable risk (PAR).

All measures of effect are based on estimates of disease risk and/or rates. They differ, and the differences between them depend on the study design used for estimating the effects (see Box 9.4). The conventional term 'relative risk' is often used in the literature as covering both risk ratio and odds ratio independent of the study design underlying the estimates. This is confusing as well as mathematically incorrect, but is acceptable for practical purposes providing certain conditions are fulfilled. Confidence of the effect estimates is given by a confidence interval, usually at a 95% level. The confidence interval covers the unknown effect measure with 95% probability. The calculation of the confidence interval for relative risk and odds ratio is given in Section 11.5 of Chapter 11. The confidence interval for incidence rate ratio is calculated below.

We examine a sample of subjects categorized into being exposed or non-exposed to a possible causal factor. All subjects are examined with regard to subsequent disease. A fraction of those exposed have become ill during the follow-up period and a fraction of those non-exposed have also become ill with the same disorder. The data can be displayed in a 2 × 2 contingency table, as shown in Table 9.3.

Table 9.3 Association between exposure and disease

	EXPOSURE		
ILLNESS	NON-EXPOSED	EXPOSED	TOTAL
Healthy	a	b	$a + b$
Affected	c	d	$c + d$
Total	$a + c$	$b + d$	n

Among the $(b + d)$ exposed there are d affected subjects. This gives a risk of $R_e = d/(b + d)$. Among the $(a + c)$ non-exposed, the risk equals $R_0 = c/(a + c)$.

The ratio between these two risks will express the risk of the exposed relative to the non-exposed. This ratio, called relative risk, is given as:

$$RR = \frac{d/(b+d)}{c/(a+c)}$$

RR is a common measure of association.

The odds ratio (OR) is another frequently used effect measure for association, but may be intuitively understood only by statisticians. Odds are the relationship between two probabilities, namely the probability for an event divided by the probability for the non-event provided that there are only two possible outcomes. Let the probability of an outcome be p and let only one alternative result emerge, e.g. that the outcome will not occur, then the probability for the non-event will be $(1 - p)$. Odds are then the ratio between p and $(1 - p)$, or $p/(1 - p)$.

In Table 9.3 $(b + d)$ exposed subjects were examined. Among these exposed subjects, d will be classified as ill or affected. The odds for being ill consequently is $d/(b + d)$ divided by $b/(b + d)$ or d/b, which is the same as the probability of being ill divided by the probability of not being ill. In the group of the $(a + c)$ non-exposed, c are counted as ill or affected and a as healthy. The odds for being ill versus healthy is c/a. The odds ratio is a comparison of these two groups by dividing the odds for both groups, giving:

$$OR = \frac{d/b}{c/a} = \frac{ad}{bc}$$

Going back to the 2 × 2 contingency table, we will see that this odds ratio is the same as multiplying the counts diagonally in the table and thereby dividing the product, resulting in ad/bc. Thus, this product is called the cross-product. The odds ratio is an effect measure that tells us how much larger the odds are for the exposed to be ill than for the non-exposed, and therefore is considered as an estimate of relative risk for disease.

Table 9.3 permits assessment of the odds for being exposed, given that the subjects are affected or healthy. There were $(c + d)$ affected and $(a + b)$ healthy subjects. Among the $(c + d)$ patients there were d who were exposed to the risk factor, whereas c were non-exposed. The odds for being exposed will be $d/(c + d)$ divided by $c/(c + d)$, or d/c, whereas in the healthy control group the odds for being exposed against non-exposed were b/a. The two groups will be compared by dividing the odds for both groups. This gives:

$$OR = \frac{d/c}{b/a} = \frac{ad}{bc}$$

which is the same as we saw when assessing odds for being ill among exposed and non-exposed.

Effect measures in cohort studies (RR and IRR)

Cohort studies yield information about risk or incidence rate in exposed and non-exposed individuals. The effect measures can be either comparisons of risks or the incidence rates, which by division give relative risks and incidence rate ratios, or by subtraction give excess risk.

Example: The results of a cohort study of 1000 men are given in a 2×2 table (Table 9.4). We know that 40 men died from coronary heart disease during a 10-year period, 30 of them smokers and 10 non-smokers.

Table 9.4 Association between dying from myocardial infarction and cigarette smoking

	CIGARETTE SMOKING		
DEATHS	NO	YES	TOTAL
No	490	470	960
Yes	10	30	40
Total	500	500	1000

Table 9.4 permits the risk of myocardial infarction among smokers and non-smokers to be estimated. The relative risk is:

$$RR = \frac{30/500}{10/500} = 3$$

The risk of dying from myocardial infarction is therefore three times greater among smokers than among non-smokers. The excess risk is calculated by subtracting the rate for the non-smokers from that of the smokers, or 30/500 − 10/500, which is 20/500, indicating that among 100 smokers there will be four more people who will die from myocardial infarction than among 100 non-smokers.

Relative risk is a better measure for the strength of the association between a risk factor and disease than excess risk, because relative risk by definition is relative compared to the underlying risk.

Assume that Table 9.4 also contains total length of follow-up time (see Table 9.5).

Table 9.5 Association between death from myocardial infarction and cigarette smoking

	CIGARETTE SMOKING		
	NO	YES	TOTAL
Number of deaths	10	30	960
Person-years	950	4850	9800

We know that the incidence rate is the number of deaths divided by total length of follow-up time, $IR = d/t$. The incidence rate among smokers (the exposed) is $IR_e = 30/4850 = 6.2/1000$ and for non-smokers is $IR_0 = 10/4950 = 2.0/1000$. We define the incidence rate ratio as $IRR = IR_e/IR_0$. We then find that IRR = 3.1. The excess risk is 4.2/1000.

Confidence interval for IRR

To estimate the confidence interval for IRR we must transform to a logarithmic scale. This is the same transformation as used for estimating a confidence interval for relative risk or odds ratio (see Sections 11.5 and 11.9 of Chapter 11). The standard error for the logarithm of the incidence rate IR_e is $1/\sqrt{d_e}$ and for the logarithm of IR_0 it is $1/\sqrt{d_0}$. The 95% confidence interval for ln(RR) is:

$$\ln\left(IRR\right) - 1.96\left(\sqrt{1/d_e + 1/d_0}\right), \quad \ln\left(IRR\right) + 1.96\left(\sqrt{1/d_e + 1/d_0}\right)$$

By transforming back to an ordinary scale, the 95% confidence interval for the incidence rate ratio is:

$$\text{IRR} \cdot \exp\left(-1.96\sqrt{1/d_e + 1/d_0}\right), \text{IRR} \cdot \exp\left(1.96\sqrt{1/d_e + 1/d_0}\right)$$

Example: In the analysis of Table 9.5, IRR = 3.1, d_e = 30 and d_0 = 10. The confidence interval is:

$$3.1 \cdot \exp\left(-1.96\sqrt{1/30 + 1/10}\right), 3.1 \cdot \exp\left(1.96\sqrt{1/30 + 1/10}\right) = (1.5, 6.3)$$

There are situations in which the excess risks may be identical, although the relative risks differ markedly. If the figures in Table 9.4 had been 20 and 40 deaths in the groups of smokers and non-smokers, the excess risk would still have been 0.04 (40/500–20/500), but the relative risk in this case would have been 2. The excess risk is the same as in Table 9.4, but the relative risk is now two-thirds of the original value.

The higher the relative risk (the farther away from 1), the stronger the assumption that the association is not caused by chance, but may be an expression of a direct causal relationship (discussed further in Section 9.13).

The odds ratio can also be estimated from cohort studies as the ratio between odds for becoming ill among the exposed and the odds for becoming ill among the non-exposed. In the example in the 2×2 contingency table above (Table 9.4) the cross-product is:

$$\text{OR} = \frac{30 \cdot 490}{10 \cdot 470} = 3.12$$

This is slightly above the relative risk estimate, but the difference is small as long as the disease incidence rate (or in this case mortality from myocardial infarction) is low. Hence, the odds ratio and relative risk are used as corresponding measures of effect. Both effect measures are further described and estimations of confidence interval are outlined in Section 11.5 of Chapter 11.

Effect measures in case–control studies (OR)

The case–control study is based on samples of patients and of unaffected control subjects. This implies that risk cannot be directly estimated based on incidence rates. Control subjects and cases are studied with regard to their exposure in the past. In the analyses, odds for exposure are estimated, and the odds ratio is used as a measure for disease risk:

$$OR = \frac{d/c}{b/a} = \frac{ad}{bc}$$

Attributable risk

Attributable risk is a measure of the proportion of the disease occurrence that can be ascribed a certain risk factor. As before, the risks among the exposed and non-exposed are denoted R_e and R_0. Then, attributable risk can be expressed by estimating excess risk as $R_e - R_0$ divided by the risk for those who are exposed to the factor, R_e, i.e. AR $= (R_e - R_0)/R_e$. This gives the proportion of the excess risk for disease that can be attributed to the exposure for the factor in question. As relative risk was defined as the risk among exposed subjects divided by the risk for the non-exposed, this can be substituted in the equation for the attributable risk, giving AR $= (RR - 1)/RR$. This is a measure of the importance of the factor in explaining the occurrence of disease among the subjects who are exposed. The association by itself is not sufficient to argue for causal inference.

The prevalence of exposure p in the total population, used to estimate the importance of a factor for the total disease occurrence, has a value between 0 and 1. The term prevalence is an epidemiological measure that has been used previously in describing the proportion of the population who are affected at a given point in time (Box 9.2). The term can also be used to indicate the proportion of a population who are exposed.

Provided that the prevalence of the exposure is known, we can set up an equation: PAR $= p(RR - 1)/(1 + p(RR - 1))$ for the *attributable risk for the population*. PAR is an expression of how large a proportion of all those who become ill in the population can be attributed a certain risk factor, given the prevalence p and relative risk RR. The equation follows from the previous definitions, but its derivation will not be taken up here.

Both AR and PAR are mathematical or algebraic assessments of the statistical association, but they provide no additional information on cause and effect. PAR is a measure of how large the problem is from a public health point of view, for instance how many of all lung cancers can be attributed to smoking. If the prevalence of smoking is set to 50% or 0.5 of the population and the relative risk for lung cancer to 10, this will give a PAR of 0.82. This implies that 82% of all lung cancer in the population can be attributed to this particular risk factor. The equation tells us that if prevalence falls, the importance of the factor as a public issue will decline even if the relative risk remains the same. Attributable risk can be used as a measure for how much one can achieve with regard to prevention or health promotion if the prevalence of the factor is reduced. Thus, the attributable risk can also be expressed or viewed as the potential for a preventive effort.

9.9 EXPERIMENTAL STUDIES AND RANDOMIZED CONTROL TRIALS

Case studies, case–control studies, cross-sectional studies and cohort studies provide valuable insights into the associations between causal factors and disease risk. However, these studies provide little information on effectiveness and efficiency or treatment or intervention where one attempts to determine whether a particular exposure actually has consequences, is harmful or is beneficial. This leads to the experimental part of epidemiology in which the main emphasis is on randomized control trials. These trials differ from cohort studies in the sense that the observers or investigators responsible for the experiment actually control the exposure status. The random allocation determines who is allocated to the respective exposure groups. These types of study are further discussed in Section 8.1 of Chapter 8.

9.10 MEASUREMENT ERROR AND SOURCES OF ERROR

All measurement incurs error, or deviation from a variable's true value. Our task is to provide the best possible estimate of this true value. The registration or the measurement process implies the use of a measurement instrument (technical apparatus or a questionnaire). The magnitude of the deviation from a variable's true value depends on the measurement instrument used and the accuracy of the instrument and the method. The deviation from the true value is called measurement error. If the variable is a categorical variable, we often speak of misclassification, as the error leads to categorizing the subject in a category other than the true one. The magnitude of the measurement error found acceptable is based on costs and resources, and is a practical question depending on the nature of the problem addressed. More accurate instruments as well as larger samples are more costly. How we cope with the problem depends on the importance attached to measurement error.

Apparatus or instrument error

The purpose of a measuring instrument is to register or measure a variable. In many situations, only indirect methods of measurement are available. This raises two questions: To what extent does what we measure reflect what we want to measure? How large is the measurement error of the apparatus or method compared to the true value of the variable? We are also interested in the repeatability of the measurement instrument, which depends not only on the instrument used, but also on the variable measured. To what extent does this variable vary within a single individual during a relatively short period? It is difficult to repeat a measurement to a greater degree of accuracy when the variation is large. The intraindividual variation may imply that we would like to use more measurements from the same individuals to achieve the most valid value possible. How many measurements will be undertaken depends on the problems that are being faced and the extent of the resources available for taking measurements. The concepts of accuracy, precision, reliability and validity are discussed in Section 4.7 of Chapter 4.

Why measurement errors occur

Questionnaires used for charting eating habits and lifestyle are crude and inaccurate instruments, whereas genetic and molecular methods are likely to be far more accurate. The sources of error lie not only in the design of the instrument and the way that questions are formulated, but also in how the instruments are used and the skills of the people performing the studies. People vary with regard to their own accuracy, and there are many possibilities for inaccurate readings and misclassifications. If the error or the deviation from the true values deviates to the same extent up or down from the true value, and we also can assume that the true value does not affect the magnitude of the error, the error is classified as random. Systematic error, in contrast, occurs when the deviations always go in a certain direction or depend on the true value of the variable. Systematic error creates problems with the interpretation of results, as it may create artificial associations or bias that consistently deviates from the true estimate.

Bias and systematic error

Bias is traditionally sorted into three categories: *selection bias, information bias* and *statistical confounding* (see also Box 4.8).

Selection bias is a deviation of the results caused by skewed selection of the participants. It arises owing to a flaw in study design, when groups to be compared have been selected in different ways. The examined participants are survivors in all cross-sectional studies and most of the case–control studies of the retrospective studies. Survivors may differ from those who died from the disorder in question and this may imply that we systematically lose information, which brings about survival bias.

Longitudinal cohort studies entail a number of problems with regard to selection: first, who takes part in the study from the start of the baseline (selective nonresponse), and secondly, the extent to which all participants can be followed up with the same intensity. If we follow people who have been particularly exposed to certain risk factors, we may assume that there is a systematic deviation in the registration of disease events or endpoints.

Information bias is a systematic deviation of the results where the disease endpoint affects the exposure data. It may arise in retrospective studies in which the issues under study are known to both the observer and the observed. Subjects with a certain disorder may be more able to retrieve more details about previous diseases, the use of pharmaceutical drugs and other habits than can the healthy control subjects.

Detection bias is systematic error that may arise whenever exposure data are used as an argument for the diagnosis.

Confounding implies that the observed situation exhibits an endpoint association with a variable other than the one under study. We then face an alternative explanation to the observed association. This may occur when there is a strong association between the variable that is measured and a confounding factor that either is not measured or is measured with a large random error. A multiple regression model may be used to correct for confounding (see Section 11.9 of Chapter 11).

Confounding can be controlled provided that the variables in question have been examined; adjusting for other forms of bias is more difficult, if not impossible.

Assume that we have registered smoking habits in a cohort and have followed the group for a long time. During the follow-up period, liver damage has been registered in some subjects. In analysing the material, we find that the smokers have a higher risk of liver damage than do the non-smokers. In analysing these associations, we ask whether this expresses the causal effect of another variable known to be associated with liver damage, that is, alcohol consumption. If this were the case, it would imply that smokers had higher alcohol consumption than non-smokers, and as alcohol is associated with liver damage, we will get the observed result. For clarification, we analyse the effect of smoking stratified on those consuming alcohol and those not consuming alcohol or adjust for the consumption of alcohol in a multivariable statistical analysis. This type of adjustment implies that we have registered the variable that we wish to control. Age is often a confounding variable, because disease risk usually increases with increasing age. It is therefore appropriate to present age-specific results, using age stratification or age-standardized estimates, or using age as a control variable in a multivariate analysis (as discussed in Section 11.9 of Chapter 11).

9.11 TESTS AND VALIDITY

Just as the validity of an observed association may be questioned, one may question the validity of tests to distinguish between affected and healthy individuals. The validity of a test may be assessed against a true standard with regard to an observed individual actually having the disease. Valid tests are needed both in clinical medicine and in preventive efforts aimed at individuals to distinguish between healthy and affected individuals or to predict disease or health damage. Their nature depends on the problem being examined.

Sensitivity and specificity

A test is used to determine whether a disorder may occur or not, so it may be used to predict disease. A test may be made using a measurement instrument, an automated analyser or a questionnaire. The disease is the truth, and the better the test succeeds in distinguishing the ill from those who have remained healthy, the greater its validity. To assess the validity of the test, we divide those who have been examined into healthy, test-positive and test-negative groups, and display the results in a 2×2 contingency table, as illustrated in Table 9.6.

Table 9.6 Association between test and disease

| | DISEASE | | |
TEST	HEALTHY	AFFECTED	TOTAL
Negative	a	b	$a + b$
Positive	c	d	$c + d$
Total	$a + c$	$b + d$	n

In Table 9.6, d of the affected subjects tested positive. The validity relevant to identifying those who are ill is therefore $d/(b + d)$, since $b + d$ is the total number of affected or ill subjects. This expresses the *sensitivity* of the test. To judge the ability of the test to identify those who are healthy, we estimate the number of test-negative, a, and relate this number to all those who are healthy, $a + c$. This expresses the *specificity* of the test. These two concepts, sensitivity and specificity, express the validity of the test or, in other words, its ability to measure what it is meant to measure: to distinguish those who became ill from those who remained healthy. Assume now that the test variable is continuous and that subjects with a specific value above a threshold test positive and those with a value below it test negative. If we move the threshold, a greater proportion of the healthy subjects will be included if it is moved upwards and a smaller proportion included if it is moved downwards. The lower the threshold, the more affected subjects will be included, the ratio $d/(b + d)$ approaches 1 and the sensitivity increases. At the same time, a larger proportion of healthy subjects will be classified test positive, so the ratio $a/(a + c)$ decreases and the specificity declines. In other words, in seeking a valid test to identify all those who are or will become ill, some validity in identifying the healthy must be sacrificed.

Predictive power of tests

Another goal of testing is to estimate the proportion of those testing positive, $c + d$, who are or will become ill. This test characteristic is called the test's predictive power. It can be the predictability, given either a positive test result (positive predictive power) or a negative test result (negative predictive power). These predictive powers are evident from the data in the rows of Table 9.6. The positive predictive power is the ratio $d/(c + d)$ and the negative predictive power is the ratio $a/(a + b)$.

Example: Consider two examples where the same test (with known specificity and sensitivity) is used for two different sample populations, A and B.

A: 2000 subjects, half ill and half healthy (Table 9.7)

B: 10 000 subjects, 1% of whom are affected (Table 9.8).

Table 9.7 Association between test and disease in a sample of 2000 subjects

TEST	HEALTHY	AFFECTED	TOTAL
	DISEASE		
Negative	990	10	1000
Positive	10	990	1000
Total	1000	1000	2000

Table 9.8 Association between test and disease in a sample of 10 000 subjects that is a subset of the general population, 1% of whom are affected

TEST	HEALTHY	AFFECTED	TOTAL
	DISEASE		
Negative	9 801	1	9 802
Positive	99	99	198
Total	9 900	100	10 000

The test was valid and distinguished well between healthy and ill with a sensitivity of 0.99 and a specificity of 0.99.

In this material 990 of the 1000 affected of the subjects who tested positive are actually affected, so the predictive power of the positive test is 0.99.

In sample B the prevalence of the disorder is 1% in the general population. The sensitivity and specificity both are 0.99, but the predictive power of a positive test is 0.50 (99/198). In the hospital material, 99 out of 100 tested positive and were correctly identified as affected or ill. In this material, the test had high predictive power. When applied in the general population, the result of the same test was that one individual was falsely found to be ill for each affected individual correctly identified. The predictive power of a test depends on the prevalence of the condition and will vary even if sensitivity and specificity remain unchanged. This implication must be taken into consideration in discussing whether one should test for a rare disorder in the general population. The consequences of a prevalence of 0.1% are that for each truly identified affected person, 10 healthy people will have test-positive answers and be misclassified as affected. The arguments for and against testing should always consider the consequences for those who are falsely identified as affected among the test positive.

Predictive power and number needed to treat

When a rare phenomenon shares characteristics with frequent phenomena, those characteristics identify the frequent phenomena more often than the rare one. An example is that within five years, an otherwise healthy person aged 60 years with a systolic blood pressure of 200 mmHg is more likely to be alive than dead. We may draw similar conclusions on the effects of various treatment efforts.

Such strong predictive powers rest in part on the masking of less likely outcomes. Indeed, an old adage advises that 'when you hear hooves in the street, don't look for a zebra'. Nonetheless, we may look for the 'zebra', as seeking out a rare phenomenon is worthwhile, particularly if it is life-threatening. Yet finding it may be like looking for a needle in a haystack. Consider, for instance, that the underlying risk of death from a certain disorder is low and that there is a treatment that can reduce this mortality, but a large number of treated patients is needed to prevent one death. This phenomenon has led to the concept of number needed to treat (NNT), which also expresses the effect measure of an effort (see Section 11.5 of Chapter 11). To what extent has the intent of a treatment been achieved? It seems reasonable to treat a group of people who have a 10 times higher risk for stroke rather than those with only a three times higher risk.

However, a low absolute risk (which it may well be with RR at 10) implies that one must treat a large number of people in order to prevent a disease or death. This is what clinicians face in efforts to reduce the risk of myocardial infarction, hip fracture, diabetes or stroke in the population of otherwise healthy subjects. Among healthy subjects, the probability of these disorders is relatively low, even when looking five to 10 years ahead. The probability for disease is low, so attempts to reduce the probability further necessitate treating or intervening upon a large number of people (e.g. to quit smoking or change their eating habits) to prevent a disease event. In short, we can only say something about the probability within a group with regard to disease risk, and little about the single individual, so we must approach the entire group to reduce the risk of disease. To attain that goal, namely that not one person succumbs to an illness during the period studied, a large number of people must be treated; but for all those except the one individual who actually needs the treatment, the treatment will have no consequences. The problem is that we cannot identify the individual who will benefit from the treatment and thereby enjoy a longer life.

9.12 CAUSES OF DISEASE

Definition and a condition

The cause of disease or health damage can be defined as the factor that alone or in concert with other factors can bring forth disease or health damage. It is evident that not everyone exposed to a particular causal factor becomes ill. The factor may increase the risk of being affected, but only rarely alone and absolutely. Risk may be defined as the probability for a certain event within a certain period, and varies between 0 and 1. Therefore, the cause of a disease can be defined as a factor that increases the probability of its occurrence. The variation of a factor in a population is a precondition for identifying it as a causal factor. That is easier said than done, because homogeneous distributions in which everyone in a population has been exposed to a factor completely obscure any association between health damage and this particular factor. Any variation that occurs with regard to the disease must be attributed to variations in other factors. A theoretical example is to assess the effect of letting everybody in a population smoke 40 cigarettes a day from the age of 15 until the age of 65 years. During this time a number of people will become ill, but as they were all smokers, the harmful effect of the tobacco would be impossible to detect. All other variables, such as diet and genetic disposition, would emerge as associated factors, either increasing risk or being protective. Homogeneous distributions of the causal factors may be a problem, since we then may be unable to detect an effect. This may be the reason why it is difficult to detect an effect of salt intake on blood pressure, or an effect of saturated fat on breast cancer, within populations, whereas ecological studies between populations show strong associations. This also implies that the factor that triggers a disease in a single individual in the population differs from the factors determining whether there is more disease in one population than in another.

The precondition for being able to identify an association is that the variation between individuals examined (interindividual variation) is larger than the variation of repeated measurements on single individuals (intraindividual variation). Thus, in overly homogeneous populations or with excessive intraindividual variations, it is impossible to identify associations. One way to avoid this problem is to examine groups of individuals known to differ with regard to the factor examined. Associations that are seen in ecological studies but not in studies where the individuals are the observation units may lead to the conclusion that there is no association between the factor examined and the disease risk. However, this conclusion may be false as the individual variation may be large. Alternatively, differences between groups may be due to factors other than the one examined. For instance, the example of salt and blood pressure may imply that populations with low blood pressure consume considerable quantities of fruit and vegetables in addition to having a low-salt diet and otherwise differ from salt consumers.

Necessary, contributing and sufficient causal factors

Causal factors of disease or health damage are said to be necessary, contributing or sufficient.

- A necessary factor must be present for a disorder to occur. Without it, the disorder will not occur. For instance, a member of the *Mycobacterium tuberculosis* family must be present for tuberculosis to occur; without it, tuberculosis does not occur. Few, if any necessary factors are deterministic in that they act alone and always cause a disorder. Indeed, many people have been exposed to the tuberculosis bacterium without contracting the disease.

- A contributing factor acts in concert with one or more necessary factors to increase the probability of an occurrence of a disorder. Alone, a contributing factor need not be necessary, as the disorder may occur without it.
- A sufficient factor usually comprises a complex of one or more necessary factors and contributing factors that together can trigger an occurrence.

The complex that is sufficient to cause a disorder consists of factors that may be considered equally important. However, the factors vary among individuals. For instance, one subject may have a genetic disposition that makes cigarette smoking more harmful (lack of special reparation enzymes?), whereas another person may be more susceptible to saturated fatty acids. This line of reasoning has consequences for planning and organizing preventive strategies. Some single factors in causal complexes may be easier to handle than others; for instance, environmental and lifestyle relationships are simpler to manipulate than genetic dispositions.

Comparisons

Causes of diseases and disorders are often studied by comparing groups or reference groups, but comparison of groups is seldom straightforward. Consider the extensively studied issue of whether contraceptive pills increase the risk of death, by causing an increase in breast cancer, uterine cancer or other disorders. When assessing this problem, one must ask what the alternative is. With whom shall the users of contraceptive pills be compared? Should they be compared to women with low sexual activity who avoid the risks associated with pregnancy and childbirth, or to women using alternative methods of birth control, or to women not using any sort of birth control? The final conclusion to your research question may depend on the choice of reference groups to whom the exposed subjects have been compared.

9.13 ASSOCIATION VERSUS CAUSALITY

When does an association express a causal relationship?

The association between a measured factor and a disease (endpoint) may express three associations relevant to predicting disease:

- The association is causal: variable A leads to disease B, such as cigarette smoking leading to bronchial carcinoma.
- The observed variable is the intermediary part of a causal chain. X, a factor so far unknown to us, has affected the organism, whereby variable A will emerge or be changed and be detectable. This is a process that will precede the disorder of event B in the causal chain. Examples are cytological changes in the cervix predicting malignant changes, or electrocardiographic changes predicting later myocardial infarction. Such factors are often called intermediary variables. They can be assessed as risk factors as they are associated with disease, but they may also be endpoints in a study where the main interest is to examine why the factors vary. Such variables may be called surrogate endpoints.
- The observed variable is associated with an alternative factor which is the real cause of disease. For instance, alcohol consumption may be associated with bronchial carcinoma, but cigarette smoking is the causal factor. Alcohol has a predictive power, but this depends on the association between alcohol and cigarettes, not that alcohol as a single factor is a cause of bronchial carcinoma.

Any of these associations can be used when predicting disease as part of a battery of tests in order to predict a single individual's risk of disease, even if there is no direct causal association. Whether an association should be accepted as a causal relationship is always a matter of opinion based on the available evidence. The questions we must ask ourselves when facing an association are:

- Is this association a random or chance finding?
- Is the association expressing a systematic error: a bias either in selection of the sample taking part in the study or because of the measurement methods being used?
- Can this association be explained by another factor or confounding variables?

Answering these three questions leads us to discuss whether the association really is causal in a biological sense. In his textbook, Sir Bradford Hill (1977) stated a number of issues that should be discussed before a conclusion regarding causality is reasonable. The list comprises demands with regard to the strength of the association expressed for instance by relative risk, consistent results from different studies (preferably also with different methods), that the causal factor is followed by the effects and that the association is understandable from a biological viewpoint. The list has been viewed as criteria that must be fulfilled, but several authors have criticized it. Hill's original intention was not that one should fulfil all the demands every time. Causality can never be proven, but is the result of discussion within the medical research community. The list, replicated in Box 9.5, is a useful instrument and review of how to think about causality within health sciences.

Box 9.5 Issues to be discussed in assessing causality
- Strength of association
- Consistency
- Specificity
- Temporality
- Biological gradient
- Biological plausibility
- Coherence
- Experiment
- Analogy

Of the nine issues listed, only the sixth has to be fulfilled with no discussion: one must be able to show that the factor in question was followed by the disease and not a consequence of the disease. The time relationship between the factor and the disease must fit in the cause and effect relationship.

REFERENCES

Hill AB (1977) A Short Textbook of Medical Statistics. Hodder & Stoughton, London.

World Health Organization (WHO) (1994) International Classification of Diseases (ICD). 10th edn, endorsed 1990 and put into use by WHO member states in 1994, available in 42 languages at www.who.int/classifications/icd/en

FURTHER READING

Kleinbaum DG (2002) ActivEpi. Springer, New York.

Kleinbaum DG et al. (1982) Epidemiologic Research: Principles and Quantitative Methods. John Wiley, New York.

Olsen J et al. (2001) Teaching Epidemiology (2nd edn). Oxford University Press, Oxford.

Rothman KJ (2002) Epidemiology: An Introduction. Oxford University Press, Oxford.

Rothman KJ, Greenland S (1998) Modern Epidemiology (2nd edn). Lippincott Williams & Wilkins, New York.

CHAPTER 10

QUALITATIVE RESEARCH

Harald Grimen and Benedicte Ingstad

10.1 QUALITATIVE VERSUS QUANTITATIVE RESEARCH

Qualitative methods enable researchers to study social and cultural phenomena. Data are acquired on individual and group experiences, perceptions, thoughts, notions, values, actions or feelings. The principal goal is to find out how people experience and interpret their own existence.

Qualitative research conditions generate data that cannot be quantified or for which quantification is pointless. These data can be described in words but not in numbers. The fundamental idea of qualitative research is that phenomena are accessible to the researcher via particular forms of communication with and observations of the subject or subjects studied. Expertise in such phenomena is decisive in understanding how a culture functions and for our understanding of why people act as they do in concrete situations. One cannot understand how a culture functions and why people act as they do without looking into how they experience and interpret the world. The world as people perceive it takes precedence in seeking an understanding of why people behave as they do. Realizing that this is so can be essential in healthcare and health policy matters.

Smoking is a good example. There is a soundly documented statistical correlation between smoking and lung cancer. Nonetheless, people do not stop smoking to the degree they should, and many young people start. Many types of antismoking campaign have not functioned as intended. Why is that so, and what are the challenges facing the stop smoking initiatives? One reason may be that many people have yet to know about the connection between smoking and cancer and still believe that smoking is safe. The antismoking campaigns primarily face an information problem. The medical approach traditionally has been that once the connection between smoking and lung cancer is made widely known, people will stop smoking. But smoking is more complex. So, a second reason may be that it intermingles in more extensive behavioural penchants, such as appearing tough, being socially popular, affecting indifference to risk, and so on. Whenever such aspects outweigh health for an individual, stopping smoking may seem unessential. The antismoking campaigns face a lifestyle problem. Lifestyles centre on ideals and identities. Who shall I be, how shall I live my life? Comprehending lifestyles presupposes an understanding of ideals, perceptions and emotions. A third reason may be that although the connection between smoking and lung cancer is well known, most may believe that lung cancer happens to others, not themselves. People tend to regard themselves as more invulnerable than they actually are. The antismoking campaigns arguably face the challenge of changing deep-rooted trends in the human perception of hazard.

Some people may start smoking or refuse to stop smoking to signal protest, such as to antismoking campaigns. Such behaviour comprises yet another problem for antismoking campaigns.

To make sense of why people continue to smoke and start smoking despite the known link between smoking and lung cancer, we must first fathom how they experience and perceive the world, what values and ideals they have, and so on. Quantitative research, such as surveys, can provide clues. But other ways of compiling data are needed to obtain a more differentiated picture of the role of smoking in the daily lives of people and groups. Qualitative research differs from quantitative research in at least three ways.

First, qualitative research is usually intensive, in contradistinction to extensive. Intensive research has few entities and relatively many variables. An example is a recent study of young, male homosexuals (Middelthon 2001) in which 22 young gay men were interviewed several times over a period of two years. The number of interview objects is typical for qualitative research. That said, a biography is the most extreme case of the qualitative approach. It is about a single person, so there is just one object, but there are myriad variables that may be relevant. An example is a recent biography of the famed Norwegian mathematician Nils Henrik Abel (Stubhaug 2000).

Extensive research has many entities and relatively few variables. The quintessential case is quantitative research conducted to find the correlation between two variables, such as between income and gender, in a representative sample of entities. However, in most quantitative studies, there are more than two but nonetheless a limited number of variables. Quantitative research focuses on width, qualitative research on depth.

Secondly, quantitative and qualitative research differ in the organization of the research process. The phases of a quantitative research study are relatively distinctive: design, hypothesis generation, data collection and analysis. The phases of a qualitative research study mingle and cannot be clearly distinguished. Much of hypothesis generation and analysis takes place as data are acquired. Moreover, qualitative research may involve several concurrent data collection techniques, such as observing and interviewing.

Thirdly, qualitative research uses tailored methods of acquiring data, such as participant observation, focus groups and qualitative interviews of various types (discussed below). These methods yield data that must be interpreted.

10.2 USING QUALITATIVE RESEARCH

The role of qualitative research remains debated, particularly in the social sciences. Quantitative and qualitative research have their strengths and weaknesses. But today, it is relatively accepted that there are at least six key uses of qualitative methods.

First, qualitative research is particularly suited to gaining understanding of people's lived worlds, that is, how people experience and view, believe and think, aspire and assess the world about them. What is it like to live with a chronic illness? How does one cope with a handicapped child? How do patients deal with various healthcare agencies? What opinions of health and illness exist in a population and how do they change?

Secondly, qualitative research is particularly suited to studying connections in complex social and cultural phenomena and to studying how they function in practice. What values do people hold and how are values bound into a system? What gives rise to the interactions between people and human behaviour in context? Qualitative studies focus on the relationship between human behaviour and the various contexts in which it manifests itself.

Thirdly, qualitative research may be used to describe distinctiveness. The major anthropological studies of the twentieth century were of this sort, for example Malinowski's study of the Trobriander society of the Pacific (Malinowski 1961), Evans-Pritchard's study of the magic of the Azande people in Sudan (Evans-Pritchard 1985), Barth's study of the rituals of the Baktaman people of New Guinea (Barth 1975) or Gluckman's study of the judicial processes of the Barotse people of Northern Rhodesia (Gluckman 1973). The aim of such studies is to describe a culture or an aspect of it as thoroughly as possible, to find out what is distinctive or unique about it.

Fourthly, qualitative methods may be used to generate hypotheses that then can be tested by other means. Qualitative research has an open character. The researcher often starts not with a concise hypothesis, but rather with an inkling that something is significant or relevant. It is precisely this open character that suits qualitative research to hypothesis generation.

Fifthly, qualitative studies can be amenable as pilot studies precursory to quantitative studies and also may be used to collect data that help to interpret statistical material.

Finally, qualitative studies may be used to test hypotheses, as one occurrence is sufficient to refute or falsify a general hypothesis. The hypothesis that all swans are white is refuted if a single black swan can be found (Popper 1974). Qualitative case studies are often suited to seek out 'black swans'. It is essential to distinguish between hypothesis testing and generalization. One may test a hypothesis without generalizing (the test may consist of falsifying a generalization). Conversely, one may generalize without testing.

10.3 WHAT QUALITATIVE RESEARCH CANNOT BE USED FOR

Most notable is that qualitative research cannot be used for generalization in the statistical sense, principally for two reasons, both illustrated by qualitative research interviews. The first reason is that the sample of entities in qualitative research interviews is usually too small and usually is not randomly selected. The sample usually comprises 15–30 people, sometimes fewer, sometimes more. The sample is selected strategically, not randomly. This means that the researcher chooses informants for reasons including their being able to provide relevant information for the topic being studied. However, to generalize in the statistical sense, one needs a sufficiently large sample selected in a randomized procedure.

The second reason can be illustrated by the difference between a survey based on questionnaires and a qualitative research interview study. In a survey study, the questions and often the alternative answers are standardized. This is done to permit comparison of answers to the same questions given by different people. For this to be possible, one must ensure that the respondents answer the same questions and that they interpret the phrasing of questions in the same way. The standardization of questions and sometimes of answers is done principally to promote greater comparability. This alone does not guarantee against bias. In surveys, bias may arise for reasons including the choice of questions asked and the manner in which questions are phrased.

In contrast, a qualitative research interview has no standardized questions. The researcher often has only an interview guide comprising keywords of themes to be addressed. The interview evolves as a conversation that can go in various directions according to who converses. In such an interview, there is usually no point in putting the same question to all respondents. The point can rather be to ask different questions, as different respondents will have unlike information and unlike views that may be worthwhile eliciting.

10.4 SAMPLES IN QUALITATIVE STUDIES

The problems involved in selecting a sample are as fundamental in qualitative studies as they are in quantitative studies. But in qualitative studies, the sample is strategic and often cannot be selected in advance of fieldwork. As for quantitative studies, qualitative studies are open to criticism of the criteria their researchers use to select the samples studied.

In qualitative research, one chooses the subjects to be interviewed. In some studies, it is advantageous to seek maximum variation in a sample to elicit as many types of information and as many different opinions as possible. There are several techniques to accomplish that. The time and place for interviews must be chosen, not least because people often say different things at different times in different places. It is one thing to interview people at their workplaces, another to interview them in their homes. In participant observation studies, choices must be made of who will be observed and where and when they will be observed. If, for instance, the aim is to study the routines and functioning of food services at a hospital, the researcher must be present when food is prepared and served. Many phenomena, such as routines, are cyclical, so they can be studied only if the researcher follows the cycle. Other phenomena are sporadic, so the researcher must find when and where they occur. Some phenomena may be so shrouded in secrecy that a researcher cannot observe them at all.

10.5 RELIABILITY AND VALIDITY

The bottom line of reliability rests on the accuracy with which research work is conducted. Validity relates to the question of whether the researcher and the instruments chosen actually measure that which the study aims to measure or illustrate that to be illustrated. At the outset, reliability and validity are separate questions. A meticulously accurate study need not illustrate what the researcher seeks to illustrate. Conversely, a study with the potential of illustrating what the researcher seeks may be so slipshod that it is worthless.

The question of reliability arises in similar ways in quantitative and qualitative research. Both types of research involve operations that require accurate conduct. For example, in quantitative research questionnaires must be accurately structured. In qualitative research, voice recordings must be accurately transcribed to text. If poorly transcribed or transcribed by different people in different ways, interviews are of marginal value. Reliability may be checked in both qualitative and quantitative studies, such as by seeing whether there is agreement in the results when different people code the same questionnaire or transcribe the same interview.

Validity arguably is not quite the same in the two types of research. In qualitative research, it is probably more correct to speak of a continual process of validation of interpretations and concepts. This process is basically that of dialogue. Consider an example. Canadian sociologist Erving Goffman characterized psychiatric hospitals as 'total institutions' (Goffman 1968). Validating that assertion entails checking whether it draws on any central aspects of psychiatric hospital organization. This can be done only through looking into whether the assertion concerns matters that involved parties' experience as an decisive or correct description of their ways of interaction. Do they recognize Goffman's description? Do they feel that the notion strikes to the centre of the organization of such hospitals? To answer such questions, the characterization must be presented to people involved and their reactions to it observed. However, such validation is an uncertain process with great leeway for interpretation. Some may recognize the characterization, others not. Moreover, many might recognize the characterization but be unwilling to so admit, as they view the concept of 'total institution' as provocative or as a threat to their self-image. This reveals an illuminating aspect of qualitative research. Qualitative research brings about interpretations of phenomena upon which those involved often may have strong and divergent opinions. The validation process also requires interpretation. If everyone involved thinks that the concept of 'total institution' is central, why do they agree? If they disagree, why? Is it because the concept does not strike a central matter? Or is it because the concept strikes a central matter but in such a way that those concerned will neither admit nor say so? If opinions differ, why do they differ? Is it because such hospitals have some of the characteristics that Goffman describes but not to the extent implied? Or is it because the researcher's use of the concept elicits latent conflicts that otherwise would not have surfaced? In validating concepts and descriptions, the researcher must continually deal with such questions. Answering requires considerable exercise of judgement and sensitivity to nuances of interpretation.

10.6 ETHICAL CHALLENGES IN QUALITATIVE RESEARCH

The rules of ethical research are in principle the same for qualitative and quantitative research. In practice they differ in the ways in which they may be exercised.

Informed consent is an example. Informed consent is a difficult theme. It assumes that the consenting persons are in full possession of their cognitive abilities so they can understand what they consent to. This makes it difficult, for example, to involve children in research. It also makes it difficult to involve mentally ill or retarded people in research. For the researcher seeking participant observation at a hospital ward, it is undeniably tempting to pretend to be a patient without divulging the research intent, with the thought in mind that both those who treat and those who are treated will modify their behaviour and statements if they know that a researcher is present. A similar argument might be advanced by a researcher seeking participant observations in a local community.

However, such an 'end justifies the means' rationale is clearly unethical. Researchers are obliged to identify themselves and state the reasons for their presence. Another aspect is that qualitative research is not always easily conducted. Even though researchers clearly state purposes at the outset of projects and honestly answer all questions about themselves and the goals of their projects, situations always arise on the way in which it is unnatural, destructive or directly provoking when the researcher must continually remind people of the research being conducted and that daily happenings comprise data. A qualitative researcher is often 'on the job' round the clock, so as an observer or a participant in interactive situations, it is easy to come into contact with people to a degree differing from that which the principal informants were briefed on in advance, concerning the purpose of a study. Several years ago, one of the authors of this chapter (B.I.) conducted fieldwork on Greenland in which participant observation was the principal data collection technique. The goal of the project was to study the impact of a development initiative – a shrimp canning factory – on the women of the town and the adjacent small communities. Data collection took place in the factory (in which all employees were informed), in the cabins where the workers and the researcher lived, and in all the social gatherings in which factory girls took part (dances, café and street corner chats, cinemas, etc.). Even though most of the residents of the town had a notion of why the researcher was there, every day new fishing boats (with potential boyfriends for the factory girls) arrived, so it would have been impossible to ensure that everyone present be kept orientated on the project and its goal.

Another problem is that qualitative data are subject to prevailing rules for safekeeping information and for constraining the time-frame in which data may be used. Data protection laws, such as those implemented across Europe, require that personal data be anonymous and be deleted within a specified time-frame (see, for instance, the relevant EU Directive on http://ec.europa.eu/justice_home/fsj/privacy/index_en.htm). This requirement is easily met by the quantitative researcher who progressively enters data on a computer. In contrast, the qualitative researcher often needs to trace matters back to the original informants (who have fictitious names) and to revert to the information in a relatively long analytical process.

In addition, there are ethical considerations that, although not specific to qualitative research, are conspicuous because the relationship between a researcher and an informant is closer in qualitative research than in quantitative research:

- the informants must feel that the research is meaningful
- responsibility for the informants: data protection, pseudonyms and camouflage when necessary
- responsibility for assistants
- responsibility to sponsors and door-openers, as in the form of consultations and feedback
- responsibility for colleagues
- own behaviour.

The requirement that research be seen as meaningful by the subjects of a study is not unique to, but arguably more important in, qualitative research. This is because the qualitative researcher comes in close contact with the study informants and consequently must be prepared to explain the purpose of the many questions asked. In qualitative fieldwork, friendship often develops between a researcher and the informants, at least while the researcher is in the field, and that responsibility is felt to be even more binding. The same is true for assistants, who often reside in the country or the local culture researched and will stay there after the fieldworker leaves. If the researcher blunders ethically, such as by inadequately anonymizing data or by releasing information that is in some way sensitive, the assistants may easily suffer subsequent repercussions.

In general, it may be said that one should foresee and attempt to guard against injury to those involved. First of all, this can be done by using pseudonyms for people and by camouflaging the place where data are collected. Sometimes these measures are inadequate, so accounts and destinies may need to be slightly rewritten. However, the rewriting cannot be so extensive that it masks the analytical points in the data being collected. Often, accounts may be made unrecognizable for outsiders but nonetheless remain recognizable to the persons interviewed. Then it is wise to go back to the informants and let them read the accounts (or have them read) so they can approve or propose changes. One of the authors of this chapter (B.I.) once wrote a paper based on in-depth interviews of three doctors on their personal illness experiences. It soon became apparent that complete anonymization was impossible, as the case histories were easily recognizable in the small professional community. Apprehensive, the researchers sent the draft manuscript to the three doctors to elicit comments (Ingstad & Christie 2001). Had writing the paper been in vain? Would they refuse to let us publish it? All three gave their permission and sent constructive comments back for revision of the draft. It is better to ask once too often than not enough times.

Informants may ask whether they will be paid for interviews. Qualitative interviews often take up to two hours or more, so it is not so strange that an informant may ask about compensation for time spent. Interviews conducted in Norway seldom involve payment to informants, as it is assumed that some mutual interest in the conversation may be built. However, for interviews in poor countries it may seem inequitable not to offer compensation for time that otherwise might be spent in farming or other necessary tasks. A solution may be to bring along coveted foods as gifts. Bringing money into the picture may create an atmosphere that invidiously affects the data collection. Some ethnic groups that are frequently visited by researchers have set up systems in which permission to conduct research must be specifically granted by an organization representing them. Each informant receives a fixed amount for an interview, and the organization derives income from a percentage of the royalties from the book or film about the project. This is the case for the San people of Kalahari.

The researcher is also responsible for sponsors and gatekeepers, who have made the research possible, both financially and practically. Research councils and funding organizations require reports and publications. Other organizations may expect feedback in the form of seminars, the opportunity to review drafts and so on. Many ministries in developing countries complain that researchers often neglect to send back reports and publications on projects that they have been given permission to conduct. Here, we all bear the responsibility for correcting such impressions. In recent years, it has become more usual to conduct research jointly with the professionals of a developing country. Partnership in projects ensures feedback of results to the country in which research is conducted and contributes to building expertise in the cooperating institution.

Responsibility to colleagues is exercised primarily in two ways: by properly citing research conducted by others and by not infringing upon the work of others. In principle, the rules for citing the work of others apply to all research. Yet, across research, the plagiarism of the ideas and results of others is an increasing problem. In qualitative research, paragraphs and even whole pages of text are copied without citing the source. It seems that analyses in the form of text are not respected to the same degree as analyses in numbers and figures. Infringing upon the work of others entails entering a field where others are conducting research in a manner that disturbs their work. There are no written rules for conduct in such cases, just unwritten practices. In quantitative research, two researchers working on similar problems seldom interfere with each other. But in qualitative research, the researcher who is established and has set up contacts in a local culture may be seriously perturbed if a colleague then comes in at the same time to study a related problem. However, qualitative researchers studying the same phenomenon over time can mutually enhance each other's work. Moreover, this brings in the aspect of time as a control of reliability and validity.

A good rule of thumb for a researcher's own behaviour in the field is that it should not offend. This, of course, also depends on the cultural context in which the research is conducted. For example, a female researcher in a Muslim culture need not necessarily wear a veil, but she should refrain from dressing in shorts and a tight-fitting top. A male researcher in a culture where men and women are segregated should forget western chat-up customs in his spare time. Some may view such advice as a constraint on a researcher's individuality, but the aim is to evolve a strategy that creates the best possible respect and confidence in relationships with potential informants. Moreover, one should not spoil the chances of others to conduct research. A researcher who behaves decently according to local codes will leave a good impression that will benefit colleagues who follow.

10.7 QUALITATIVE DATA COLLECTION

In qualitative research, the connotation of 'method' is broader than in quantitative research. The word designates not just the technique of collecting data, but also the collection process that comprises a theoretical approximation to a problem approach, the data collection itself and the running analysis conducted during and after qualitative data are collected. This section looks at this process, which comprises the techniques or approaches that may be used to collect qualitative research data. The list of possible methods may be long and limited only by imagination. Here, only the principal and most used methods are mentioned:

- participant observation
- qualitative interviews
- narratives
- multisited ethnography
- life-mode interviews
- grounded theory
- rapid rural assessment and participatory rural assessment.

In addition, there are related and supplementary approaches that are beyond the scope of this chapter:

- historical methods
- linguistic methods
- photographic methods
- collections of objects.

Participant observation

Participant observation is mentioned first, as in a way it is the parent of all qualitative research methods, devised by the old ethnographers who conducted long, often repeated fieldwork in isolated and well-delimited societies. Through detailed observations of what people did, together with participation in their daily doings and conversations to elicit their explanations of the meaning content of happenings and activities, the classic ethnographers 'got under the skin' of the culture studied and thereby enabled understanding of how the people interpreted their existence. This perspective has become known as emic (from *phonemic* in linguistics). At the same time, the ethnographer attempted to elicit connections and patterns that perhaps were not always clear to the individual actor. This perspective has become known as etic (from *phonetic* in linguistics).

Fieldwork is paramount in participant observation. Ideally, fieldwork should be of long duration, at least one year. The fieldworker was expected to live with the local culture and learn its language so well as not to need an interpreter. Today, few researchers can finance a research stay of more than a year's duration or often less, and the use of interpreters has become more accepted and customary.

More recently, participant observations have been used to study milieu or groups that to a degree are delimited, such as youth gangs, hospital departments, local cultures in industrial countries or rural communities in developing countries.

The importance of understanding culture is clearly prominent in participant observation, both at home and abroad, but applies equally well to all qualitative social (medical) research. Culture, understood as a network of meanings with which people surround themselves, provides the basis for understanding and interpreting qualitative data.

Qualitative interviews

Now, there are hardly any more traditional, delimited cultures in which a researcher can observe and participate in almost all communal life. Consequently, the possibilities for using traditional participant observation are limited, although participant observation is still used in delimited milieu such as local subcultures and hospital departments. At once, the newer, more complex cultural constellations create new challenges for the qualitative researcher. For example, one will seldom study the carriers of a particular medical diagnosis by putting down roots in a single local culture. The potential informants often are spread over a far larger area, probably with no more than one or two in each local culture. Another difficulty may be that modern, stressed people are often uninterested in letting others into their personal spheres.

These challenges can be faced by resorting to other supplementary methods, such as the qualitative interview. It can take many forms: a completely open and unstructured interview not planned in advance, a semistructured conversation planned in advance, or a more open query included in an otherwise structured (quantitative) questionnaire.

The open interview can be spontaneous in the follow-up of a particular happening, or it may be planned and agreed in advance with the informant. The principle of the interview is that it is a freely flowing conversation on a selected topic in which the researcher seeks to attain the trust and confidence of the person interviewed. A useful approach is for the researcher to assume the role of a pupil who seeks to learn. The informant then is given the impression of being the expert on the topic at hand, namely his or her own life. But that approach may fail if the informant becomes preoccupied with 'correct' answers, which are a poor point of departure in a qualitative interview.

However, the interview is more steered by the researcher, who preferably comes with a keyword list based on the topic to be touched upon during the conversation. The list is drawn up both to ensure that essential themes are not forgotten and to ensure that all interviews conducted are thematically fairly alike. However, within this frame, the conversation should flow as naturally as possible, and themes need not necessarily be taken up in the same order in all interviews. Some may prefer to formulate part of the questions in advance, but most have found that this constrains the conversation and makes it more like filling out a questionnaire. A workable rule for a semistructured interview is to start with something meaningful to the interviewed person in relation to the agreement made. If, for example, the interview concerns living with a particular diagnosis, it is wise to begin by asking the informant about how and when the affliction started, rather than posing the standard questions on address, occupation, income and so on, which can be put aside until later in the conversation. In this manner, it is easier to awaken the informant's interest in the interview, which is beneficial for the rest of the conversation.

Qualitative interviews may be voice recorded or may be written down retrospectively. A voice recording has the advantage that it accurately captures what has been said, but also the disadvantage that the amount of data collected is large, and transcription can be excessively time consuming and expensive when done by others. An interview respectively written by the researcher should, as much as possible, be put to paper immediately after the interview. If time goes by, or perhaps there are several interviews before any are written down, recording accuracy suffers.

A structured questionnaire, primarily intended for quantitative research, may also be combined with more open questions for which the informant's answers are written down in as much detail as possible. Some computer applications, such as the Statistical Package for the Social Sciences (SPSS), support insertion of such texts. However, the big question then is how such statements are to be treated afterwards. They are not suited for quantification unless they are subsequently subjected to extensive systematization and coding, and for qualitative research alone, they often are out of context and overly brief, and consequently of little use. They are best suited as short capsules accentuating results expressed quantitatively, such as when a general dissatisfaction or satisfaction with a treatment is illustrated with a pertinent quote from one of the informants.

Many fields have qualitative interview traditions. General practitioners, psychiatrists and psychologists all use interviews as one of their more useful tools. However, a research interview is distinguished from a classical clinical interview in that, to a far greater degree, the control over and definition of what information is vital is up to the person interviewed. It is decisive that the conversation elicits the interviewee's concept of reality, not clinical findings that may help a specialist to reach a diagnosis. To a greater degree than ever, general practitioners and psychiatrists now seek to elicit their patients' experiences with their own cases, and also integrate these in research. Psychologist Steinar Kvale is among the leaders in the development of the qualitative interview as a research tool. In a landmark text (Kvale 1996), he sets forth the theoretical basis for and the practical aspects of interviewing in research. The principal message of the book is that social interaction results in knowledge production.

Narratives

Narratives are elicited in interviews aimed to prompt informants to give accounts of their careers. Triggered by the researcher's question, the story is usually angled towards a particular theme, such as in eliciting stories of migrations or the development of living conditions for a particular social group. As for qualitative interviews in general, it is pertinent to interpretation to be aware that a career communicated in an interview is not a documentation of historical fact, but rather a story that bears the stamp of the context in which it is told, of the relationship between the interviewer and the interviewee, and of the presentation of the self that the informant wishes to give.

Narratives most often designate shorter stories of limited time span, such as those of medical cases. Narratives are subject to the same caveats on interpretation as are qualitative interviews in general. They are formed by the time and place, by the relationship between the interviewer and the interviewee, and by the interviewee's wishes for self-presentation.

Multisited ethnography

The challenges of the complex, modern society have caused anthropologist George E. Marcus to launch multisited ethnography (Marcus 1995), which is not a method of collecting data, but rather a statement of the ways of collecting data that capture exactly the complexities of culture that reflect the diminishing localization of social phenomena. Marcus recommends one or more of the following strategies for data collection:

- Follow people
- Follow an object
- Follow a metaphor
- Follow a story
- Follow a life
- Follow a conflict.

Studies of migration are typical of those in which people are followed in various connections over time. The migration studied may be within a bounded region, such as one inhabited by nomads, or may comprise following people who relocate over greater parts of the globe, as do the refugees and the workers of today's increasingly global labour market. One might also envision studies that follow individuals in various social contexts in modern cultures, such as patients' meetings with primary care doctors, hospitals, workplaces and social benefit offices. A study of this type was conducted by Torunn A. Sørheim on a Pakistani immigrant family's encounters with Norwegian healthcare and social services when they have a disabled child (Sørheim 2004).

An object, such as a gift, money or a work of art, may be followed to see how it enters various types of transaction and value creations. An interesting example is a global network for studies of organ donations led by Nancy Scheper-Hughes of the University of California at Berkeley. The studies literally follow human organs, donors and recipients round the world to reveal the global trade in organs (Scheper-Hughes & Wacquant 2002).

An example of a metaphor or concept followed in various contexts is 'health'. One of the authors of this chapter (B.I.) and Per Fugelli interviewed people in several local communities in Norway to find what they picture as 'good health'. We also conducted a similar study among the San people of Kalahari (Ingstad & Fugelli 2006). One of the many fascinating finds of these studies is the high regard that Norwegians as well as the San people have for nature as a source of good health. In the USA, Emily Martin has studied conceptions of the immune system by following them in various connections, via interviews, in art and literature, in studies of texts and so on (Martin 1994).

In following a story one may, for example, become interested in how a happening is related differently by different individuals involved in it. These differences can themselves provide leads to identify important institutions, power structures, relations and phenomena in a culture. Unni Wikan used this method in her study of the Fadime 'honour killing' case in Sweden (Wikan 2004).

Finally, we will discuss an old anthropological strategy that used to be called the extended case method, that is, the following of a dramatic happening or conflict. Descriptions of mining communities in Zambia in the 1960s that followed conflicts revealed the various actors' tribal relationships and principal agenda. More recently, the method has been used in follow-ups of court cases (e.g. Wikan 2004). In 2002 in Uppsala, Sweden, a young Kurdish woman, Fadime Sahindal, was murdered by her father. The reason given was that she had found a Swedish boyfriend and wished to live her life in the modern Swedish manner. The family felt dishonoured. Unni Wikan followed the murder trial of Fadime's father and wrote a book and several articles on the case. A main theme in the book is the question of why the father acted as he did. Put simply, the principle is that one unravels from an incident to all involved actors and charts their interests, strategies and values in relation to the relevant drama.

Life-mode interviews

Psychologist Hanne Haavind and colleagues have developed a method that seeks to compensate for the difficulty of conducting participant observations over time in modern, busy families (Jensen 2005). At the outset, the method is a study of how time is spent, but goes further in seeking to elicit a family's mode of life, hence the name 'life-mode interviews'. Haavind and colleagues have developed methods for studies of the daily lives and social interactions of ordinary Norwegian families with five-year-old children.

The method entails interviewing family members (assembled or individually, depending on the approach of the project) on the course of the previous day, from the first sign of life in the morning until the last family member goes to bed in the evening. A record is made of what each family member did, individually and together, with entries at short intervals (10–15 min); it is amazing what people can recall of the previous day once they get started. Aside from asking 'what did you do yesterday?', for each sequence of happenings one can ask 'was it a usual day or happening?' and 'was that as it should be?' In this manner, the researcher can correct for unusual days and also include the normative aspect. However, with so detailed descriptions of use of time, it often turns out that the informer(s) are exhausted or impatient midway in the description of the day, so then it is best to stop and come back later in a fresh interview that begins where the first stopped. Hence, for each interview object or family, yesterday consists of two half-days. It can be advantageous if the interview is not concerned with a particular day, but rather with the life mode of the family in the course of a day.

Grounded theory

Grounded theory is a method of collecting and analysing qualitative data that was developed in the 1960s by sociologists Glaser and Strauss (1999). Their aim was to give qualitative methods and analyses a more systematic character than previously had been the case in much of the research conducted in the field.

Among other reasons, this was done to meet the quantitative research requirement of stringency. Observations of interactions are accurately recorded and coded with no predetermined theories. Instead, the theories arise from the data. Researchers of other qualitative professional traditions (including the authors of this chapter) maintain that such a starting point is illusory and not even desirable. As professionals, we always start with ideas concerning the research we conduct, and indeed our theoretical ballast is a useful tool in the collection and analysis of data (see Section 10.9). Regardless of standpoint, grounded theory is instructive for others in its meticulous recording of interactions observed in the field. Grounded theory probably has been most used by nursing researchers and to a degree also by doctors. This is natural considering that the relevant studies are often of domains that are easily followed (such as hospital departments) where almost all social actions that take place may be recorded.

Rapid rural assessment and participatory rural assessment

Rapid rural assessment (RRA) and participatory rural assessment (PRA) are methods developed for rapid collection of qualitative data, especially for use in planning and evaluation. The principal difference between the two methods is that RRA has less participation of the population involved than does PRA. Both RRA and PRA build on simple techniques that can be used with local resources, preferably with active involvement of communities to which they are applied. RRA and PRA have been used principally in developing countries, but also in developed countries. In Norway, PRA has been used in the Askvoll municipality to assess the population's views of healthcare (Mæland & Haglund 1999). Askvoll is a typical western Norwegian coastal community. It is on an island and has a population of a few thousand, most of whom live by fishing, farming and a little trading. It is probably typical of many small local communities around the world (see Box 10.1).

Even though RRA and PRA were not specifically designed for qualitative research but are more for applications in project planning and conduct, there is one exception, namely focus groups. The focus group technique is used extensively, partly in pre-projects to chart a new field before a larger (often quantitative) study starts, partly to carry on themes that have come up in larger enquiries and partly as a

---

Box 10.1 Participatory rural assessment – an example

Eelco Boonstra and colleagues wish to develop methods for local mobilization and charting of the healthcare priorities of individuals. One of several methods used was to hold popular meetings in the municipal wards. The meetings, arranged by a local volunteer organization, called in all households in a ward. In the discussions of the meetings, participants were encouraged to speak out on problems concerning health and well-being in their local communities and any problems that the communities had been or would be exposed to. The meetings ended with elections of local project groups to carry out the measures that had been agreed upon. The sum of experiences from the meetings showed that the people had a broad understanding on what could promote healthcare in their municipality. The principal sectors that people wished to prioritize were measures for the elderly, measures for well-being across generations, measures for physical activities, study circles, volunteer centres and cultural activities.

(Mæland & Haglund 1999)

---

solo method. In short, as in the RRA/PRA method, a focus group is a group of people brought together and encouraged to discuss a specific topic. One or more moderators ('facilitators') direct the discussion, ensure that all have their say and put forward additional questions to keep the discussion going. The discussion is voice recorded for subsequent transcription, or an assistant takes detailed written notes. The size and composition of a group depend on the theme researched. Usually there are no more than 10 people in a group, and often it is suitable to have fewer than 10. If, for instance, a study aims to study nursing habits, the group will bring together mothers who are nursing their babies, but it also may be illuminating to have a focus group of fathers, a group of grandmothers, a group of mothers who have finished nursing, and so on.

In analysing data collected at a focus group meeting, be aware that the forum affects the statement. Statements made in a public group discussion often reflect what is commonly accepted and believed more than what people actually do and think. For example, in a focus group discussion in local community in southern

Africa, the question of views on functionally disabled people often elicits answers that they are looked down upon, excluded and even hidden away. Yet, in the same local community, observations made of disabled people or conversations with them and their families often reveal a different picture: most families with disabled members manage as well as they can, and most disabled people have accepted places in households and in the community. Hence, data from focus group discussions can easily mislead if the influence of the public nature of the group is neglected and the data are processed in the same manner as are data from individual interviews.

The authors of this chapter advise against preferential use of focus group discussions to generate qualitative data. We maintain that the method is best suited to preliminary studies conducted before other methods (quantitative or qualitative) are used or for in-depth field enquiries on subjects already familiar, such as through a quantitative study, a poll or long-term participant observations.

10.8 TRIANGULATION

Triangulation in research refers to the use of several theories, data sources, methods or researchers in studying a single phenomenon. Multimethod triangulation combines two or more research methods. The goal may be for the methods to supplement each other or for their combination to contribute to enhancing the validity and reliability of an investigation. The term triangulation is derived from land surveying, where it refers to the use of sequences of triangles to map out an area.

In principle, qualitative and quantitative methods can be triangulated in two ways. First, a pattern noticed in the data of a quantitative investigation can be qualitatively studied in depth. Secondly, the findings of a qualitative investigation in a delimited geographical area or limited sample may in a subsequent quantitative study be tested to explore their applicability to a larger and more representative sample of a population.

In practice, almost all forms of qualitative research include an element of triangulation. For example, participant observations often may be supplemented with semistructured interview conversations with key informants or with data gleaned from searches in archives and public documents. Likewise, in qualitative interview conversations, researchers keep their eyes and ears open as much as possible to gain an impression of the informants' surroundings. Small happenings during an interview may also influence the interpretation of the data collected.

Triangulation may take place in many ways, as illustrated by the example from Botswana in Box 10.2 (Ingstad et al. 1992). At the outset, the project was qualitative, yet it used several qualitative methods, to validate and supplement each other. The multidisciplinary and multinational composition of the research team may also be viewed as a form of triangulation. A few years after the end of the project, the same institutions in Norway and Botswana cooperated in a national survey of the elderly, based on a representative sample of the population. As this book goes to press, these data have yet to be fully processed and analysed, but thus far the findings support those of the in-depth study described in Box 10.2.

Box 10.2 Triangulation: combining quantitative methods – an example

In 1990–1995, researchers at the Institute of General Practice and Community Medicine of the University of Oslo conducted a study of the elderly in Botswana in cooperation with the University of Botswana. The project, entitled 'Care for the elderly – care by the elderly', aimed to find out how the rapid development of Botswana since it gained independence had affected the elderly, both those needing care and those providing care. In planning the project, the researchers faced the decision of whether it should be quantitative or qualitative, that is, whether it should be focus on breadth or depth. The qualitative approach was chosen because the principal aim was to collect data on how the elderly experienced their situations. How did they view the changes that had taken place, and how did they feel that they best could be aided in their prevailing situation? From that, the research team selected a village assumed to be fairly representative of villages across the country and so close to the capital to be in many ways affected by the modernization process. All people over 60 were interviewed by a multinational and multidisciplinary research team of doctors, nurses and anthropologists. The methods used included a medical checkup, a simple health test, a health interview, a household interview, a depth

---

interview of key people, focus group discussions, drawings and essays from school children and participant observation. Upon completion of the project, a seminar was held in the capital, and authorities, planners, donor organization representatives and village people were invited. The project results were presented. The National Gallery featured an exhibition of project photos, and a video was shown to the participants. The attention that the seminar attracted contributed to concern for the elderly being put on the political agenda and to an old-age pension for everyone over 65 being introduced in Botswana a couple of years later.

10.9 ANALYSES OF QUALITATIVE DATA

The data resulting from a qualitative research process are of many sorts:

- personal diaries
- systematic project logbooks
- retrospectively written interviews
- voice recordings of interviews or focus groups
- narratives written by the informants themselves
- documents, literature sources and newspaper clippings
- videos, films, still photos, drawings
- objects.

The project logbook is a pivotal data record. In it, researchers record the phases of the research process, observations made, questions asked, answers given by people named, and their own recollections and reactions in the course of the work. A logbook should be kept regardless of the strategy chosen for qualitative data collection. It allows going back to review one's own notes and refreshing initial reactions at a later stage at which answers may have been found to questions that were initially puzzling. In the first phase of fieldwork, the researcher is a sponge for absorbing all impressions. Later, when the fieldwork has become more routine, it can be illuminating to recall that initial curiosity.

If so desired, the logbook may be divided in two, a diary for more personal notes and a more systematic project logbook organized by themes. The analysis of qualitative data differs from that of quantitative in that it continues and takes place throughout the research process. The continually ongoing analyses necessitated in qualitative research help to guide the evolution of data collection. In contrast, in quantitative research, analysis is a more explicit phase that takes place after the data collection phase. To a greater degree, analysis comprises finding suitable statistical measures relevant to the data acquired. Later, the resulting coefficients are interpreted, but upon finishing a qualitative investigation, one is left with myriad data and some rudimentary attempts made to analyse it during the course of the fieldwork. The sheer amount of data can be overwhelming. Consider, for instance, that a one-hour qualitative research interview may result in 20–25 pages of transcribed text. So 30 interviews will result in 600–750 pages of text. In addition, participant observations may generate hundreds of pages of notes in various forms, from keywords and single sentences to longer text passages. The records generated in focus groups will add more text. The analysis aims to create order in this wealth of material, in view of the approach of the project. This may be done in many ways. Here we discuss two, concept formation and narrative.

A greater part of the analysis of qualitative data comprises finding rich, interesting concepts. As sociologist Erik Olin Wright points out, concepts are made, not discovered (Wright 1985). They are made by researchers, not found in finished form in data collected. The essential hallmark of a rich concept is that it opens vistas that would have gone unseen without it. An interesting concept affords a new, unexpected perspective that otherwise would not have existed. Ervin Goffman's concept of a 'total institution' is a good example. In studying psychiatric hospitals, Goffman saw that they tended to take over the lives of patients, erecting formidable barriers to interaction with outsiders. To describe the phenomenon of institutionalization he coined the term 'total institution'. Using that concept, he drew attention to surprising likenesses between such hospitals and other institutions with similar characteristics, such as prisons. Goffman's concept also sheds light on another aspect of a good concept, namely that it arranges and systematizes data. A good concept provides a vantage point from which the way the pieces of a puzzle fit together can be seen. It arranges, if not all the data collected, in any case, a greater part of them. The starting point for creating a rich concept often lies in the themes or viewpoints that recur in the collected data. Therefore, the first step in creating worthy concepts is often to search out recurrent themes in the collected data. Yet there is no straight method for creating a concept. Most social scientists test concepts taken from the existing literature on their own collected data. A few innovate and create concepts that stand in the history of science.

A good concept must also provide grounds for comparison. Was it only psychiatric hospitals that exhibited the characteristics that Goffman emphasized, or are they found in other major social institutions? Such comparisons enable us to find out how widespread a concept may be.

With narratives, the researcher tries to arrange collected data in stories. A story has a beginning, a middle and an end, and it has a plot. The plot is the essence, it drives the story, says what it is about. The plot and the story's time-frame can arrange the collected data to appear as a meaningful whole. Narrative analysis can be used in various ways. Qualitative research interviews often are amenable to stories, as the respondents are free to speak as they please and often are encouraged by the interviewer to tell a story. Stories also are well suited to comparison. For example, interview data may reveal that many or most of the respondents have the same stories, similar stories or dissimilar stories of similar inner structure. Ingrid Hanssen used this analytical technique in analysing interviews with Norwegian nurses and their ways of relating to patients from other cultures (Hanssen 2002). A story is a means of conveying experiences. Similar stories from many respondents imply that they have the same basic experiences. But stories need not be used only in analysing interviews.

Social anthropologist Ellen Kristvik used the story approach (Kristvik 2002) in studying the lives of prostitutes in Nepal, basing the analysis in part on interviews and in part on observations and being with them. The dissertation is constructed as a network of stories of the lives of individual prostitutes. Parts of the stories comprise extracts from interviews. Other parts consist of information that Kristvik gleaned through informal conversations and participant observation. Together, the network of the stories of individual lives give a shocking picture of the lives of a particular group of prostitutes in Nepal.

Many of the classics of qualitative works in social anthropology are built up of stories. An example is Bronislaw Malinowski's brilliant account of the Kula exchange of the Trobiand Islands of the South Pacific (Malinowski 1961). It comprised stories based on two years of participant observation, arranged so that they described all the steps leading to the high point of the ceremonial exchange of necklaces and armbands between groups. He focused principally on describing canoe building, production of the valuable barter objects, and so on. The reader goes along with each step towards the ceremonial exchange, and through the structures of the stories, gains an overview of the entire ceremonial exchange system. One gains a magnificent view of a large exchange system that even the people taking part in it did not fully comprehend. Nonetheless, Malinowski's analytical technique was not new. It is part of the historian's standard repertoire of analytical techniques, from ancient times, when historians described wars and battles. Aspects such as these underscore that the analysis of qualitative data is not an automatic process, but rather one that demands considerable intellectual effort.

Computer applications for qualitative research are to a considerable degree contrived for grounded theory data, but may also be used for other qualitative data. These applications do not analyse data, but rather help to arrange and locate data. They replace the file cards, scissors and glue of days past, but do not replace keeping logbooks. However, the texts entered as logbook notes may be systematized using such applications.

10.10 RELEASING QUALITATIVE DATA

According to Kvale (1996), there are five possible stages of a qualitative research process:

1. the antipositivist enthusiasm phase
2. the interview-quoting phase
3. the working phase of silence
4. the aggressive phase of silence
5. the final phase of exhaustion.

In the antipositivist enthusiasm stage, a conscious choice has been made of an approximation that closely approaches what is germane and seeks opinions and experiences, not quantifiable objective data. As the fieldwork progresses and the researcher still is enthusiastic, the interview-quoting phase begins, in which the researcher yearns to share new findings with others, of course within the constraints of confidence. Later, after writing starts and those nearest begin to tire of accounts of fieldwork, the researcher enters the working phase of silence. The duration of silence depends on the scope of the data and on the researcher's energy and writing skill, but invariably is longer than expected. The researcher then enters the aggressive phase of silence, where even friendly queries on the progress of the work are taken as personal insults. It is hoped that few ever reach the final phase of exhaustion, in which one gives up the entire project. But all researchers, regardless of method, inevitably occasionally find themselves in situations where expectations must be adjusted according to the art of the possible. The best must not be the enemy of the good.

Qualitative data are analysed continuously. So, from the first observations or interviews, one cannot distinguish finds and analysis in releases in the same manner that they are distinguished in medical journals. This may be a barrier to publishing. Fortunately, with time, Social Science and Medicine and other journals have been launched to overcome that barrier.

The presentation may, but need not necessarily follow the traditional sequence of introduction, theory, method, etc. One may, for example, start with a biographical sketch or a description of a situation that draws the reader into the context studied. One should aim for readable presentation, not use unnecessary foreign words, and put forth theory that clarifies, not fogs the material for the reader. This is particularly the case in writing for medical readers, who often are more concerned with applicability than with theoretical constructions. The distinction between good qualitative research and good writing rests in part on theory being the foundation of research. But at the same time, it is best that qualitative research reads well, and the approaches of good writing are close to those of qualitative research.

Qualitative research has come to stay in medicine, and with time is likely to gain greater acceptance as more medical professionals start using qualitative methods.

REFERENCES

Barth F (1975) Knowledge and Ritual among the Baktaman of New Guinea. Universitetsforlaget, Oslo; Yale University Press, New Haven, CT.

Evans-Pritchard EE (1985) Witchcraft, Oracles and Magic among the Azande. Clarendon Press, Oxford.

Glaser B, Strauss A (1999) The Discovery of Grounded Theory. Aldine de Gruyter, New York.

Gluckman M (1973) The Judicial Process among the Barotse of Northern Rhodesia. Manchester University Press, Manchester.

Goffman E (1968) Asylums. Penguin Books, London.

Hanssen I (2002) Facing differentness: an empirical inquiry into challenges in intercultural nursing. Doctoral dissertation, University of Oslo.

Ingstad B, Christie VM (2001) Encounters with illness: the perspective of the sick doctor. Anthropology and Medicine, 8: 2–3.

Ingstad B, Fugelli P (2006) Our health was better in the time of Queen Elizabeth. The importance of land to the health perception of the Botswana San. Senri. Ethnological Studies: 70: 61–79.

Ingstad BF et al. (1992) Care for the elderly – care by the elderly. The role of elderly women in a changing Tswana society. Journal of Cross-Cultural Gerontology 7: 379–398.

Jensen TK (2005) The interpretation of child sexual abuse. Culture and Psychology 11: 5.

Kristvik E (2002) Nepali sex workers: narratives of violence and agency. Doctoral dissertation, University of Oslo.

Kvale S (1996) InterViews: An Introduction to Qualitative Research Interviewing. Sage, London.

Mæland JG, Haglund BJA (1999) Health promotion development in the Nordic and related countries. In: Bracht N (ed.) Health Promotion at the Community Level. New Advances. Sage, Newbury Park, pp. 178–197.

Malinowski B (1961) Argonauts of the Western Pacific. EP Dutton, New York.

Marcus GE (1995) Ethnography in/of the world system: the emergence of a multi-sited ethnography. Annual Review of Anthropology 24: 95–117.

Martin E (1994) Flexible bodies. Tracking Immunity in American Culture from the Days of Polio to the Age of AIDS. Beacon Press, Boston.

Middelthon AL (2001) Being young and gay in the context of HIV: a qualitative study among young Norwegian gay men. Doctoral dissertation, University of Oslo.

Popper KR (1974) The Logic of Scientific Discovery. Hutchinson, London.

Scheper-Hughes N, Wacquant L (eds) (2002) Comodifying Bodies. Sage, London.

Sørheim TA (2004) The impact of gender on parental response to children with disability. A study of Pakistani families in Norway. In: Trausdadottir R, Kristiansen K (eds) Gender and Disability in Nordic Countries. Studentliteratur, Lund.

Stubhaug A (2000) Niels Henrik Abel and his Times: Called Too Soon by Flames Afar. Springer, Berlin.

Wikan U (2004) Deadly Distrust: Honor Killings and Swedish Multiculturalism. In: Hardin R (ed.) Distrust. Russel Sage, New York.

Wright EO (1985) Classes. Verso, London.

FURTHER READING

Denzin NK, Lincoln YS (2005) The Sage Handbook of Qualitative Research. Sage, Thousand Oaks, CA.

Denzin NK, Lincoln YS (2005) Collecting and Interpreting Qualitative Materials. Sage, Thousand Oaks, CA.

STATISTICAL ISSUES

Petter Laake, Thore Egeland and Eva Skovlund

11.1 INTRODUCTION

This chapter covers the statistical analysis of categorical and numerical data (see Section 4.6 of Chapter 4 for descriptions of types of data). For categorical data, contingency tables and logistic regression are used to analyse association between variables. The binomial distribution is briefly described, as it is essential to understanding the assumptions and analyses related to categorical data. For continuous numerical data, statistical analyses of data that can be described by normal distributions as well as by non-normal distributions will be discussed. In the latter case, non-parametric methods will be examined. The choice of a statistical method rests on understanding the distributional assumptions made. The normal distribution is particularly important, because many methods depend on it. So this is discussed in depth. Towards the end of the chapter more advanced regression models are discussed and exemplified.

The focus will be on the effect measures, the associated effect estimate, on confidence intervals and P-values. These concepts are essential for the statistical analyses of categorical and continuous data, as summarized in Box 11.1.

Several examples of analyses of data files are presented. The data files are accessible at http://books.elsevier.com/9780123738745, in both standard text format and

> ### Box 11.1 Main concepts
> - The effect measure is derived from the outcome variable. It is the magnitude that provides the statistical description of interest in a study.
> - The null hypothesis, H_0, is the hypothesis that we seek to reject. A null hypothesis is tested by computing a test statistic and the associated P-value. The P-value is the calculated probability for the observed result or results that are even more contradictory to the null hypothesis, *when the null hypothesis is correct.*
> - A 95% confidence interval includes the unknown true value with 95% probability.

SPSS (Statistical Package for the Social Sciences) format. The StaTable program (Cytel 1998) is used to compute P-values and percentiles.

11.2 EFFECT MEASURE, HYPOTHESIS TESTING AND CONFIDENCE INTERVAL

The main concepts involved in statistical analyses are summarized in Box 11.1.

Concepts

Most studies entail the assessment and measurement of a health outcome. The outcome may be the result of a treatment in a randomized controlled trial or in an epidemiological study, such as a study of the effect of smoking during pregnancy on birth weight. Accordingly, first an effect measure must be chosen. For the case of the effect of smoking during pregnancy, the effect measure could be chosen as the difference in average birth weights of babies born to non-smoking and smoking mothers. Then, the effect measure must be estimated, giving us the effect estimate. Clearly, the effect measure chosen and the way it is estimated depend on the study design and type of data for the variables that measure the outcome. A confidence interval, most often with a 95% confidence level, is used to assess uncertainty. A 95% confidence interval includes the unknown true value with 95% probability. Professional journals increasingly require that the effect measure be stated along with its confidence interval to enable assessment of the magnitude and significance of an effect.

The approach to a study is formulated in its research hypothesis. For example, a research hypothesis may be that the outcomes of two treatments differ or that there is an association between smoking during pregnancy and birth weight of babies. Then, a testable statistical hypothesis, the null hypothesis, is formulated. An example might be that two normal distributions have the same mean, in which case the null hypothesis is 'no difference'; the normal distribution is presented in greater detail in Section 11.6. The P-value provides the basis for rejecting the null hypothesis and consequently accepting an alternative hypothesis, which is the research hypothesis. Computation of the P-value is discussed below.

A research paper usually should include three relevant measures: the effect estimate, the 95% confidence interval for the effect measure and the P-value for the statistical hypothesis related to it. In the following, these concepts are discussed in greater detail. Then, for selected situations, the choice of effect measure will be explained, as well as the computation of the associated confidence interval and P-value.

Null hypothesis, alternative hypothesis and P-values

An approach to a problem including computation of P-values is inextricably linked to the null hypothesis H_0 and the alternative hypothesis H_1. Hypothesis testing always calls for a null hypothesis, which is *the postulate that we seek to reject*. Assume, for instance, that we wish to determine whether the effects of two drugs differ. The null hypothesis is then that the two drugs have equal effects. The effect measure is then zero, and the alternative hypothesis is that the effects of the two drugs differ. The alternative hypothesis may be one-sided or two-sided. It is one-sided if changes are bound to be in one direction. It is two-sided if we are interested in changes to either side, such as either positive or negative effects. Two-sided alternative hypotheses are the most common in medical studies.

A null hypothesis is tested by computing a test statistic. In statistical analyses, we wish to learn whether an effect is large compared to random variation, expressed as standard error. Consequently, a test statistic is selected based on

$z = $ *effect estimate/standard error.* This is the standard form for all parametric tests. The *P*-value is the calculated probability of occurrence of an observed result (z_{obs}) as well as other results even more contradictory to the null hypothesis, *when the null hypothesis is correct.* Computation of the *P*-value for a one-sided alternative hypothesis differs from that for a two-sided alternative hypothesis. For a two-sided hypothesis, the *P*-value is normally twice the *P*-value of a one-sided hypothesis. Most often, one decides that a *P*-value less than 5% indicates a statistically significant result, so that the null hypothesis may be rejected. A significant result may be interpreted as a strong indicator of a real difference. In contrast, a high *P*-value, such as greater than 5%, is not synonymous with the null hypothesis being correct.

Confidence interval

An effect measure is estimated in an experiment. Its form depends on the design and on the statistical model used. The estimation of the effect measure is associated with uncertainty. In turn, uncertainty is quantified by the confidence interval, which expresses in a percentage, usually 95%, the probability that it includes the true, but unknown value of the effect measure. If the effect measure is a *difference*, the value of 0 corresponds to 'no difference', such as no difference between the effects of two treatments. If a 95% confidence interval does not include the value of 0, the *P*-value of a corresponding hypothesis test will be less than 0.05. Accordingly, the confidence interval may be used to draw conclusions on an effect. If, however, we consider a ratio, such as the relative risk (Section 11.4), the value 1 corresponds to 'no effect' and our conclusions will be based on whether the 95% confidence interval contains the value 1 or not. The three main results listed in Box 11.2 summarize the statistical analysis.

Box 11.2 Presentation of the main statistical results
- Effect estimate
- Uncertainty of the effect estimate, reflected in the 95% confidence interval
- Results of one or more hypothesis tests, expressed in *P*-values

---

11.3 BERNOULLI TRIAL

A sequence of n independent trials is called a Bernoulli trial provided that exactly one of two outcomes, A or $\bar{A}$, can occur in each trial. In all trials the probability is the same for the outcomes, namely $P(A) = p$ and $P(\bar{A}) = 1 - p$. Births in a maternity ward comprise an example of a Bernoulli trial. For example, outcome A can indicate the birth of a girl.

In n Bernoulli trials, the variable X representing the number of A outcomes will follow a binomial distribution. The true, unknown probability is denoted p (see Box 11.3). The notation p circumflex, $\hat{p}$, designates the estimator of the unknown probability, as is standard in the statistical literature.

Box 11.3 Estimation, standard error and confidence interval for p

The estimator of the unknown probability p is:

$\hat{p} =$ proportion in the sample with property A

The standard error is estimated by:

$$SE(\hat{p}) = \sqrt{\frac{\hat{p}(1 - \hat{p})}{n}}$$

The binomial distribution can be approximated by a normal distribution in which the confidence interval for p is given as

$$\hat{p} - 1.96SE(\hat{p}), \ \hat{p} + 1.96SE(\hat{p})$$

The key equations for the estimator $\hat{p}$ and the standard error of $\hat{p}$ result from the properties of the binomial distribution.

Example: The Framingham Heart Study is a large cohort study aimed to study the risk factors for coronary heart disease. A random selection of 500 people from data on 1615 people aged 31–65 years, reported by Carroll et al. (1995) and given at http://www.stat.tamu.edu/~carroll/data.php is made available in the

FRAMINGHAM.SAV data file at http://books.elsevier.com/9780123738745. Here, the variables of coronary heart disease (FIRSTCHD) will be studied as a categorical variable with two mutually exclusive categories, presence or absence of coronary heart disease. Consequently, we choose to present the univariate distribution as a frequency distribution.

In Table 11.1, let outcome $A = $ *Evidence of coronary heart disease (CHD)*. The assumptions for a Bernoulli trial are then fulfilled since this is a random sample, and the (unknown) probability for coronary heart disease is estimated as $\hat{p} = 37/500 = 0.074$. Consequently, the proportion of coronary heart disease is estimated to be 0.074 or 7.4%. $\text{SE}(\hat{p}) = \sqrt{\hat{p}(1-\hat{p})/n} = 0.0117$, which gives a confidence interval of $(0.051, 0.097)$. A different sample would have resulted in an estimate differing from 0.074. The uncertainty in the estimation of p is given by the confidence interval above and calculated as explained in Box 11.3. More precisely, we can say that we are 95% certain that the true value falls within a 95% confidence interval.

Table 11.1 Frequency distribution of coronary heart disease (CHD)

EVIDENCE OF CHD	COUNT	PERCENTAGE	CUMULATIVE %
No evidence	463	92.6	92.6
Evidence	37	7.4	100.0
Total	500	100.0	

First of all, we are interested not only in the frequency distribution of the categorical variables, but in the bivariate association between two categorical variables as they are presented in contingency tables; Table 11.2 is one example. Associations are presented between an outcome variable (dependent variable) and an explanatory variable (there are many synonyms, including exposure variable, independent variable and covariate). There is no generally accepted practice for presenting such tables. We choose to present the contingency table with the outcome variable in the rows and the explanatory variable in the columns. The association between the variables of coronary heart disease (FIRSTCHD) and elevated systolic blood pressure (HYPERTENSION) can be presented as in Table 11.2. In this case, elevated blood pressure is defined as a pressure over 120 mmHg.

Table 11.2 Association between coronary heart disease (CHD) and elevated systolic blood pressure

EVIDENCE OF CHD		ELEVATED SYSTOLIC BLOOD PRESSURE		TOTAL
		NO	YES	
No evidence	Count	151	312	463
		96.8%	90.7%	92.6%
Evidence	Count	5	32	37
		3.2%	9.3%	7.4%
Total	Count	156	344	500
		100.0%	100.0%	100.0%

The aim is to study the effect of systolic blood pressure on coronary heart disease, so the percentages are presented within each category of systolic blood pressures. As can be seen in Table 11.2, 9.3%, or a proportion of 0.093 with a systolic blood pressure of more than 120 mmHg had coronary heart disease. The corresponding proportion among those without elevated blood pressure is 0.032.

The effect measure is based on comparing the risk of coronary heart disease among men with systolic blood pressure over and under 120 mmHg. Of the many ways this can be done, medical statistics most often use *relative risk* (RR) and *odds ratio* (OR) as measures. The *absolute risk reduction* (ARR) or the equivalent *risk difference* (RD) is less often used, although it might be just as natural. Regardless of the effect measure chosen, attention is focused on its associated confidence interval. Moreover, the significance of the association is assessed by statistical testing. Here, Pearson's chi-squared test statistic will be used to determine whether there is significant association between the variables.

This presentation includes several equations, so that computations may be performed on a calculator or manually, although a statistical program, such as SPSS, more readily provides the results. We have chosen to do this for several reasons. In general, detailed knowledge can give deeper understanding. Moreover, in this area, relatively simple equations may be used in many cases. Yet another and perhaps more vital reason is that one should be able to compute OR and RR manually, as the results provided by statistical software depend on how the variables are coded. For example, reversing the coding of the HYPERTENSION blood pressure variable might change OR from 2 to 1/2. The conclusion drawn from the analysis remains unchanged, but its formulation must match the coding or the chosen frame of reference. Simple computations are therefore vital, as they provide an independent check on the results produced by a statistical program. For some examples we have chosen small datasets to make manual computations easy.

In this chapter, the important equations for calculating the confidence interval and the P-values of statistical hypotheses are based on an approximation between the binomial distribution and the normal distribution. The approximation is quite good when np and $n(1 - p)$ both exceed 5, which means that the expected number of observations in the cells of the table is greater than 5. When the expected number of observations is less than 5, an exact method is used to compute the correct confidence interval and P-values. Fisher's exact test is one such test for *2 × 2 contingency tables*. Table 11.2 is an example of such a table. More generally, if there are r rows and c columns, the corresponding table will be an $r \times c$ contingency table. As an alternative to exact methods, Monte Carlo methods may be used to compute the confidence interval and P-values. These methods are not discussed in this chapter; for details, see statistical textbooks, such as those listed in the reference list to this chapter. Exact and Monte Carlo computations can be performed using statistical programs, such as SPSS ('crosstabs/exact') or STATXACT.

11.4 COMPARING TWO PROPORTIONS

Table 11.2 may be used to study the association between FIRSTCHD and HYPERTENSION. In analysing associations the observations are considered as outcomes of Bernoulli trials. There is one Bernoulli trial for individuals with blood pressures below 120 mmHg ($n = 156$) and one for those above ($n = 344$). Our goal is to determine whether the probabilities for coronary heart disease differ for the classifications of below and above 120 mmHg systolic blood pressure.

In general, we assume that we have two trials, with the probability of property A given by $p_0 = P(A)$ for trial 1 and $p_1 = P(A)$ for trial 2. We find the estimators for p_0 and p_1, respectively, and assess whether $p_0 \neq p_1$. The research hypothesis includes a description of the association between the outcome variable (property A) and the explanatory variable (trial). Falsification of the null hypothesis of $p_0 = p_1$ implies that an association between the outcome variable and the explanatory variable exists. In the example at hand (Table 11.2), this means that there is an association between the outcome variable FIRSTCHD and the explanatory variable HYPERTENSION.

In general, the association between the outcome and the exposure is studied as it appears in Table 11.3. Here the outcome variable is ILLNESS, with the categories of Healthy and Affected, and the explanatory variable is EXPOSURE, with the categories of Non-exposed and Exposed. Let $p_0 = P(Affected)$ be the unknown probability among the non-exposed and $p_1 = P(Affected)$ among the exposed. The association between outcome and explanatory variable is given in Table 11.3.

Table 11.3 Association between the outcome variable and the explanatory variable

		EXPLANATORY VARIABLE	
		NON-EXPOSED	EXPOSED
Outcome variable	Healthy	$1 - p_0$	$1 - p_1$
	Affected	p_0	p_1
		1	1

The null hypothesis of no association between ILLNESS and EXPOSURE is tested:

$H_0: p_0 = p_1$ against the alternative $H_A: p_0 \neq p_1$

The statistical test is based on the observations in the cross-table of ILLNESS and EXPOSURE and has the form illustrated in Table 11.4. The comparison of two binomial trials is summarized in Box 11.4.

Table 11.4 Observed association between the outcome variable and the explanatory variable

		EXPLANATORY VARIABLE		TOTAL
		NON-EXPOSED	EXPOSED	
Outcome variable	Healthy	a	b	$a + b$
	Affected	c	d	$c + d$
	Total	$a + c$	$b + d$	n

Box 11.4 Comparison of two binomial trials

The two unknown probabilities are estimated as:

$$\hat{p}_0 = \frac{c}{a+c} \quad \text{and} \quad \hat{p}_1 = \frac{d}{b+d}$$

From the theory of Bernoulli trials it follows that when $p_0 = p_1$, the standard error of the estimator is:

$$SE(\hat{p}) = \sqrt{\frac{c+d}{n}\frac{a+b}{n}\left(\frac{1}{b+d} + \frac{1}{a+c}\right)}$$

The normal distribution approximates the binomial distribution, so it follows that:

$$z = \frac{\dfrac{d}{b+d} - \dfrac{c}{a+c}}{\sqrt{\dfrac{c+d}{n}\dfrac{a+b}{n}\left(\dfrac{1}{b+d} + \dfrac{1}{a+c}\right)}}$$

is approximately standard normally distributed (i.e. the location is 0 and the standard deviation is 1). Note that the size of z measures the difference between p_0 and p_1, since we compare the difference with the random variation given by the standard error. Again, the test statistic $z = $ *effect estimate/ standard error* provides the basis for testing. The observed value z_{obs} is used as the starting point for testing the hypothesis H_0 and H_0 is rejected if $2P(z \geq z_{obs}) \leq 0.05$. The normal distribution in StaTable is used (Cytel 1998).

Example: In Table 11.2 for the association between FIRSTCHD and HYPERTENSION,

$$\hat{p}_1 = \frac{32}{344} = 0.093 \quad \text{and} \quad \hat{p}_0 = \frac{5}{156} = 0.032.$$

Moreover, $\dfrac{c+d}{n} = \dfrac{37}{500} = 0.074$ and $\dfrac{a+b}{n} = \dfrac{463}{500} = 0.926$,

which gives $\mathrm{SE}(\hat{p}) = \sqrt{0.074 \cdot 0.926 \left(\dfrac{1}{344} + \dfrac{1}{156} \right)}$.

Consequently, $z = \dfrac{0.093 - 0.032}{\sqrt{0.074 \cdot 0.926 \left(\dfrac{1}{344} + \dfrac{1}{156} \right)}} = 2.41$.

From the StaTable (Cytel 1998) for the normal distribution, $2P(z \geqslant 2.41) = 0.016$, which means that there is a statistically significant association ($p \geqslant 0.016$) between FIRSTCHD and HYPERTENSION.

The following will show that the P-value corresponding to the z-statistic is identical to the chi-squared test statistic, and in practice the more general Pearson's chi-squared test statistic is used to test hypotheses in two binomial trials.

Chi-squared test

The idea of the Pearson's test statistic is to compare the observed number of observations in each cell with the expected. Let the number of observations in cell (i, j) be n_{ij} and the expected number of observations be e_{ij}. Then the test statistic in Box 11.4 may be written:

$$z = \frac{\dfrac{d}{b+d} - \dfrac{c}{a+c}}{\sqrt{\dfrac{c+d}{n} \dfrac{a+b}{n} \left(\dfrac{1}{b+d} + \dfrac{1}{a+c} \right)}} = \sqrt{\frac{n(bc - ad)^2}{(a+b)(a+c)(b+d)(c+d)}}$$

$$= \sqrt{\sum_{cells} \frac{(n_{ij} - e_{ij})^2}{e_{ij}}}$$

The last expression is the square root of Pearson's chi-squared test statistic that is used to test the assertion that ILLNESS is independent of EXPOSURE. This extends to the general cases (see Box 11.5).

> ## Box 11.5 Pearson's chi-squared test statistic
>
> A test statistic for the null hypothesis of no association can be based on the difference between the observed and expected cell counts. The following test statistic is used:
>
> $$\chi^2 = \sum_{\text{cells}} \frac{(\text{observed} - \text{expected})^2}{\text{expected}} = \sum_{\text{cells}} \frac{(O - E)^2}{E}$$
>
> In the above equation, O means observed and E means expected. This test statistic is Pearson's chi-squared test statistic. The χ^2-distribution depends on the number of cells in the table, which sets the number of degrees of freedom as $(r - 1)(c - 1)$, where r is the number of rows and c is the number of columns.
>
> Now, let the observed χ^2-value be χ^2_{obs}. H_0 is rejected if $P(\chi^2 \geqslant \chi^2_{obs})$ $\leqslant 0.05$. The P-value for an observed χ^2-value corresponding to $(r - 1)$ $(c - 1)$ degrees of freedom is found from StaTable (Cytel 1998).

Example: In the FRAMINGHAM.SAV data file, the association between FIRSTCHD and HYPERTENSION was tested with tests for two Bernoulli trials. The result was $z = 2.41$ with a P-value of $2P(Z \geqslant 2.41) = 0.016$. In this case, $r = c = 2$ and so there is $(r - 1)(c - 1) = (2 - 1)(2 - 1) = 1$ degree of freedom. The chi-squared test results in the figures given in Table 11.5.

Table 11.5 Pearson's test statistic for the association between coronary heart disease and elevated systolic blood pressure

	VALUE	DF	ASYMP. SIG. (2-SIDED)
Pearson chi-square	5.823	1	0.016
N of valid cases	500		

In Table 11.5 DF means degrees of freedom, and ASYMP. SIG. means that the calculation of the P-value is based on the approximation to a normal distribution.

The observed Pearson's chi-squared test statistic (5.82) is equal to the square of the z-value resulting from the case of two Bernoulli trials, and the P-values are the same in both cases. There is 1 degree of freedom, as $r = c = 2$. Accordingly, it can be concluded that we may use the chi-squared test in all cases to find P-values for associations in tables, including 2×2 tables.

11.5 MEASURES OF ASSOCIATION IN 2×2 TABLES

The measures of association between illness and exposure are based on the occurrences of illness among the exposed and non-exposed, and various measures are defined below. An effect measure can be defined by examining either the relationship between risks (RR or OR) or the difference between them (ARR). As noted earlier, ARR is identical to RD. Let $\widehat{RR}$, $\widehat{OR}$ and $\widehat{ARR}$ denote the estimated effect measures based on the observed tables.

Absolute risk reduction and its confidence interval

The absolute risk reduction (ARR; see Box 11.6) is defined as:

$$ARR = p_1 - p_0$$

Note that two exposure groups are compared, so that the risk difference expresses the difference between the exposed and the non-exposed.

Box 11.6 Absolute risk reduction

The absolute risk reduction is estimated by:

$$\widehat{ARR} = \hat{p}_1 - \hat{p}_0 = d / (b + d) - c/(a + c)$$

The standard error of $\widehat{ARR}$ is found from:

$$SE(\widehat{ARR}) = \sqrt{\frac{\hat{p}_1(1 - \hat{p}_1)}{b + d} + \frac{\hat{p}_0(1 - \hat{p}_0)}{a + c}}$$

The above equations evolve from the theory of Bernoulli trials (Section 11.3). In the customary manner, the confidence interval for ARR is found from:

$$\widehat{ARR} - 1.96\, SE(\widehat{ARR}),\ \widehat{ARR} + 1.96\, SE(\widehat{ARR})$$

Example: When we want an estimate of the absolute reduction of risk of coronary heart disease among people with elevated compared to those with normal systolic blood pressure, the above equations give an estimated risk reduction of $\widehat{\text{ARR}} = 0.093 - 0.032 = 0.061$ or 6.1%. The standard error of this estimate is

$$\text{SE}(\widehat{\text{ARR}}) = \sqrt{\frac{0.093(1 - 0.093)}{344} + \frac{0.032(1 - 0.032)}{156}} = 0.021,$$

which in turn yields a confidence interval for risk reduction of (0.020, 0.102). The confidence interval does not include zero, so we know that the *P*-value for testing the null hypothesis will be less than 0.05.

In randomized controlled trials, it may sometimes be useful to report the results using the number needed to treat (NNT), which is the reciprocal of the absolute risk reduction, that is, 1/ARR. It determines the number of patients who need to be treated to prevent one bad outcome. A large treatment effect leads to a small NNT.

Relative risk and its confidence interval

The relative risk (RR; see Box 11.7) is defined as the ratio of the probabilities of being affected among exposed and non-exposed:

$$\text{RR} = p_1/p_0$$

The relative risk is estimated based on the observations in Table 11.4 as:

$$\widehat{\text{RR}} = \frac{d / (b + d)}{c / (a + c)}$$

A transformation is required to calculate the confidence interval for RR. This is because the estimated relative risk does not follow a normal distribution, but is skewed. In this case a logarithmic transformation is used.

Box 11.7 Relative risk

The relative risk is estimated as:

$$\widehat{RR} = \frac{d / (b+d)}{c / (a+c)}$$

The standard error of $\ln \widehat{RR}$ may be calculated as:

$$SE(\ln \widehat{RR}) = \sqrt{\frac{1}{d} - \frac{1}{b+d} + \frac{1}{c} - \frac{1}{a+c}}$$

The confidence interval of ln RR is as usual provided by the theory of the normal distribution, namely by its interval:

$$\ln \widehat{RR} \pm 1.96 \, SE(\ln \widehat{RR})$$

The inverse transformation back to $\widehat{RR}$ via the exponential function results in the confidence interval for RR being:

$$e^{\{\ln \widehat{RR} - 1.96 \, SE(\ln \widehat{RR})\}}, \quad e^{\{\ln \widehat{RR} + 1.96 \, SE(\ln \widehat{RR})\}}$$

Example: Again, we analyse the association between FIRSTCHD and HYPER-TENSION in Table 11.2 by using relative risk as the effect measure. The estimated RR is:

$$\widehat{RR} = \frac{32 / 344}{5 / 156} = 2.90$$

and the standard error is:

$$SE(\ln \widehat{RR}) = \sqrt{\frac{1}{32} - \frac{1}{344} + \frac{1}{5} - \frac{1}{156}} = 0.471$$

Hence, using the above equation, the confidence interval for RR becomes:

$$(e^{\ln 2.90 - 1.96 \cdot 0.471}, e^{\ln 2.90 + 1.96 \cdot 0.471}) = (1.15, 7.31)$$

As the relative risk is calculated as 2.90, this result implies that high elevated systolic blood pressure raises the risk of coronary heart disease by almost a factor of 3, and that the true, unknown value of the risk ratio in this case is likely to be within the interval $(1.15, 7.31)$.

SPSS may be used to compute the RR and the confidence interval. The results are shown in Table 11.6. The result agrees with that computed manually, and the confidence interval of RR is again $(1.15, 7.31)$.

Table 11.6 Relative risk (RR) for the association between coronary heart disease and elevated systolic blood pressure

	VALUE	95% CI	
		LOWER	UPPER
Relative risk for evidence of CHD	2.902	1.153	7.307
N of valid cases	500		

Odds ratio and the confidence interval

As its name implies, the odds ratio (OR) is the ratio between two expressions of odds. In turn, odds may be defined as the ratio of two probabilities, the probability of being affected to the probability of being healthy. Odds are calculated both for exposed people and for non-exposed people. From Table 11.3, the odds for exposed people are $p_1/(1 - p_1)$, while the odds for non-exposed are $p_0/(1 - p_0)$.

The odds ratio is the ratio of the exposed to the non-exposed, that is:

$$OR = \frac{p_1/(1 - p_1)}{p_0/(1 - p_0)} = \frac{p_1(1 - p_0)}{p_0(1 - p_1)}$$

Note that this is the cross-product in Table 11.3. Thus, OR is estimated by calculating the cross-products in Table 11.4. The result is an estimated odds ratio of:

$$\widehat{OR} = \frac{ad}{bc}$$

As for the risk ratio, in computing the confidence interval for OR, a logarithmic transformation is used (Box 11.8).

Box 11.8 Odds ratio

The odds ratio is estimated as:

$$\widehat{OR} = \frac{ad}{bc}$$

The standard error of $\ln \widehat{OR}$ is:

$$SE(\ln \widehat{OR}) = \sqrt{\frac{1}{a} + \frac{1}{b} + \frac{1}{c} + \frac{1}{d}}$$

The confidence interval of ln OR is as usual provided by the theory of the normal distribution, namely by its interval:

$$\ln \widehat{OR} \pm 1.96 \, SE(\ln \widehat{OR})$$

The inverse transformation back to $\widehat{OR}$ via the exponential function results in the confidence interval for OR being:

$$e^{\{\ln \widehat{OR} - 1.96 \; SE(\ln \widehat{OR})\}}, \; e^{\{\ln \widehat{OR} + 1.96 \; SE(\ln \widehat{OR})\}}$$

Example: In the table of association between FIRSTCHD and HYPERTENSION, the estimate of OR is:

$$\widehat{OR} = \frac{151 \cdot 32}{312 \cdot 5} = 3.10$$

and the standard error of the estimate is:

$$SE(\ln \widehat{OR}) = \sqrt{\frac{1}{151} + \frac{1}{312} + \frac{1}{5} + \frac{1}{32}} = 0.491$$

327

By entering terms and inverse transforming, the confidence interval for OR is:

$$(e^{\ln 3.10 - 1.96 \cdot 0.491}, e^{\ln 3.10 + 1.96 \cdot 0.491}) = (1.18, 8.11)$$

This means that excessive systolic blood pressure increases the odds of coronary heart disease by a factor of three to one and that the confidence interval is $(1.18, 8.11)$.

This corresponds with the result produced by SPSS in Table 11.7

Table 11.7 Odds ratio (OR) between coronary heart disease and elevated systolic blood pressure

	VALUE	95% CI	
		LOWER	UPPER
Odds ratio for evidence of CHD	3.097	1.183	8.108
N of valid cases	500		

Thus far, the focus has been on the computational aspects of the three effect measures, absolute risk reduction, relative risk and odds ratio. The first stands apart from the other two because it expresses an absolute difference. Although much has been written about absolute and relative measures of effects, confusion prevails about these concepts. In particular, for a rare event, it might be confusing to report a relative effect instead of an absolute effect. Assume that an event occurs among 2% of untreated patients and among 1% of treated patients. The absolute risk reduction is 1%, which is read as one percentage point. The relative risk indicates considerable effect, since often treatment halves the risk of the event. The relative risk then is an apparently large number, despite representing a small change of just 1%. Apparent disagreements on the magnitudes of effects can often be ascribed to differences in effect measures. Regardless of the effect measure chosen, to be germane, the changes must be assessed relative to initial values, and the changes must be of scientific interest.

Likewise, much has been written about the choice between RR and OR, in part with divergent opinions. Often, the effect measure is spoken of as relative risk, although the odds ratio actually is computed. This is unfortunate, particularly as it can create unnecessary confusion. However, the two measures yield approximately the same results for rare events. The analysis of Table 11.2 in this case yields an RR of 2.91 and an OR of 3.10. The probability of coronary heart disease is small, so the results for RR and for OR coincide. The estimated probability of illness can be estimated in cohort studies and cross-sectional studies. Hence, RR normally is used in such studies. In case–control studies, samples are taken among patients and among healthy controls. As the ratio between the numbers of patients and controls is given, one cannot estimate the probability of illness, just the probability of being exposed. Then, OR is used as an effect measure. However, this classification is oversimplified, as OR is often used, regardless of the type of trial involved. The odds ratio permits simpler statistical analyses in more complex models, such as in logistic regression (see Section 11.9). Hence, there are computational reasons for the extensive use of the odds ratio. However, interpretation of RR may be simpler, so many prefer to use RR whenever possible.

11.6 NORMAL DISTRIBUTION

The normal distribution plays a key role in the applications of statistical methods. The distribution is symmetrical, and its form is described by its location (expectation) and its variation (standard deviation). The location is denoted by the symbol μ and the variation by the symbol σ; σ is called the standard deviation and σ^2 the variance. A normal distribution of X with location μ and variance σ^2 is written $X \sim N(\mu, \sigma^2)$. The shape of the normal distribution is determined by its location on the x-axis (μ) and the width of the distribution (σ).

In principle, an infinite number of values is possible when variables are continuous, so it is no longer possible to compute the probabilities of individual outcomes. Instead, one may compute the probability for a variable lying within an interval or being less than a given value. For example, for variable X and a particular measured value a, one may compute $P(X \leq a)$. For this purpose, the concept of probability density is introduced. The probability density is represented by a curve and areas under the curve correspond to probabilities: the probability of the variable X being less than a is denoted $P(X \leq a)$ and equals the area to the left of a. The entire area under the curve is unity. Then, a is denoted as the 90th percentile if $P(X \leq a) = 0.90$ (=90%). The 25th percentile is often called the lower quartile, the 50th percentile the median and the 75th percentile the upper quartile. The interquartile distance is the difference between the lower and upper quartiles, and it covers 50% of the distribution.

Many statistical methods are based on the assumption that the data follow a normal distribution. This assumption is checked using a normality plot that establishes whether the data (approximately) follow a normal distribution.

Now we look more closely at estimating the location of a distribution with a confidence interval, when the data follow a normal distribution. In a statistical experiment comprising n observations, as of n persons, the data $X_1, X_2, ..., X_n$ follow a normal distribution having a location μ and a variance σ^2. The mean of the observations, $\bar{X}$, is calculated. The distribution of $\bar{X}$ will also be normal, with a location at μ but with a variance of σ^2/n, and therefore $\bar{X} \sim N(\mu, \sigma^2/n)$. This means that the distribution of the average of the observations has the same location, but with a variance of $1/n$ of that of a single observation. The empirical variance is given by

$$s^2 = \frac{1}{n-1} \sum_{i=1}^{n} (x_i - \bar{x})^2$$

and s is used as the estimator for the standard deviation. Accordingly, the location of the normal distribution is estimated as the mean of the observations, and the standard error of the mean (SEM) may be written

$$SE(\bar{X}) = \frac{s}{\sqrt{n}} = SEM.$$

This gives the relationship between the standard error of the mean and the standard deviation.

If we are interested in the distribution of the individual data points, as might be the case in assessing the uncertainty of measurement methods or in describing the variation in populations, we will be concerned with the magnitude of s. However, if we are interested in the location of the distribution the SEM is of interest, because it estimates the uncertainty of the mean.

Note that the confidence interval (Box 11.9) is approximate, as it is based on the normal distribution and not on the t-distribution, which is described in Section 11.7. For small samples ($n < 50$) the confidence interval based on the t-distribution (Section 11.7) is more accurate.

> **Box 11.9 Confidence interval of the mean**
>
> $\bar{X}$ is the estimate of μ, which is the true, unknown location of the distribution.
>
> Because $\bar{X}$ follows a normal distribution, an approximate 95% confidence interval for the location of the distribution is
>
> $$\bar{X} - 1.96 \text{ SEM}, \bar{X} + 1.96 \text{ SEM}$$
>
> Here, SEM is multiplied by -1.96 and 1.96, which point to the 2.5 and 97.5 percentiles of the normal distribution. This ensures that the confidence interval includes the location of the distribution with 95% probability. The percentiles in the normal distribution can be found by using the StaTable application (Cytel 1998).

Example: A descriptive analysis of the mean systolic blood pressure (MEANSBT) of the FRAMINGHAM.SAV data file as it is produced by SPSS, is presented in Table 11.8.

Table 11.8 Descriptive analysis of mean systolic blood pressure (SBP)

MEAN SBP	STATISTIC	SE
Mean	131.3020	0.86293
95% confidence interval for mean		
Lower bound	129.6066	
Upper bound	132.9974	
5% trimmed mean	129.8744	
Median	128.0000	
Variance	372.327	
Standard deviation	19.29579	
Minimum	77.50	
Maximum	226.00	
Range	148.50	
Interquartile range	21.50	

SE: standard error

In this table, the estimated location of the distribution is at 131.3, the estimated variance is 372.3 and the estimated standard deviation is 19.3. The standard error is 0.86; this is the SEM. The study comprises $n = 500$ data points, so SEM $=$ $19.3/\sqrt{500} = 0.86$ and the approximate confidence interval is (129.6, 133.0). This closely matches the values listed in Table 11.8. The confidence interval computed this way may deviate from that computed in tables when n is not large, because tables are based on t-distributions.

11.7 COMPARISON OF MEANS

The t-tests are often used to compare means. They were first proposed in 1908 by W.S. Gosset, who wrote under the pseudonym 'Student', so the approach is often referred to as 'Student's t-tests'. One type of t-test is used to draw conclusions on the mean of one group of observations (one sample), and another type of t-test is used to compare the means of two groups of observations (two samples). Pairs of observations, such as two observations of the same patient, may be examined by treating them as a single sample of differences between two values.

t-distribution

Assume a data set comprising n data randomly selected from a normal distribution located at μ with a standard deviation σ. Let $\bar{X}$ be the mean, and let SEM $= s/\sqrt{n}$ be the standard error of $\bar{X}$. The test statistic is based on *effect estimate/standard error*, so in this case $t_{obs} = \bar{X}/\text{SEM}$. We seek $P(t \geq t_{obs})$, and to find it we need to know the distribution of t. The test statistic t has a (Student's) t-distribution with $n - 1$ degrees of freedom. There are many t-distributions. The t-distribution selected depends on the number of observations, or more precisely the number of degrees of freedom. It is known that $\bar{X}$ follows a normal distribution and that the difference between the normal distribution and the t-distribution is that for the latter the standard deviation (σ) is regarded as unknown and is estimated by s. The two distributions may differ. The t-distribution is symmetrical and bell-shaped, as is the normal distribution, but it has heavier tails. As the number of observations increases, the t-distribution more closely resembles the normal distribution. The difference between the two distributions is negligible for about 100 or more degrees of freedom.

The StaTable application (Cytel 1998) may be used to find P-values for a given t_{obs} as well as percentiles in the t-distribution for various degrees of freedom.

The 97.5 percentile in a t-distribution with 50 degrees of freedom is 2.009. Degrees of freedom of 100 result in a percentile of 1.98. As that is nearly 1.96, the 97.5 percentile of the normal distribution, in this case the t-distribution can be approximated by a normal distribution.

Two examples of use of the t-test – the one-sample test and the two-sample test – are now presented. The one-sample test is used for pairs of observations.

Paired t-test

A principal application of the one-sample t-test is for pairs of observations, such as the results of cross-over trials (see Chapter 8) or the comparison of observed values before and after a specified treatment. For instance, the null hypothesis to be tested can be:

H_0: the mean before treatment equals the mean after treatment.

The alternative hypothesis can then be:

H_A: the means before and after treatment differ.

In practice, the difference (D) is calculated between two observations of each patient and the result tested to see whether it is compatible with the assertion that the difference is zero (Box 11.10). The null hypothesis can then be expressed as:

H_0: the location of the distribution of D is 0.

Box 11.10 Paired t-test

The test statistic is based on the mean of the differences between the observations for the individual patients, denoted by $\bar{D}$ instead of $\bar{X}$ to indicate clearly that we are dealing with a paired situation. Under the null hypothesis, the mean value of the difference is 0, and the test statistic may be written:

$$t = \frac{\bar{D}}{\text{SEM}} = \frac{\bar{D}}{s / \sqrt{n}}$$

This test statistic is t-distributed with $n - 1$ degrees of freedom. For a particular value of the test statistic, t_{obs}, the null hypothesis may be rejected if the P-value is $2P(t \geq t_{obs}) \leq 0.05$. StaTable (Cytel 1998) or statistics programs, such as SPSS, may be used to compute this P-value.

Example: Studies have demonstrated that some calcium antagonists may interact with constituents of grapefruit juice. Apparently, intake of grapefruit juice leads to increased drug concentration in plasma. If this is the case, the result may be a more pronounced antihypertensive effect and a greater risk of adverse effects.

Assume that a trial is conducted with $n = 9$ volunteers who on one occasion take the drug with water and on another occasion take it with a measured quantity of grapefruit juice. The maximum plasma concentration of the drug is c_{max}, measured in nmol/l. The results shown in Table 11.9 are from a hypothetical trial that nonetheless is plausible as it reflects previously published results. The CMAX.SAV data file is accessible at http:/books.elsevier.com/9780123738745

Table 11.9 Plasma concentration of calcium antagonist taken with and without grapefruit juice

PERSON	CALCIUM ANTAGONIST + WATER (c_{MAX} NMOL/L)	CALCIUM ANTAGONIST + GRAPEFRUIT JUICE
1	22.5	31.8
2	8.2	12.4
3	11.3	15.7
4	5.9	4.2
5	20.7	23.1
6	19.3	18.5
7	10.4	12.0
8	12.1	14.9
9	18.3	21.2

The observed calcium antagonist concentrations c_{max} with and without intake of grapefruit juice are compared for each person by taking their differences, as shown in Table 11.10.

Table 11.10 Drug concentration c_{max} differences, intake with grapefruit juice minus intake without

PERSON	1	2	3	4	5	6	7	8	9
Difference	9.3	4.2	4.4	-1.7	2.4	-0.8	1.6	2.8	2.9

The mean difference is $\bar{d} = 2.79$ and the standard deviation is

$$s = \sqrt{\frac{1}{n-1}\sum_{i=1}^{n}(d_i - \bar{d})^2} = 3.19$$

We test whether the mean difference is compatible with the null hypothesis by computing

$$t = \frac{2.79}{3.19/\sqrt{9}} = 2.62.$$

StaTable (Cytel 1998) is used to find the P-value. As $n = 9$, there are $n - 1 = 8$ degrees of freedom. For a t-value of 2.62 and eight degrees of freedom, the P-value is 0.03, so we can reject the null hypothesis and conclude that simultaneous intake of grapefruit juice increases the maximum concentration of calcium antagonist in plasma.

This problem can easily be analysed using SPSS with a paired t-test. The results are shown in Tables 11.11 and 11.12.

Table 11.11 Descriptive analysis of c_{max}

PAIRS OF OBS.	MEAN	N	SD	SEM
Grapefruit juice	17.08889	9	7.843539	2.614516
Water	14.30000	9	5.979339	1.993113

SD: standard deviation; SEM: standard error of the mean

Table 11.12 Single sample t-test for c_{max}

PAIRS	PAIRED DIFFERENCES					T	DF	SIG. (2-TAILED)
	MEAN	SD	SEM	95% CI OF THE DIFFERENCE				
				LOWER	UPPER			
Grapefruit juice–water	2.78889	3.193526	1.064509	0.33413	5.24365	2.620	8	0.031

From Table 11.12, the mean difference is 2.79 and the standard error is 1.06. This gives a t-statistic of 2.62 (by dividing the mean by the standard error), with a corresponding P-value of 0.03. Note that in SPSS, the P-value is computed for a two-sided alternative.

Box 11.11 Effect measure for paired samples

The mean difference is the natural effect measure in this case. Hence, we must find the 95% confidence interval for the true difference, as given by the equation:

$$\bar{D} - t_{0.975,n-1} s / \sqrt{n}, \bar{D} + t_{0.975,n-1} s / \sqrt{n}$$

which can be rewritten:

$$\bar{D} - t_{0.975,n-1} \text{SEM}, \bar{D} + t_{0.975,n-1} \text{SEM}$$

where $\bar{D}$ is the mean difference, s the standard deviation and, as before, n is the number of patients. The constant $t_{0.975,n-1}$ is the 97.5 percentile in the t-distribution having $n - 1$ degrees of freedom, as found using StaTable (Cytel 1998).

Example: In the previous example, the mean difference is $\bar{d} = 2.79$. As there are 8 degrees of freedom, $t_{0.975,8} = 2.306$. A 95% confidence interval then is calculated to be

$$\left(2.79 - 2.306 \cdot \frac{3.19}{\sqrt{9}}, 2.79 + 2.306 \cdot \frac{3.19}{\sqrt{9}}\right) = (0.33, 5.24).$$

The interval does not include zero and consequently corresponds to a hypothesis test that results in rejection of the null hypothesis. As seen in Table 11.12, this confidence interval is the same as that found using SPSS.

Two independent samples

One of the most common statistical analyses is the comparison of two independent groups of observations (see Box 11.12). In these cases, we wish to find whether there is a difference in the locations of the distributions of the two groups. Let the locations of the two distributions be μ_1 and μ_2. It is assumed that the observations of the two groups are independent and follow normal distributions. It is also assumed that the standard deviations of the two groups are the same and, as usual, equal to σ. Now, one may test the null hypothesis:

H_0: there is no difference between the groups, that is: $\mu_1 = \mu_2$

against the two-sided alternative

H_A: the groups differ, that is: $\mu_1 \neq \mu_2$

Box 11.12 Testing in the two-sample situation

Assume that the two groups respectively comprise n_1 and n_2 observations. The means of the groups are $\bar{X}_1$ and $\bar{X}_2$. The test statistic, which is the ratio between the effect estimate and its standard error, is

$$t = \frac{\bar{X}_1 - \bar{X}_2}{s\sqrt{\dfrac{1}{n_1} + \dfrac{1}{n_2}}}$$

where s is the estimate of the pooled standard deviation of the two groups, expressed by

$$s = \sqrt{\frac{s_1^2(n_1 - 1) + s_2^2(n_2 - 1)}{n_1 + n_2 - 2}}$$

Example: We wish to determine whether there is a difference between the measured ratios of free and total prostate-specific antigen (PSA) in patients with cancer of the prostate (CAP) and those with benign prostate hyperplasia (BPH). The ratios of free to total PSA are measured in a small sample consisting of $n_1 = 10$ CAP patients and $n_2 = 9$ BPH patients. The data, as listed in Table 11.13, are from the PSA.SAV study file at http:/books.elsevier.com/9780123738745

Table 11.13 Percentage free/total PSA in patients with cancer of the prostate (CAP) and with benign prostate hyperplasia (BPH)

CAP	BPH
1.4	25.1
36.0	27.0
27.3	20.6
11.0	18.0
2.3	34.1
14.9	40.3
13.9	22.5
11.6	21.1
7.7	8.8
10.7	

Now, we test the null hypothesis:

H_0: there is no difference between the mean of the free/total PSA of the two groups, that is: $\mu_1 = \mu_2$

against the two-sided alternative

H_A: there is a difference between the mean of the free/total PSA of the two groups, that is: $\mu_1 \neq \mu_2$

In this example, the mean free/total PSA is $\bar{x}_1 = 13.7$ for CAP patients and $\bar{x}_2 = 24.2$ for BPH patients. The standard deviations are estimated as $s_1 = 10.65$ and $s_2 = 9.13$, respectively. An estimate of the pooled standard deviation is

$$s = \sqrt{\frac{s_1^2(n_1 - 1) + s_2^2(n_2 - 1)}{n_1 + n_2 - 2}} = \sqrt{\frac{10.65^2 \cdot 9 + 9.13^2 \cdot 8}{10 + 9 - 2}} = 9.97$$

and

$$t = \frac{13.7 - 24.2}{9.97\sqrt{\dfrac{1}{10} + \dfrac{1}{9}}} = -2.29$$

There are $10 + 9 - 2 = 17$ degrees of freedom, and StaTable (Cytel 1998) is used to find that $p = 0.04$.

SPSS may also be used to analyse the data, as shown in Tables 11.14 and 11.15.

In the analysis, the two groups are assumed to have the same variance, which is credible on the basis of the descriptive analysis of Table 11.14. The results are thus read from the first row of Table 11.15, 'Equal variances assumed', and the P-value (for the two-sided alternative) is found to be $p = 0.035$. Accordingly, it may be concluded that the ratio between free and total PSA in serum is higher in patients with BPH than those with CAP. There is, however, more information in

Table 11.14 Descriptive analysis of free/total prostate-specific antigen (PSA)

PSA	N	MEAN	SD	SEM
CAP	19	13.6800	10.65456	3.36927
BPH	9	24.1667	9.13044	3.04348

CAP: cancer of the prostate; BPH: benign prostate hyperplasia.

Table 11.15 Two-sample *t*-test for free/total PSA

PSA	LEVENE'S TEST FOR EQUALITY OF VARIANCES		*T*-TEST FOR EQUALITY OF MEANS					95% CI OF THE DIFFERENCE	
	F	SIG.	T	DF	SIG. (2-TAILED)	MEAN DIFFERENCE	SEM	LOWER	UPPER
Equal variances assumed	0.079	0.783	−2.290	17	0.035	−10.4867	4.57925	−20.14803	−0.82530
Equal variances not assumed			−2.310	16.969	0.034	−10.4867	4.54035	−20.06728	−0.90605

Table 11.5. The assumption of equal variances is assessed by Levene's test, given by F (Levene 1960). In this case the resulting P-value is 0.783 and we do not reject the hypothesis of equal variances. This further justifies reading of results from the line 'Equal variances assumed'. If the P-value from Levene's test is below 0.05, one should use the line 'Equal variances not assumed'.

Box 11.13 Effect measure and confidence interval

The effect measure for the difference between the two groups is:

$$\bar{X}_1 - \bar{X}_2$$

This is a plausible effect measure as it estimates the difference between the means of the distributions, $\mu_1 - \mu_2$.

A 95% confidence interval for the difference is:

$$\bar{X}_1 - \bar{X}_2 - t_{0.975,\, n_1 + n_2 - 2} \cdot s \sqrt{\frac{1}{n_1} + \frac{1}{n_2}},$$

$$\bar{X}_1 - \bar{X}_2 + t_{0.975,\, n_1 + n_2 - 2} \cdot s \sqrt{\frac{1}{n_1} + \frac{1}{n_2}}$$

The $t_{0.975, n_1 + n_2 - 2}$ constant is found from StaTable (Cytel 1998) for a t-distribution with $(n_1 + n_2 - 2)$ degrees of freedom.

Example: In the PSA example above, the difference between the means is -10.5 and the standard error of the mean difference is 4.58. The constant is $t_{0.975,17} = 2.79$. Using the equations in Box 11.13, a 95% confidence interval for the true difference is $(-20.15, -0.83)$.

11.8 NON-PARAMETRIC METHODS

The approaches described in Section 11.7 are based on the assumption that the data follow a normal distribution and for the two-sample situation on the assumption that the samples have the same variance. Assumptions of data following a normal distribution are verified using a normality plot. The assumption of two samples having the same variance can be checked by a descriptive analysis or more formally by Levene's test.

Two alternative methods are available for cases in which the data do not follow a normal distribution. A set of data may be transformed so that it (approximately) follows a normal distribution, or non-parametric methods may be used. The need for transformation arises whenever the tails of a distribution are heavier than those of a normal distribution. A logarithmic transformation is most often used. Non-parametric methods are also used, as explained below. Note, however, that non-parametric methods are as intolerant as the *t*-test to deviations from the assumption of equal variance.

Effect measure in non-parametric situations

The mean of a set of observations is a sensible estimator of the location of a distribution whenever the data completely or approximately follow a normal distribution. However, the mean is excessively influenced by extreme observations. The median is a more appropriate effect measure whenever a distribution is skewed or the data contain extreme observations. This is because the median is not influenced by extreme observations. For a symmetrical distribution, the mean and the median are the same.

Non-parametric methods should be used whenever the observations do not approximately follow a normal distribution. Often, such methods are used only for statistical testing and not for estimating the effect measure. That said, we will show how the confidence interval may be found for median differences in non-parametric situations.

Wilcoxon's signed rank test

Wilcoxon's signed rank test is the most commonly used test for paired comparisons (see Box 11.14).

> ## Box 11.14 Wilcoxon's test for paired comparison
> - Calculate the difference for each pair of observations.
> - The absolute values of the paired differences are arranged in ascending rank order.
> - The ranks are added up for the positive (or the negative) differences.
> - Find the P-values from Table 1 at http:/books.elsevier.com/ 9780123738745

Example: Consider the same example as that of the one-sample *t*-test in Tables 11.9 and 11.10.

The absolute values of the paired differences are arranged in ascending rank order. Thereafter, rank values are added up for the positive (or the negative) differences. Accordingly, the test statistic is denoted T^+ or T^-. The differences ordered in ascending absolute value are shown in Table 11.16.

Table 11.16 Ranked differences in c_{max}, intake of drug with grapefruit juice minus intake without

Absolute difference	0.8	1.6	1.7	2.4	2.8	2.9	4.2	4.4	9.3
Sign	−	+	−	+	+	+	+	+	+
Rank	1	2	3	4	5	6	7	8	9

The sum of the ranks for negative differences is $t^- = 1 + 3 = 4$, and the sum of the rank values for positive differences is $t^+ = 2 + 4 + 5 + 6 + 7 + 8 + 9 = 41$. Let T^- be the test statistic. The P-value is found from Table 1 at www.textbooks.elsevier.com, or from tables in statistics texts, such as Altman (1991) or Rosner (2005). The table for $n = 9$ shows that the lower limit is 5 and the upper limit is 40. As our test statistic falls outside this range, it may be concluded that the effect of grapefruit juice is statistically significant ($p < 0.05$).

When there are more than 15 pairs, the rank sum is approximated well by a normal distribution. The expectation and standard deviation of T^- are

$$E(T^-) = \frac{n(n+1)}{4} \quad \text{and} \quad SD(T^-) = \sqrt{\frac{n(n+1)(2n+1)}{24}}$$

In the example $E(T^-) = 22.5$ and $SD(T^-) = 8.44$. The test statistic is equal to the difference between the observed value (that is 4) and the expected value (22.5) divided by the standard deviation (which is 8.44):

$$z = \frac{4 - 22.5}{8.44} = -2.192$$

which gives a two-sided P-value of 0.028.

The data also may be analysed in SPSS. The results are shown in Tables 11.17 and 11.18. The results in Table 11.18 correspond to those of manual computation, both exact and approximate (Asymp. sig.).

Table 11.17 Sums of rank values for the positive and negative differences of c_{max}

WATER–GRAPEFRUIT JUICE	N	MEAN RANK	SUM OF RANKS
Negative ranks	7[a]	5.86	41.00
Positive ranks	2[b]	2.00	4.00
Ties	0[c]		
Total	9		

[a] Water $<$ grapefruit juice; [b] water $>$ grapefruit juice; [c] grapefruit juice $=$ water

Table 11.18 Test statistic and P-value for Wilcoxon's test for paired comparison

	WATER–GRAPEFRUIT JUICE
Z	-2.192[a]
Asymp. sig. (2-tailed)	0.028

[a] Based on positive ranks

Here, the non-parametric test led to the same conclusion as that of the t-test. This suggests that the observations (that is, the observed differences) approximately follow a normal distribution.

In this case, the information in the data are in the differences, the d_is. We wish to find a confidence interval for the median of the d_is. There are several ways to compute the median and the confidence interval (see Box 11.15). Here, the Hodges–Lehmann procedure is used. First, the differences are ranked by magnitude.

Then the means $w_{ij} = (d_i + d_j)/2$, $i \leqslant j$ are calculated. In the above example, there are nine differences that result in 45 such means: w_{11}, w_{12}, w_{22}, w_{13}, w_{23}, w_{33}, w_{14}, ..., w_{59}, w_{69}, w_{79}, w_{89}, w_{99}. Finally, the median of these means is calculated. An effect estimate of 2.8 is found, which in this example agrees with the median of the differences. However, that will not be the case when there are identical observations, or ties, among the pairs, which result in zero differences. Accordingly, the previous method is preferable.

Example: The confidence interval is found from Table 2. We rank all 45 observations. For $n = 9$ observations, the critical values are 6 and 40, so if we choose the value of rank ordered mean number 6, corresponding to $w_{33} = (0.40 + 0.40)/2 = 0.40$, and number 40, corresponding to $w_{49} = (2.4 + 9.3)/2 = 5.85$, we find an approximate 95% confidence interval of (0.40, 5.85).

Box 11.15 Effect measure and confidence interval
- Calculate the difference for each pair of observations and sort them in increasing order.
- Calculate the mean of the differences as explained in the text.
- Arrange the paired means in increasing order.
- Calculate the median of the means (the effect estimate).
- Calculate the 95% confidence interval in the effect estimate, such as from Table 2 at http:/books.elsevier.com/9780123738745

Two independent samples: the Wilcoxon–Mann–Whitney test

This test has many names, including Wilcoxon's two-sample test, Wilcoxon's rank sum test and Mann–Whitney U-test (see Box 11.16). The null hypothesis to be tested is

H_0: the observations of the two groups are from the same distribution

and the alternative hypothesis is:

H_A: the observations of the two groups are from two distributions that have the same shape, but there is a shift in location.

345

> **Box 11.16** Wilcoxon–Mann–Whitney test for two independent samples
> - Rank observations for both groups together.
> - Calculate the rank sum for one of the groups.
> - Find the P-value from Table 3 at http:/books.elsevier.com/9780123738745

The Wilcoxon–Mann–Whitney test was made for cases when there is a pure shift in location between the distributions. While this setting is ideal, the test may also be used in more general cases.

To look more closely at the test, let the number of patients in group 1 be n_1 and in group 2 n_2. The total number of patients then is $n = n_1 + n_2$. The observations can be plotted in ranked order so the groups to which they belong are obvious. Thereafter, rank values are assigned to the observations. Similar sums of the ranks (W) for the two groups indicate that the null hypothesis is true. A rank sum for one of the groups that is considerably higher or lower than anticipated implies that the alternative hypothesis is true. The critical values depend on the choice of significance level and probability distribution of W under the null hypothesis.

Example: A trial to examine the effect of psychological treatment of noise sensitivity included 22 migraine patients. Half of the patients were given psychological treatment (relaxation training) and half comprised a control group. Each patient listened to an audio tone that progressively increased in volume to the level of perceived discomfort. That level was registered. A low score indicates that a patient tolerates little noise. The results of the study are shown in Table 11.19. The data are replicated from Daly et al. (1995), and are found in the SCORE.SAV file.

The sums of the ranks for the two groups differ considerably (Table 11.20). There is a trend towards lower scores in the control group than in the treated group. If the null hypothesis is true, the rank sums of the two groups should be similar because the groups are equally large. Assume that the control group rank sum of W_1 is chosen as the test statistic. The observed rank sum is $w_1 = 90$. Is this sum sufficiently small to reject the null hypothesis? The answer can be found in Table 3 or in statistical texts such as Altman (1991) or Rosner (2005). We enter the table

Table 11.19 Sound level (score) perceived as uncomfortable

TREATED	CONTROLS
5.70	2.80
5.63	2.20
4.83	1.20
3.40	1.20
15.20	0.43
1.40	1.78
4.03	11.50
6.94	0.64
0.88	0.95
2.00	0.58
1.56	0.83

for $n_1 = n_2 = 11$ and find a lower limit of 96 and an upper limit of 157. Our test statistic falls outside this range, so $p < 0.05$ and H_0 is rejected.

An alternative approach to finding P-values is to approximate the rank sum distribution with a normal distribution. Usually exact tables cover values only up to $n_1 = n_2 = 12$, so an approximation to a normal distribution must be used for higher values, unless relevant statistical software is used. The approximation is considered good when n_1 and $n_2 \geq 10$. The expected rank sum under H_0 is

$$E(W_1) = \frac{n_1(n_1 + n_2 + 1)}{2} = 126.5$$

and the standard deviation is

$$SD(W_1) = \sqrt{\frac{n_1 \cdot n_2(n_1 + n_2 + 1)}{12}} = 15.23 \,.$$

When we divide the difference between the observed and expected rank sums by the standard deviation, we find a test statistic

$$z = \frac{90 - 126.5}{15.23} = -2.397 \,.$$

347

Table 11.20 Ranked values of the sound levels perceived as uncomfortable

CONTROL GROUP		PSYCHOLOGICALLY TREATED	
RANK	SOUND LEVEL	RANK	SOUND LEVEL
1	0.43		
2	0.58		
3	0.64		
4	0.83		
		5	0.88
6	0.95		
7.5[a]	1.20		
7.5[a]	1.20		
		9	1.40
		10	1.56
11	1.78		
		12	2.00
13	2.20		
14	2.80		
		15	3.40
		16	4.03
		17	4.83
		18	5.63
		19	5.70
		20	6.94
21	11.50		
		22	15.20
W1 = 90		W2 = 163	

[a] An average rank value is assigned to coincident observations.

This gives $p = 0.017$, which is found in the StaTable (Cytel 1998) for the normal distribution.

SPSS also may be used to perform this analysis, in which case the results are as shown in Tables 11.21 and 11.22. These results correspond to those above.

Assume that the application concerns two independent groups, one treated and one untreated. We wish to find the median difference between the two groups. First, all possible pairwise combinations of results in the treated group minus results in the untreated group are calculated. Then, the differences are

Table 11.21 Rank sums for the control group and the treatment group

	TREATMENT	N	MEAN RANK	SUM OF RANKS
Score	Control	11	8.18	90.00
	Treatment	11	14.82	163.00
	Total	22		

Table 11.22 Test statistic and P-value for Wilcoxon–Mann–Whitney test

	SCORE
Mann–Whitney U	24.000
Wilcoxon W	90.000
Z	-2.397
Asymp. sig. (2-tailed)	0.017

ranked and the median difference is found (Box 11.17). The confidence interval can be found from Table 4.

Box 11.17 Effect measure and confidence interval
- Calculate the difference for all possible pairs of observations, one observation from each group.
- Rank the differences in ascending order.
- Calculate the median of the differences (the effect estimate).
- Calculate the 95% confidence interval for the effect estimate, such as from Table 4 at http:/books.elsevier.com/9780123738745

Example: Considering the same case as that of the previous example, we will calculate $11 \times 11 = 121$ differences. We rank all the calculated differences and find their median to be 2.57. As for Wilcoxon's test for paired comparison, a table is used to find the confidence interval. The starting point is Table 4. For $n_1 = n_2 = 11$, the lower limit observation is number 31 and the upper limit observation is number 91. Then a 95% confidence interval is $(0.36, 4.50)$.

11.9 REGRESSION ANALYSIS

Regression analysis generalizes studies of association between one outcome variable and one or more exposure variables. For an overview, see Table 4.2 in Chapter 4. Here we will look closer at linear, logistic and Poisson regression analyses, for the cases when the outcome variable is measured via continuous, nominal or count data, respectively. In regression analysis, variables are commonly denoted as dependent and explanatory variables, rather than outcome and exposure or independent variables.

Linear regression

Linear regression analysis comprises the study of association between two variables assumed to be continuous (see Table 4.2). The relationship is assumed to be linear, that is, a straight line in the slope-intercept form, where x is the explanatory and y the dependent variable: $y = \alpha + \beta x$.

Observed data are regarded to be variables fitting the slope-intercept equation, although the observations have an additional randomness or 'noise'. This is expressed by adding a term to the slope-intercept equation:

$$Y = \alpha + \beta x + \varepsilon$$

where α is the intercept, β the slope and ε the randomness or 'noise'. In this interpretation, the observations do not lie on a straight line, but rather deviate from it owing to individual variations.

Again consider the FRAMINGHAM.SAV data as an example. Now, we will look at the relation between systolic blood pressure, cholesterol level, age and smoking. The relation between systolic blood pressure (MEANSBT) and cholesterol level (CHOLESTEROL) is shown in Figure 11.1. The relation between the variables is shown in the plotted points, and a regression line has been drawn through them. For a given value of x, there is a y on the regression line, given by the equation $y = \alpha + \beta x$. This value is called the predicted value. The difference between the predicted and the corresponding observed value is called the residual. As can be seen from the figure, the residuals are relatively large. This means that the

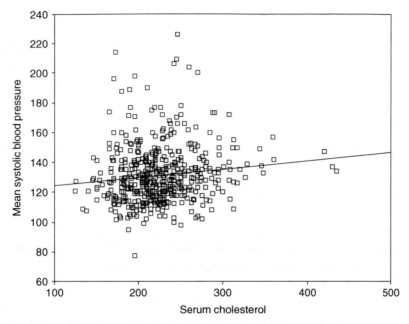

Figure 11.1 Association between systolic blood pressure and cholesterol level

randomness or 'noise' is considerable, which in turn indicates that the individual variations are large and raises the question as to whether variables other than the independent variable chosen may influence the dependent variables.

Regression coefficient

Now we will find the line that best fits the observations. We choose a linear fit.

> ### Box 11.18 Least squares regression line
> There are n pairs of observations of (x_i, Y_i) and $\bar{x}$ and $\bar{Y}$ denote the respective means of the xs and the Ys. The intercept α and slope β are estimated by
>
> $$\hat{\alpha} = \bar{Y} - \hat{\beta}\bar{x}$$

and

$$\hat{\beta} = \frac{\sum_{i=1}^{n}(x_i - \overline{x})(Y_i - \overline{Y})}{\sum_{i=1}^{n}(x_i - \overline{x})^2}$$

$\hat{\alpha}$ is called the constant, and $\hat{\beta}$ is called the regression coefficient. Both $\hat{\alpha}$ and $\hat{\beta}$ may be calculated manually, but they are far more easily calculated using a statistical program package.

See Box 11.18. Note that $\hat{\alpha}$ and $\hat{\beta}$ represent estimates of the unknown parameters α and β from the data, and the interpretation is discussed below.

Example: We continue using the FRAMINGHAM.SAV data in the regression analysis. The least squares method is used to find the line that provides the best fit to the observations shown in Figure 11.1. It entails analysing the association between MEANSBT as the dependent variable and CHOLESTEROL as the explanatory variable in a regression analysis. The results are given in Table 11.23.

Table 11.23 Regression analysis of systolic blood pressure with respect to cholesterol level

MODEL	UNSTANDARDIZED COEFFICIENTS	
	B	SE
Constant	118.566	4.652
Cholesterol	5.659E–02	0.020

The regression analysis yields $\hat{\alpha}$ (constant) and the regression coefficient $\hat{\beta}$ (cholesterol level). In the regression analysis for CHOLESTEROL, $\hat{\alpha} = 118.6$ and $\hat{\beta} = 0.057$. An interpretation of the constant term $\hat{\alpha}$ is usually of minor interest in a regression analysis. This is because the value of y when $x = 0$, in this example when CHOLESTEROL $= 0$, is irrelevant in the analysis. The analysis indicates that MEANSBT changes by 0.057 units when CHOLESTEROL changes by 1 unit. Scales of variables must be remembered in comparing regression coefficients with respect to each other, as a one-unit change in a particular variable may differ from a one unit change in another variable.

Regression coefficient testing and confidence interval

The P-values and confidence intervals resulting from regression analyses require observations to be independent and 'errors' to have the same variance σ^2 (which must be estimated). If there is no association between the dependent and the independent variables, that is $\beta = 0$, the variance is estimated from the variations around the mean, that is

$$s_y^2 = \frac{1}{n-1} \sum_{i=1}^{n} (Y_i - \bar{Y})^2.$$

This is the same measure of variance as in Section 11.6. If association exists between the variables, that is the regression coefficient β is no longer zero, the estimate of the variance is based on variations around the regression line instead of variations from the mean, that is,

$$s_0^2 = \frac{1}{n-1} \sum_{i=1}^{n} (Y_i - \hat{\alpha} - \hat{\beta} x_i)^2$$

is now used as the estimator for σ^2. In this case, the denominator is $n - 2$ instead of $n - 1$; again, this is of marginal practical importance, so it is not discussed further.

Box 11.19 Confidence interval of regression coefficient

The estimator for the regression coefficient β is given in Box 11.18. The standard error of the estimated regression coefficient is:

$$SE(\hat{\beta}) = \frac{s}{\sqrt{\sum_{i=1}^{n} (x_i - \bar{x})^2}}$$

This standard error is essential, as its magnitude determines the length of the confidence interval. It is tabulated in the SE column of Table 11.23.

The confidence interval of the regression coefficient β is as usual:

$$\hat{\beta} - t_{0.975,n-2}SE(\hat{\beta}), \hat{\beta} + t_{0.975,n-2}SE(\hat{\beta})$$

The value of $t_{0.975,n-2}$ is found from StaTable (Cytel 1998) in its t-distribution with $(n-2)$ degrees of freedom. If the number of observations is large, $t_{0.975,n-2}$ approximates 1.96.

Example: The FRAMINGHAM.SAV data comprise $n = 500$ observations. The standard error of the regression coefficient between MEANSBT and CHOLES-TEROL was found above. We use $t_{0.975,498}$ from StaTable (Cytel 1998) to calculate the confidence interval for the regression coefficient. But as there are so many degrees of freedom, the percentile for the t-distribution and the percentile for the normal distribution coincide, that is, $t_{0.975,498} = 1.96$. Hence the 95% confidence interval for the regression coefficient of MEANSBT with regard to CHO-LESTEROL is (0.017, 0.097), based on Box 11.19 and Table 11.24.

Table 11.24 Regression analysis of systolic blood pressure with respect to cholesterol level

MODEL	REGRESSION COEFFICIENTS			SIG.	95% CI FOR B	
	B	SE	T		LOWER BOUND	UPPER BOUND
Constant	118.566	4.652	25.488	0.000	109.426	127.706
Cholesterol	5.659E–02	0.020	2.786	0.006	0.017	0.097

When the regression coefficient $\beta = 0$, the relationship between the dependent variable y and the explanatory variable will be given by a straight line parallel to the x-axis. This means that there is no association between y and x, while $\beta > 0$ indicates a positive association and $\beta < 0$ indicates a negative association.

Hence, we wish to find whether there is an association between y and x. So, we put forth the null hypothesis

$H_0: \beta = 0$

and the alternative hypothesis

$H_1: \beta \neq 0$

Box 11.20 Test of the hypothesis that the regression coefficient is zero

The hypothesis that there is no relation between the dependent and independent variables is tested with the test statistic

$$t = \frac{\hat{\beta}}{SE(\hat{\beta})}$$

The observed value of t is t_{obs}. H_0 is rejected if $2P(t \geq t_{obs}) \leq 0.05$.

Returning to the example of the FRAMINGHAM.SAV data file, the relationship between systolic blood pressure (MEANSBT) and cholesterol level (CHOLESTEROL) can be summarized as in Table 11.24.

In Table 11.24, the confidence interval is $(0.017, 0.097)$, which agrees with the manual calculation above. Moreover, the t-value is 2.79, with a corresponding P-value of $p = 0.006$, so Table 11.24 provides the information needed in this case.

The above example illustrates a relationship between the dependent variable and one explanatory variable. But most often, a single explanatory variable is an oversimplification in a regression model, as usually several explanatory variables are needed. Other models may include explanatory variables such as gender, age and other demographic variables. An observed association between the dependent variable and an explanatory variable may be due to covariation with other explanatory variables. Even in cases focusing on the effect of a specific exposure variable, a regression model must also include other relevant explanatory variables. These other explanatory variables may be regarded as control or confounding variables used to verify the effect of the variables of interest.

Regardless of the aim of a regression analysis, the procedure for extending a regression model to several explanatory variables is known as a multiple regression model. The estimated regression coefficients are again calculated using the least squares method. However, the manual computations are complicated and will consequently not be discussed further here. Hence, the description is confined by writing the relation between the dependent variable Y and the p explanatory variables as

$$Y = \alpha + \beta_1 x_1 + \beta_2 x_2 + \cdots + \beta_p x_p + \varepsilon$$

and the estimated regression coefficients as $\hat{\beta}_1, \hat{\beta}_2, \ldots, \hat{\beta}_p$. Let two sets of explanatory variables be $(x_{11}, x_{21}, \ldots, x_{p1})$ and $(x_{12}, x_{22}, \ldots, x_{p2})$. Hence, the difference in y for the sets of explanatory variables is $\beta_1(x_{12} - x_{11}) + \beta_2(x_{22} - x_{21}) + \cdots + \beta_p(x_{p2} - x_{p1})$. Consequently, β_1 is interpreted as the change of the dependent variable caused by a unit change of x_1 when $x_2, x_3, \ldots, x_p$ are held constant. In other words, β_1 is the effect of x_1 when we have controlled for $x_2, x_3, \ldots, x_p$. The other regression coefficients are interpreted in the same manner.

Example: Returning to the example of the FRAMINGHAM.SAV data file, with systolic blood pressure as the dependent variable and CHOLESTEROL, AGE and SMOKE as the explanatory variables (which in this case might be called exposure variables), we find the results given in Table 11.25.

Table 11.25 Regression analysis of systolic blood pressure with respect to cholesterol level, age and smoking status

MODEL	UNSTANDARDIZED COEFFICIENTS			SIG.	95% CI FOR B	
	B	SE	T		LOWER	UPPER
Constant	94.809	6.435	14.733	0.000	82.166	107.453
Cholesterol	5.628E–02	0.020	2.864	0.004	0.018	0.095
Age	0.562	0.096	5.872	0.000	0.374	0.749
Smoking status	−2.374	1.932	−1.229	0.220	−6.170	1.422

From Table 11.25, the conclusions of the multiple regression model are by and large the same as those of the simple regression analysis for cholesterol level. When controlled for age and smoking, the effect of the cholesterol level changes negligibly, from 0.057 to 0.056.

The analysis of variance (ANOVA) may be regarded as a variety of regression analysis. ANOVA is used when the explanatory variable is categorical. Regression analysis may also be used in cases with categorical explanatory variables. So with categorical explanatory variables, one may choose either ANOVA or regression analysis.

Logistic regression

Whenever the dependent variable is categorical, the plot of the relation between the dependent variable and the explanatory variable differs from plots exhibiting the relations between two continuous variables. FIRSTCHD is the categorical variable in the FRAMINGHAM.SAV file. FIRSTCHD is plotted against MEANSBP in Figure 11.2.

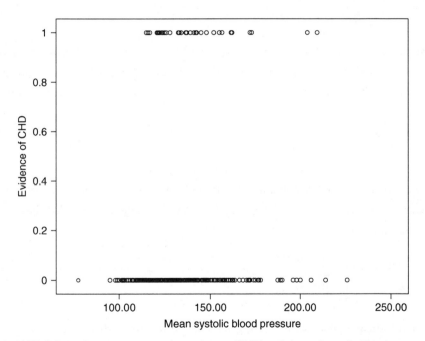

Figure 11.2 Relation between coronary heart disease (CHD) and elevated systolic blood pressure

The plot may indicate that the probability of coronary heart disease increases with increasing blood pressure, as there is a tendency toward more 1s at higher blood pressures, but this may admittedly be hard to see from the figure.

The relationship with a dependent variable that has only the values 0 or 1 is expressed as the probability of the dependent variable having the value 1 in a

357

non-linear function of the explanatory variable. The function usually used to express this relation is the logistic (which is the root of the term 'logistic regression analysis'):

$$\Pr(Y = 1 \mid x) = p(x) = \frac{e^{\alpha + \beta x}}{1 + e^{\alpha + \beta x}}$$

To interpret the values of α and β, we transform the equation for $p(x)$, and calculate the odds $p(x)/(1 - p(x))$. Hence,

$$\ln \frac{p(x)}{1 - p(x)} = \alpha + \beta x$$

which is called log-odds, or logit. As the log-odds is linear in α and β, they are readily interpreted. The interpretation of α is the log-odds for $x = 0$. But this interpretation is of little practical importance, so its value is almost never used. Because

$$\ln \frac{p(x) / (1 - p(x))}{p(x - 1) / 1 - p(x - 1)} = \beta$$

the value of β will tell us how much the log-odds for $y = 1$ changes when x changes by one unit. It is more amenable to interpret β as a change in odds. From the above equation,

$$\frac{p(x) / (1 - p(x - 1))}{p(x - 1) / (1 - p(x - 1))} = e^{\beta}$$

which indicates that when x changes by one unit, the odds change by e^{β} units. The value of β is the effect measure of a logistic regression analysis, as it assesses the degree of association between the explanatory variable and the dependent variable. The value of e^{β} is used to interpret the relationship.

Example: We use logistic regression to analyse the association between FIRSTCHD and MEANSBP of the FRAMINGHAM.SAV data. The results are shown in Table 11.26.

> ## Box 11.21 Confidence interval for the regression coefficients
>
> There are n pairs of observations of (x_i, Y_i). Values of α and β may be found so the data best suits the logistic function, maximizing the likelihood for what is observed. This is called the maximum likelihood method. A program for logistic regression, such as SPSS, provides maximum likelihood estimates $\hat{\alpha}$ and $\hat{\beta}$ and the associated standard errors $\text{SE}(\hat{\alpha})$ and $\text{SE}(\hat{\beta})$.
>
> The confidence interval for β is found in the customary manner by associating the estimated $\hat{\beta}$ with the standard error $\text{SE}(\hat{\beta})$. The estimator $\hat{\beta}$ very nearly follows a normal distribution, provided there are sufficient data, so in the usual manner, we have an approximate 95% confidence interval for β:
>
> $$\hat{\beta} - 1.96\text{SE}(\hat{\beta}), \hat{\beta} + 1.96\text{SE}(\hat{\beta})$$
>
> Through transforming via an exponential function we find a confidence interval for e^{β} of
>
> $$e^{\{\hat{\beta} - 1.96\text{SE}(\hat{\beta})\}}, e^{\{\hat{\beta} + 1.96\text{SE}(\hat{\beta})\}}$$

Table 11.26 Logistic regression of coronary heart disease with respect to systolic blood pressure (SBP)

MODEL	B	SE	WALD	DF	SIG.	EXP(B)	95% CI FOR EXP(B)	
							LOWER	UPPER
Mean SBP	0.024	0.007	11.630	1	0.001	1.025	1.010	1.039
Constant	−5.846	1.021	32.817	1	0.000	0.003		

Note that EXP(B) in Table 11.26 is the odds ratio we use for intepretation of the results.

Here, the regression coefficient is $\hat{\beta} = 0.024$, $e^{\beta} = 1.025$ and the confidence interval for e^{β} is $(1.0130, 1.039)$. This means that the odds for coronary heart disease increase by a factor of 1.025 for each millimetre increase in systolic blood pressure. We see that the 95% confidence interval in this case is calculated to be $(1.010, 1.039)$.

Now we wish to test the null hypothesis that there is no association between y and x. So, we put forth a null hypothesis H_0 that $\beta = 0$, against the alternative hypothesis H_A that $\beta \neq 0$ (Box 11.22).

Box 11.22 Test of the hypothesis that the regression coefficient is zero

The test of the hypothesis of no association is conducted in the usual way by comparing the effect estimate with that of the standard error. The following test statistic is used:

$$t = \frac{\hat{\beta}}{SE(\hat{\beta})}$$

The observed value of t is called t_{obs}. H_0 is rejected if $2P(t > t_{obs}) \leq 0.05$.

Note that some applications, such as SPSS, calculate not t, but rather

$$t^2 = \left(\frac{\hat{\beta}}{SE(\hat{\beta})} \right)^2$$

which is called the Wald statistic. Whether t^2 is used instead of t is immaterial, with respect to calculating the P-values.

Example: We revert to the results of Table 11.26 and find that

$$t_{obs}^2 = \left(\frac{\hat{\beta}}{SE(\hat{\beta})} \right)^2 = (0.024 / 0.007)^2 = 11.76$$

which agrees with the result of the above analysis (save a small rounding-off error). We find a P-value of less than 0.001.

As discussed above, an observed association between the dependent variable and an explanatory variable might be due to association with other explanatory variables. The effect of one explanatory variable, controlled for other variables, may be studied using multiple logistic regression. Assume in general that there are p explanatory variables and that the logistic regression model may be expressed

$$\Pr(Y = 1 \mid x_1, x_2, \ldots, x_p) = p(x_1, x_2, \ldots, x_p) = \frac{e^{\alpha + \beta_1 x_1 + \beta_2 x_2 + \cdots + \beta_p x_p}}{1 + e^{\alpha + \beta_1 x_1 + \beta_2 x_2 + \cdots + \beta_p x_p}}$$

This expression is equivalent to

$$\ln\left(\frac{p(x_1, x_2, \ldots, x_p)}{1 - p(x_1, x_2, \ldots, x_p)}\right) = \alpha + \beta_1 x_1 + \beta_2 x_2 + \cdots + \beta_p x_p$$

For two measurements of the explanatory variable, $(x_{11}, x_{21}, \ldots, x_{p1})$ and $(x_{12}, x_{22}, \ldots, x_{p2})$, the difference in the log-odds is $\beta_1(x_{12} - x_{11}) + \beta_2(x_{22} - x_{21}) + \cdots + \beta_p(x_{p2} - x_{p1})$. Consequently, β_1 is interpreted as the change of the log odds caused by one unit change in x_1 when $x_2, x_3, \ldots, x_p$ are held constant. It is often said that β_1 is the effect of x_1 when we have controlled for $x_2, x_3, \ldots, x_p$. The other regression coefficients are interpreted in the same manner. Again, the effects are interpreted via changes in odds, not via log-odds. When x_1 changes by 1 unit and the other explanatory variables are held constant, the odds for $y = 1$ change by e^{β_1} units. The results for the other explanatory variables are similar.

Example: Using the FRAMINGHAM.SAV data file, with coronary heart disease as the dependent variable and systolic blood pressure (MEANSBT), cholesterol level (CHOLESTEROL), age (AGE) and smoking (SMOKE) as the explanatory variables, we find the results tabulated in Table 11.27.

The effect of systolic blood pressure is approximately the same for the multiple as for the simple logistic regression analysis, but there is also an effect of cholesterol level and age.

Table 11.27 Logistic regression of coronary heart disease with respect to systolic blood pressure (SBP), cholesterol level, age and smoking

MODEL	B	SE	WALD	DF	SIG.	EXP(B)	95% CI FOR EXP(B)	
							LOWER	UPPER
Mean SBP	0.017	0.008	4.771	1	0.029	1.017	1.002	1.032
Cholesterol	0.008	0.004	4.169	1	0.041	1.008	1.000	1.015
Age	0.067	0.022	8.909	1	0.003	1.069	1.023	1.116
Smoking status	0.408	0.432	0.892	1	0.345	1.504	0.645	3.510
Constant	-10.097	1.720	34.474	1	0.000	0.000		

Poisson regression

In clinical and epidemiological studies the effect measure is often based on the incidence rate, as discussed in Chapter 9. A useful measure is the ratio of the incidence rates among the exposed and the non-exposed, IRR. A feasible way to study an incidence rate is to assume that the total number of cases follows a Poisson distribution. Now let Y be the number of cases, I the incidence rate and t the length of follow-up time or observation time. Then we write that $Y \sim$ Poisson $(I \cdot t)$ to signify that Y follows a Poisson distribution with mean $I \cdot t$. The estimator for the incidence rate I is $\hat{I} = Y/t$. This is the estimator discussed in Chapter 9. As for the relative risk and the odds ratio, we have to transform to a logarithmic scale to estimate the standard error. The standard error of $\ln \hat{I}$ is $1/\sqrt{y}$. If there are two groups, one exposed and one non-exposed, with incidence rates $\hat{I}_e = y_e/t_e$ and $\hat{I}_0 = y_0/t_0$, the confidence interval for IRR $= I_e/I_0$ is as stated in Section 9.8.

As mentioned for linear and logistic regression, we may wish to control for several explanatory variables. When the explanatory variables are categorical, the incidence rate is constant or nearly so within each category, and the Poisson model is the most suited analytical tool. In general, there are p explanatory variables, x_1, $x_2, ..., x_p$, and the Poisson regression model then may be expressed by

$$Y \sim \text{Poisson} (\lambda(x_1, x_2, ..., x_p) \cdot t)$$

where $\lambda(x_1, x_2, ..., x_p) = \exp(\alpha + \beta_1 x_1 + \beta_2 x_2 + \cdots + \beta_p x_p)$ is the incidence rate, and t is the total length of follow-up time. As usual, α is the constant term and

$\beta_1, \beta_2, \ldots, \beta_p$, are the regression coefficients of the regression model. Consider two people with explanatory variables $(x_{11}, x_{21}, \ldots, x_{p1})$ and $(x_{12}, x_{22}, \ldots, x_{p2})$. The IRR is then

$$\frac{\lambda(x_{12}, x_{22}, \ldots, x_{p2})}{\lambda(x_{11}, x_{21}, \ldots, x_{p1})} = e^{\beta_1(x_{12}-x_{11})+\beta_2(x_{22}-x_{21})+\cdots+\beta_p(x_{p2}-x_{p1})}$$

and $e^{\beta_1(x_{12}-x_{11})}$ is the change in incidence rate for a change in x_1. When x_1 changes by one unit and the others are held constant, the change in incidence is e^{β_1}.

An application program for general linear models, such as SPSS, can be used to find estimates and confidence intervals for the βs.

Example: Table 9.5 may be analysed using Poisson regression. Let smoking be the explanatory variable (with two categories). Using a generalized linear model for analysis with one explanatory variable, namely smoking, the results are as shown in Table 11.28.

Table 11.28 Poisson regression analysis of the incidence rate relative to smoking

MODEL	ESTIMATE	SE	Z	SIG.	95% CI	
					LOWER BOUND	UPPER BOUND
Constant	−6.205	0.316	−19.622	0.000	−6.824	−5.585
Smoking	1.119	0.365	3.065	0.002	0.403	1.835

The results of the analysis are on a logarithmic scale, so the estimated incidence rate is $\widehat{\text{IRR}} = \exp(1.119) = 3.1$ and CI = $(\exp(0.403), \exp(1.835)) = (1.5, 6.3)$. These results correspond to those calculated manually in Section 9.8.

REFERENCES

Altman DG (1991) Practical Statistics for Medical Research. Chapman & Hall, London.

Carroll RJ et al. (1995) Framingham Data. Texas A&M University Study, College Station, TX.

Cytel, StaTable, free online application that provides access to the 25 most frequently used statistical distributions, at www.cytel.com/Products/StaTable/Default.asp

Daly F et al. (1995) Elements of Statistics. Addison-Wesley. Data quoted with permission of the Open University.

Levene H (1960) In Contributions to Probability and Statistics: Essays in Honor of Harold Hotelling. Stanford University Press, Stanford, CA, pp. 278–292.

Rosner B (2005) Fundamentals of Biostatistics (6th edn). Duxbury Press, Pacific Grove, CA.

FURTHER READING

Armitage P et al. (2002) Statistical Methods in Medical Research (4th edn). Blackwell, London.

Lang TA, Secic M (1997) How to Report Statistics in Medicine. Annotated Guidelines for Authors, Editors, and Reviewers. American College of Physicians, Philadelphia.

EVIDENCE-BASED PRACTICE AND CRITICAL APPRAISAL OF SYSTEMATIC REVIEWS

Liv Merete Reinar and Peter M. Bradley

12.1 INTRODUCTION

Evidence-based practice or evidence-based healthcare (EBHC) is an approach to decision making which combines research information, clinical expertise and users' values and preferences. It can be defined in this way:

> 'Evidence Based Medicine (or health care) sees clinical expertise as the ability to integrate patient circumstances, research evidence, and patient preferences to help patients arrive at optimal diagnostic and treatment decisions'.
>
> (Guyatt et al. 2004)

Evidence-based or evidence-informed practice emphasizes the importance of research-based evidence in decision making. However, it is not the intention that research evidence should replace professional expertise and experience or deny patients' choices. On the contrary: evidence-based practice suggests that only by using clinical expertise and listening to patients can research results be applied in clinical practice or policy making.

The five steps classically described in EBHC are listed in Box 12.1.

> **Box 12.1 The five steps in EBHC**
> 1. Define a clearly defined clinical or research question.
> 2. Search for the evidence. Search the literature for relevant articles.
> 3. Critically appraise the evidence for its validity and usefulness.
> 4. Implement useful findings in clinical practice.
> 5. Evaluate the practice or what you have decided to do when appropriate.

This chapter focuses on the third and fourth steps of Box 12.1. They are about the critical appraisal of research literature and about using the results in real life. That said, there are many reasons why clinicians do not act on research evidence. Research on implementation of research in practice suggests the most common barriers are that: clinicians do not have time, there are insufficient incentives to change behaviour, clinicians do not have the right skills or knowledge about EBHC, they lack support from superiors and they are unable to access appropriate, valid research evidence. Ironically, the fact that thousands of research papers are published each week can also be a barrier for many clinicians, who experience an 'information overflow'.

Reviews can be a good way to make information more accessible and to address the information overflow. Reviews sum up the results from several research papers on the same topic and therefore make the findings easier to access. However, publishing a review is not enough to implement evidence-informed practice. To influence healthcare workers' behaviour, measures are needed to ensure that the results of reviews are made available, disseminated and implemented to the appropriate target audience within the health service.

12.2 SYSTEMATIC REVIEWS

The aim of a *review* is to combine the results from individual studies in one paper or report. Many reviews do not use a systematic method for combing their results and may give an unbalanced picture of all research results for one particular issue or question. In *systematic reviews* the authors use an explicit and systematic method to search for, critically appraise and combine the results from individual studies on the topic of interest. As illustrated by Peter Morgan, former Scientific Editor, Canadian Medical Association Journal:

'The medical literature can be compared to a jungle. It is fast growing, full of dead wood, sprinkled with hidden treasure and infested with spiders and snakes.'

This chapter will concentrate on systematic reviews based on controlled studies that answer questions about the effectiveness of interventions in healthcare. In other words, reviews that aim to answer questions such as 'Does this treatment work, or does this prevention intervention or this rehabilitation intervention work? What is the effect of this treatment? Does it cause any harm?' The majority of available systematic reviews, including those in the Cochrane Library, www.thecochranelibrary.com (see Section 5.8 of Chapter 5), review these kinds of questions.

The process that helps you to decide whether or not research is valid is called *critical appraisal*. It has been described as follows. Critical appraisal is 'the evaluation of a piece of completed research, usually a published paper in a professional journal, by means of structured checklists that address issues of study design and methodology, data analysis, presentation of results, and relevance of conclusions' (Greenhalgh 2006). There are three main reasons for rejecting a research paper:

- The authors fail to tell you what the paper is about. They have not stated a clear research question.
- The authors might have used a research method that leads to *bias* (a systematic error or deviation in results or inferences from the truth). If the results are biased owing to how the authors conducted their piece of research you cannot trust the results. The methodological quality of a piece of research is called the *internal validity* (see Section 4.9 of Chapter 4).
- The results from it are not relevant and applicable for your practice, your population or your own piece of research. The way in which the results from a study are used in practice is called the *external validity* or generalizability of the research.

In the rest of this chapter, a paper is critically appraised and one example from a systematic review (Herxheimer & Petrie 2002) is used.

12.3 CRITICAL APPRAISAL OF A SYSTEMATIC REVIEW: ONE EXAMPLE

This section, will critically appraise a systematic review (Herxheimer & Petrie 2002) using a checklist that can help to appraise the quality of a review. The questions are taken from the checklist developed by the Critical Appraisal Skills Programme (CASP), called '10 questions to help you make sense of reviews', downloadable free in a PDF file from www.phru.nhs.uk/casp/critical_appraisal_tools.htm. This is, in turn, based on a checklist from Users' Guides to the Medical Literature, Choosing Evidence Worksheet (1E Summarizing the evidence) available from http://www. usersguides.org

Please read the scenario in Box 12.2 before moving on.

> ### Box 12.2 Scenario for melatonin and jet lag
>
> Some friends have contacted you. They are going away to a conference in Australia from Europe and have read that taking melatonin might help to prevent and treat jet lag. You have heard this too and as you are travelling internationally across time zones this year you are interested to know more about it. In the Cochrane Library you find a review (Herxheimer & Petrie 2002) that apparently may provide answers to two relevant questions:
>
> 1. Is melatonin effective in preventing and treating jet lag?
> 2. Should you use melatonin if you were travelling long distance?

If you have access to the Cochrane Library please find the review and take the time to read the paper now. When you have done so, answer the two questions in Box 12.2. After you have been through the whole chapter you can address the questions again and see whether you have changed your mind.

Critically appraising a paper consists of four main stages. First, you should consider whether the review is worth spending time on. The two first questions we will consider are therefore called screening questions (CASP questions 1 and 2):

- Did the review ask a clearly focused question?
- Did the review include the right type of studies?

Only if you can answer yes to both these it is worth continuing to read the review more closely. Secondly, you need to judge whether the results of the review are valid by considering the scientific methodology of the review (CASP questions 3–5). After this, you should look at the results of the review (CASP questions 6 and 7). Finally, look at CASP questions 8–10 to see whether the results can realistically be applied in the real world.

Screening questions

1. *Did the review ask a clearly focused question?*
Consider whether the question is 'focused' in terms of:

- the population studied
- the intervention given or exposure
- the outcomes considered.

For any research paper we read, we need to know the question addressed. A paper on the effectiveness of an intervention should define a focused question in terms of the study population, the intervention and the outcomes considered.

In the melatonin paper the answers to the question on population, intervention and outcomes are stated in the *Objectives* and in the following paragraph, *Criteria for considering studies for this review*. In the Objectives it states that the aim is:

'… to evaluate whether melatonin taken by mouth can prevent or alleviate jet lag associated with air travel across several time zones. The review also examines the evidence for the effectiveness of different dosage regimens, in particular the timing of doses in relation to the flight and to the day–night cycle at the destination.'

The inclusion criteria for this systematic review are stated under *Criteria for considering studies for this review*; studies were included if they compared oral melatonin with (1) non-active treatment (placebo) or (2) other medication, taken before, during and/or after medication.

More detail is given under the heading *Types of participants*. Participants were included if they were airline passengers, airline staff or military personnel. Oral melatonin (the *intervention*) could be taken before, during and/or after travel. The primary *outcome* was the self-reported rating of jet lag. The review authors also stated that they were interested in the following secondary outcomes: fatigue, day-time tiredness, onset of sleep at destination, onset and quality of sleep, psychological functioning, duration of return to normal, and measures indicating the phase of circadian rhythms.

It seems that a clear question was addressed by the review and it is worthwhile continuing.

2. Did the review include the right type of study?
Consider whether the included studies:

- address the review's question
- have an appropriate study design.

As stated earlier, a review can in theory include studies of all kinds of study design. However, the studies included must be appropriate to answer the type of question addressed. Box 12.3 categorizes those commonly asked questions and suggests the appropriate type of study design to answer them. More than one study design can be used for addressing any of the question categories and Box 12.3 is only meant for guidance (see also Table 5.4 in Chapter 5).

Box 12.3 Which study design is appropriate for different types of study questions?

Question	Preferred study design
Why do some get a disease while others stay well? (causation)	
What is the probability that the problem will resolve by itself?	Studies on the course of a condition (prognosis)
	• Cohort studies
	• Case–control studies
How can we decide whether someone has a specific condition?	Studies on the accuracy of diagnostic tests (diagnosis)
Could this problem or condition be prevented?	

What should I do to treat this condition? Is there anything we can do to speed up recovery?	Studies on the effects of intervention • Randomized controlled trials
How does the person feel or experience having this condition?	Studies on experiences • Qualitative studies
How many people have this condition?	Studies on prevalence • Surveys
Is this intervention or exposure harmful?	Studies on harm and aetiology • Cohort studies • Case–control studies (Herbert et al. 2005)

To consider whether Herxheimer and Petrie had included the right sort of studies in their paper, we need to refer once more to the review's inclusion and exclusion criteria. Under *Types of studies* it specifies that randomized controlled trials were included. The same information is also given in the abstract under *Selection criteria*. They also looked for reports of suspected adverse effects of melatonin that were not reported in the randomized trials, but these were not included in the main analysis.

Randomization in practice means that the participants in the trial all have an equal chance of receiving active or control treatments. This happens regardless of any external factor, e.g. age, gender, motivation to participate in the study or previous experience of long distance travel. In other words, successful randomization ensures that the participants' characteristics and other factors that might influence the results or outcomes are equally distributed between the groups, and the result of the study is not influenced by the extent to which these factors are known to the researcher.

Non-randomized controlled trials can lead to *bias* as the participants' characteristics and other factors are not equally distributed between the groups. This imbalance might in itself influence the results of the trial. For example, in a non-randomized trial with 100 participants where the researchers were comparing melatonin with placebo, 50 of the participants could have previous serious symptoms of jet lag. The researchers could, knowingly or unknowingly, have allocated more of these participants to the placebo group.

Detailed questions about internal validity

3. Did the reviewers try to identify all relevant studies?
Consider:

- which bibliographic databases were used
- whether there was a follow-up from reference lists
- whether there was personal contact with experts
- whether the reviewers searched for unpublished studies
- whether the reviewers searched for non-English-language studies.

One of the main goals in a systematic review is to give the reader a full and balanced overview on all available research on one issue. It is therefore necessary that systematic reviews are based on a thorough search of the literature and give a balanced overview of the studies found. This is a big challenge. Medline, one of the biggest bibliographic databases within medical research, indexes more than 4900 journals, but still includes less than a fifth of the total number of journals relevant to medicine and healthcare. In addition, many journals are not indexed in databases and many studies are never published. It can therefore be problematic if systematic reviews only concentrate on studies indexed in the biggest available databases, a phenomenon known as *publication bias*.

In Herxheimer and Petrie's review the details of the search strategy are given in the section *Search strategy for identification of studies*. The authors looked for published and unpublished studies by searching databases and by contacting authors of relevant studies. The databases searched were: the Cochrane Controlled Trials Register (now called Cochrane Central), Medline, Embase and Psychlit. They searched the Science Citation Index to identify trials that had cited the studies. They handsearched two journals (Aviation, Space and Environmental Medicine, and Sleep, from 1986 to 1999) and they also considered conference abstracts published there. Further searches were performed in the citation lists of relevant studies for relevant trials. The authors also state which search terms they used (melatonin, jet lag, jet-lag, aviation, air travel, airtravel).

Moreover, the authors searched for reports on adverse effects of melatonin that were not reported in the studies retrieved in the above searches. It seems that the authors searched for literature up to 1999. The review was updated in 2003 and further studies were included then, but the authors do not state how they searched for literature at that point. Generally speaking, it seems that the authors did a good job in identifying relevant studies.

4. Did the reviewers assess the quality of the included studies?
Consider:

- whether a clear, predetermined strategy was used to determine which studies were included, look for:
  - a scoring system
  - more than one assessor.

The authors of a systematic review should first exclude the primary studies that did not meet the inclusion criteria as stated in the protocol. The authors should then critically appraise all the studies they have left in terms of the internal and the external validity of the individual study, and the results.

Since critical appraisal is partly a subjective process, it is good practice that this selection is done independently by two people. Reviewers should use a valid checklist, which has been piloted and truly assesses the quality of the study. There are many checklists available. Prevalidated checklists can be used together with locally developed checklists that address questions of particular importance to the review. More details on checklists and critical appraisal of randomized trials are included in the bibliography at the end of this chapter.

In the systematic review on the effects of melatonin the authors state in the methods section that they undertook a quality assessment of the included studies:

'Quality assessment: allocation concealment and blinding was looked for, described and evaluated. Methods of measurement used in the trials are described ant their relevance, validity and reproducibility discussed.'

The authors used standard criteria to critically appraise randomized controlled trials, related to 'allocation concealment' and 'blinding'. They also considered whether researchers had measured the outcomes in the trials using valid methods.

5. *If the results of the studies have been combined, was it reasonable to do so?* Consider whether:

- the results of each study are clearly displayed
- the results were similar from study to study (look for tests of heterogeneity)
- the reasons for any variations in results are discussed.

The authors of a systematic review can decide whether to describe the results from included studies as a narrative, describing each study in turn. Alternatively, they can numerically combine the results for each outcome using a *meta-analysis*. A meta-analysis should only be performed if the individual studies are similar enough (*homogeneous*) to make 'clinical sense' (i.e. similar populations, interventions and outcomes are considered).

Whether studies are sufficiently homogeneous can be assessed using two main methods: first, by assessing individual studies' characteristics, and secondly, by using statistical tests. The first method leads to a subjective and clinical judgement of whether studies are similar enough to combine in a meta-analysis. The second method identifies potential *heterogeneity* (for example, unexplained variation in the results from the individual studies), that is, subgroups of studies where results seem to differ from the overall results and require further investigation.

Herxheimer and Petrie performed a meta-analysis of four of the 10 trials they had included. The authors' justification for combining these results and excluding others was that these four papers had used a global visual analogue jet lag score to measure the outcomes. Three of the trials used a dosage of 5 mg melatonin and one study used 8 mg. There was also some difference in the way the participants were instructed to use the melatonin. However, the results were consistent in the four trials, so it seemed reasonable to combine the results from the four studies in a meta-analysis, as illustrated in Figure 12.1.

The authors performed a statistical test for heterogeneity which suggested whether it was reasonable (from a statistical point of view) to combine the results.

Review: Melatonin for the prevention and treatment of jet lag
Comparison: 01 Melatonin versus placebo: Eastward flights
Outcome: 01 Global jet-lag ratings

Study	Melatonin		Control		Weighted Mean Difference (Fixed) 95% CI	Weight (%)	Weighted Mean Difference (Fixed) 95% CI
	N	Mean(SD)	N	Mean(SD)			
87 Arendt	8	11.31 (9.34)	9	55.20 (38.24)		11.1	−43.89 [−69.70, −18.08]
88 Arendt	29	21.40 (19.39)	30	39.20 (30.67)		43.5	−17.80 [−30.85, −4.75]
91 Nickelsen	18	52.00 (25.00)	18	66.00 (21.00)		32.6	−14.00 [−29.08, 1.08]
92 Claustrat	15	34.53 (30.89)	15	52.79 (36.15)		12.8	−18.26 [−42.32, 5.80]
Total (95% CI)	70		72			100.0	−19.52 [−28.13, −10.92]

Test for heterogeneity chi-square = 4.02 df = 3 p = 0.26 I = 25.3%
Test for overall effect z = 4.45 p < 0.00001

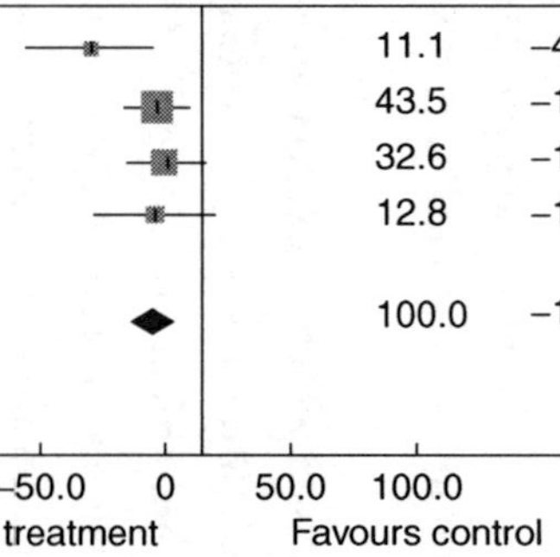

−100.0 −50.0 0 50.0 100.0
Favours treatment   Favours control

Figure 12.1 Meta-analysis of comparison melatonin versus placebo (*Source:* The Cochrane Library: Cochrane Reviews; John Wiley & Sons)

Further information on statistical methods on testing for heterogeneity is provided by Egger et al. (2001) and Higgins & Green (2005).

6. How are the results presented and what is the main result?
Consider:

- how the results are expressed (e.g. effect measures such as odds ratio, relative risk, etc.)
- how large the effect is and how meaningful it is
- how you would sum up the bottom-line result of the review in one sentence.

There are many ways in which to express the results from individual studies and meta-analyses. Roughly, there are two types of data. The first type is called *categorical* (*nomimal* and *ordinal*) outcome data, and the simplest case is when you can answer a question with a simple 'yes' or 'no'. For example, has the jet lag got better or not? Such data are also called *dichotomous*. The other main category of data is *continuous* data that measure relative changes, for example measuring improvement of jet lag on a visual scale (see also Section 4.6 of Chapter 4).

When dichotomous outcomes are used, the effect measure can be expressed as a ratio or an absolute difference. A ratio expresses the risk of something happening against it not happening, e.g. *incidence rate ratio*, *relative risk* or *risk ratio*, *odds ratio*. Alternatively, the effect measure is an absolute risk reduction. An absolute difference is used when one result is subtracted from another, e.g. *absolute risk reduction*. The most usual way to compare continuous data in a meta-analysis is based on subtraction of one result from another. The results can be calculated as *weighted mean difference* (if the outcomes were measured in the same way in the individual studies) or *standardized weighted mean difference* (if the outcomes were measured differently in the individual studies). You can read more about how to express results in Chapter 11 in this book and in the Cochrane Handbook.

The overall results in the meta-analysis for melatonin versus placebo are calculated as a weighted mean difference by combining the data from all the individual studies to give an overall score, represented as a diamond in Figure 12.1. There are two statistical methods for calculating effects, using fixed and random effects models. Here the meta-analysis has used a 'fixed effect'. The individual studies are *weighted*, a method of giving more weight to bigger studies (studies with most participants). To read more about fixed and random effects and analysis, see Egger et al. (2001).

The overall result from the meta-analysis on melatonin versus placebo is still less than 0, indicating that the result favours the use of melatonin to treat jet lag. It is worth noting that when interpreting such results, you need to bear in mind what type of outcome is being measured. If the outcome had been the opposite of that considered in the jet lag studies, i.e. how many people had *not* responded to therapy, the weighted mean difference would have been greater than 0.

7. How precise are these results?
Consider:

- whether a confidence interval is reported: would your decision about whether or not to use this intervention be the same at the upper confidence limit as at the lower confidence limit?
- whether a *P*-value is reported where confidence intervals are unavailable.

In the comparison above (melatonin versus placebo, Figure 12.1) the uncertainty in the mean difference is shown as a *confidence interval* (CI). The confidence interval takes into account whether the positive effect of melatonin on jet lag could be due to random events (chance) rather than any real effect. Roughly speaking, there is a probability of 95% that the effect of melatonin lies in the range represented by the confidence interval limits.

When you look at the forest plot in Figure 12.1 you can see the confidence intervals for each trial as horizontal lines going through the point estimate. In two of the trials the confidence interval crosses the vertical line (line of no significance) and numerically crosses zero. This means that we do not know whether the treatment has had a beneficial or detrimental effect on jet lag. This type of result is known as a *non-significant* result. However, when the results were pooled to obtain an overall result, the confidence interval does not cross the line of no significance. This means that the overall review result statistically shows a clear reduction in the severity of jet lag symptoms. It is worth noting that the trials included in the meta-analysis recruited relatively few participants, varying from 17 to 59. In forest plots, small trials are seen to be most at risk of chance events as they have the biggest confidence intervals. In contrast, the overall result from the meta-analysis has the shortest confidence interval and hence the most precise result. It represents a larger 'virtual' study than any of the component studies.

8. Can the results be applied to the local population?
Consider whether:

- the population sample covered by the review could be different from your population in ways that would produce different results
- your local setting differs much from that of the review
- you can provide the same intervention in your setting.

The systematic review addresses the effect of oral melatonin (in different dosages) for alleviating jet lag after air travel across several time zones. When considering questions on external validity on research as in question 8 above, it is an advantage that the authors have summarized 10 different studies that all show similar results. The studies were all performed on participants travelling across six or more time zones mostly to or from Europe or to or from the USA, New Zealand or Australia. The studies covered both eastward and westward flying. The participants were 'healthy volunteers'. Both men and women were included and the age range covered was 20–68 years. There is limited additional information about the participants in the tables of included studies given in the review.

You might consider that the population sample is similar to yourself or your friends, but be aware that we have no information on the ethnic origins, social class or level of education of the participants. The majority of passengers seem to have been Europeans, Americans, Australians or New Zealanders. Your local setting may be different. Note also that there are no studies that considered children or old people. Nonetheless, jet lag is a common condition for people who travel across time zones and generally the review of 10 studies suggests that melatonin could help many passengers.

Quite apart from this, you need to consider whether melatonin is available in your setting. Melatonin is not licensed and cannot be easily obtained in most countries. This is an obvious barrier to implementing the review's findings.

9. Were all important outcomes considered?

Consider outcomes from the point of view of the:

- individual
- policy makers and professionals
- family/carers
- wider community.

The primary outcome considered was a self-reported rating of jet lag. The authors also looked for fatigue, daytime tiredness, onset of sleep at destination, quality of sleep, psychological functioning, duration of return to normal and measures indicating the phase of circadian rhythms as secondary outcomes. Nine of the 10 randomized controlled trials reported potential side-effects, but only three looked for symptoms systematically. Some of the symptoms reported in these trials were symptoms of jet lag (daytime sleepiness, dizziness, headache and loss of appetite). The authors discuss the possibilities of side-effects on various groups of people. This should help you to make an informed decision on the effect of melatonin. The possibility of side-effects needs to be balanced against the compelling evidence for the positive effect on jet lag symptoms.

10. Should policy or practice change as a result of the evidence contained in this review?

Consider:

- whether any benefit reported outweighs any harm and/or cost; if this information is not reported can it be filled in from elsewhere?

The information on the effect of and the side-effects of melatonin is not complete. The individual studies have not reported all relevant outcomes, although the main question on subjective rating of jet lag symptoms is well covered. Information from systematic reviews will inevitably be limited as the authors can only base the recommendation on currently existing studies. We have to consider whether the information with which we were presented is sufficient to make an informed decision. Note that the review was last updated in 2003. It might be worthwhile to check whether any new randomized trials on melatonin for jet lag have been published since then.

Turn back to the scenario and see whether you can decide on the questions presented therein:

1. Does melatonin help for jet lag?
2. Would you use melatonin if you were travelling long distance?

Have you changed your mind after having read the review, systematically using the checklist? If so, why have you changed your mind?

12.4 SUMMARY

Systematic reviews can provide comprehensive, transparent and valid information that can help to answer specific clinical questions. They are useful tools for anyone who feels overwhelmed by research information or who wants to invest precious time efficiently by reading high-quality, critically appraised literature. By appraising the individual studies included in reviews, authors have selected available high-quality studies on a particular topic and excluded lower quality research papers. By using a quantitative method of combining results from individual studies in a meta-analysis, a more precise and convincing conclusion can be presented than from individual studies alone.

Unfortunately, systematic reviews are of variable quality. However, by critically appraising them using accepted criteria, poor reviews can be identified and discarded. Poor research evidence can promote ineffective and even harmful interventions. Reviews of low methodological quality should therefore not be used to support healthcare decisions, whether by the public, clinicians and policy makers. In contrast, systematic reviews of high methodological quality can help decision makers, even when reviews highlight the lack of available evidence. Quite apart from this, the findings from systematic reviews need to be adapted to take account of local settings, the availability of resources, users' and patients' experience, ethics and values, before evidence is applied to practice.

REFERENCES

Egger M et al. (2001) Systematic Reviews in Health Care: Meta-Analysis in Context. BMJ Publishing Group, London.

Greenhalgh T (2006) How to Read a Paper: The Basics of Evidence Based Medicine (3rd edn). BMJ Books, Blackwell Publishing, Oxford.

Guyatt GH et al. (2004) Evidence based medicine has come a long way and the second decade will be as exciting as the first. BMJ 329: 990–991.

Herbert R et al. (2005) Practical Evidence-Based Physiotherapy. Elsevier, Butterworth-Heinemann, Oxford.

Herxheimer A, Petrie KJ (2002) Melatonin for the prevention and treatment of jet lag. Cochrane Database of Systematic Reviews (Issue 2).

Higgins JPT, Green S (eds) (2005) Cochrane Handbook for Systematic Reviews of Interventions 4.2.5 [updated May 2005]. In: The Cochrane Library (Issue 3). John Wiley, Chichester.

FURTHER READING

Guyatt G, Drummond R (2002) Users' Guides to the Medical Literature: A Manual for Evidence-Based Clinical Practice. American Medical Association, Chicago.

CHAPTER 13

SCIENTIFIC COMMUNICATION

Haakon Breien Benestad

13.1 INTRODUCTION

Communication, from the Latin word spelled the same way, means to share a message having content, not just send it out and hope that it is understood. As a sender, you must know a bit about your receivers, about their wants as well as their knowledge of and ability to understand your message. Know your audience. Indeed, textbook authors, newspaper journalists and actors send differing messages. Similar but smaller contrasts exist within disciplines, as scientific papers, poster presentations and professional lectures differ.

13.2 THE SCIENTIFIC PAPER

First things

There is only one way to learn how to write good factual prose: practise, again and again, throughout life. In addition, read good professional literature and learn from others. It also helps to study texts on scientific writing and on improving writing skills (some are listed under Further Reading at the end of this chapter). But there is no substitute for trying and trying again, and with each trial, eliciting and heeding the critique of colleagues. These private referees can be your advisers.

You also may include colleagues working in specialities other than yours, as they will be able more easily to spot flaws in logic or statements to which you may be blind. It is wise, and should be natural, to have friends among colleagues who mutually help each other in the writing process.

Once you have chosen the audience for your paper, you can choose a suitable journal. That is easier said than done. On the one hand, you may hope to publish in a major international journal, but on the other hand, you may prefer to avoid the delays of successive rejections by Science, Nature and Cell. If your work is not worthy of international publication, it is not likely to be good enough to be part of a biomedical dissertation either.

Carefully read the journal's 'Instructions to authors' (usually downloadable from its website) as well as papers in a recent issue (either the electronic or paper versions, to which your library may subscribe). Thereafter, you preferably should summarize your results, the basic data for your paper, in figures and tables, in the format suiting the journal chosen. Prepare several versions; you may have spent years in your experimental work, so why not spend a few days to present your data so that it is educational?

Then it is time to write. Start by thinking thoroughly through the paper, and continue by compiling an outline that logically connects the sections of it. I have found it best to start with Materials and Methods and then follow with Results, then Discussion or Introduction, and finally Abstract, with the title page. Some prefer to start with Results. That makes it easier to gauge the extent of the method description. Finally, in incidental order: Keywords, Acknowledgements, References and Legends to figures. For tidiness, the sections are discussed below in their customary order of appearance in a published paper. Some journals have dispensed with the Introduction, Methods, Results and Discussion (IMRAD) recipe, but in one way or another, the paper must cover these aspects.

Ideally, you should write when you have the time, peace and desire to write. Some have a writing block and must force themselves to start, as by setting aside a couple of hours in the morning for writing and never using these hours for other pursuits. A colleague maintains: 'Don't think that you can do any other demanding work in a productive writing period. For most of us, writing is so tiring that we work at the limit of our abilities.' And: 'Force yourself to stop writing when you are well on the way with a point or a paragraph.' That may be torturous, he admits, but it is best to quit before you are empty, so you have something to start on the next day. 'The points don't vanish.'

Another colleague has remarked that it is not shameful to put in a full stop. A reader-friendly sentence seldom is more than 22–25 words long. If you have several things in mind, say them one at a time. So, I maintain, it is wise to read through your manuscript to check sentence length, in addition to practising the rule of 'Use your ears': reading the manuscript aloud to yourself to check sentence flow and rhythm. Do not catch the noun ailment. An example: 'This chapter is a summary of current concepts …' is better written: 'This chapter summarizes current concepts …'. And ensure that the (main) verb comes early in the sentence, so that what you have to say can be more easily understood. Fear not expressing yourself concretely. It may be daring to change 'It is clear that much additional work will be required before a complete understanding …' to 'I do not understand it …', but the example is illustrative. Do not waste words and letters; do not write 'effectuate an alteration' instead of 'change', or 'produce an inhibitory effect' instead of 'inhibit'. Nigh empty phrases such as 'it is interesting (important) to note that' or 'it should be remembered that' often are best deleted. Finally, do not be afraid of the personal tone. It is natural to write in the passive voice in Materials and Methods, but otherwise you can write 'I/we found that …'. 'It is the opinion of the present author …' can be 'I think …'.

A summary of commonplace grammatical difficulties is given in Box 13.1, and manuscript structure is summarized in Box 13.2.

Box 13.1 Things to pursue in manuscript revision

- Use of nouns when verbs would be better
- Overuse of prepositions

For example, 'There had been marked *changes in* the presentation related *to* the data accumulated as a consequence *of study of* the results of treatment in cancer *of* the breast' can be rewritten, 'The surgeon markedly *changed* her presentation after she had *studied* the results of *treated* breast cancers'.

- Overuse of the verb 'to be'

  'This short chapter *is* an excellent summary of ...' is better written, 'This short chapter excellently summarizes ...'.

- Empty phrasing

  '*The fact is that (It is clear that)* our findings falsify generally held hypotheses ...' perhaps could be simplified to 'Indeed, our findings falsify ...'.

- Impersonal 'it'

  '*It is* possible that ...' can be 'Perhaps, ...' or 'Possibly, ...'.

- Present participle with no subject (dangling participles)

  'Assaying the cytokine concentrations, the results showed that ...' should be rephrased, as the results did not measure the cytokine: 'Assaying the cytokine concentrations, we found that ...'.

- Ordinary, convoluted constructions

  'The experimental cell cultures had a very rapid growth rate, *compared to* the controls' can be written, '... grew more rapidly *than* the controls'.

  'The results for the test group were comparable to those for the control group' perhaps should be '... similar to those ...'.

- Long sentences – full stops far apart

Box 13.2 IMRAD structure

Title

Include the *keywords* of the paper. Avoid superfluities and ponderous, verbose titles. Preferably state the type of work, the experimental animals handled and the main method used.

Abstract

1. *Why, how and what was done.*
2. *Results.*
3. *Conclusions,* or inferences of the results.

The summary/abstract must not be too long. Maximum 200–250 words; see Instructions to authors. Include only what you are certain about, that is, *avoid doubt.* The summary should be *informative, not indicative* (do not write statements such as: 'The findings will be discussed in relation to the … hypothesis').

After the abstract: up to 10 *keywords, index words,* preferably from MeSH.

Introduction

1. Presentation in two or three sentences, setting the topic in perspective.
2. Background, perhaps as a historical overview, including only what is needed to lead to the next point. Past tense. Refer readily to overview articles. It should be clear what the references point to.
3. *Approach, purpose, perhaps hypotheses to be tested.* Precise. Put forth why the work is important.
4. *Maybe* clarify the work progression and the results.
   Summary: You should answer the question 'Why?'

Materials and Methods

Order: usually according to importance, or in logical order, starting with experimental animals or test subjects (and approval by relevant ethical committees): species, number, gender, weights/ages, food and drink. Maybe designate the time of day of the various interventions or operations. Maybe italicized headings (see 'Instructions to authors' for journal chosen). *Manufacturer's* name, address and maybe batch number for biological materials. Donor's name: 'provided by Dr Hansen …'. For a method *previously described: cite!* For modifications: add what and why different.

Maybe a subsection: 'Methodological considerations' with methodological discussions. The experimental set-up must be *described so the trials can be replicated* by competent researchers in the field at hand: Materials and Methods is a 'recipe book'. Finish by giving an account of the statistical methods under the heading *Statistics*.

Summary: You should answer the question 'How?'

Results

1. Give a *brief* description of the findings, in logical order.
2. *Do not repeat numbers listed in tables or appearing in figures, but describe trends and courses* (such as with 'Table ...' or 'Figure ...' in parentheses). Preferably state *how large* the changes are, such as: '... about 25% larger than the control'. No real discussion in the Results section, but *some interpretations and explanations* are often necessary.
3. You might start *Results* with a documentation of certain terms (validation of methods or study set-up).
4. State the *uncertainty of the localization parameter and/or the effect measure* [mean ± standard error of the mean (SEM), median with a 95% confidence interval]. Now and then the *variation width* of the measurements [standard deviation (SD), coefficient of variation (CV) or quartile interval] is stated, but often is better placed in Materials and Methods. State the *P-values, number of replicated measurements, number of trials*, preferably in connection with tables/figures.
5. Data normalization, that is conversion of data to percentages of median or mean in a control group, such as to permit combining the results of different trials without including the interexperimental variation, *can* be taken too far. With the same result of five trials, one can now and then describe the course of one experiment and then say that the other four gave '... in principle the same result ...'; or maybe, 'Atypical result of two trials ...', and the reason for it.

Summary: You should answer the question 'What did you find?'

Discussion

1. *Briefly point out the principal findings, in relation to the approach* of the Introduction.
2. *Interpreting* results: preferably several possibilities (references), preferably presented with diminishing probability. How reliable are the results? Do they agree with the findings of others? Transfer value or generalizability?
3. Why have *no others found* this? Accurate references!
4. Maybe something on the *conditions* for a successful result.
5. *Unanswered questions and future tasks.*
6. *Ending* preferably in the form of *Conclusions* in which you summarize what is most important, put forth your own opinions, and preferably include item 5.

Summary: You should answer the question 'What do the findings mean?'

Title

The title should be short and specific. It may be divided into two parts, the second of which narrows the first. If possible, state the principal find or conclusion in the title. It also helps to indicate the experimental animals used in an animal study or the subjects used in clinical or other trials on humans.

Some examples might be: 'Dendritic cells as effector cells: gamma-interferon activation of murine dendritic cells triggers oxygen-dependent inhibition of *Toxoplasma gondii* replication'. 'Minocycline inhibits cytochrome *c* release and delays progression of amyotrophic lateral sclerosis in mice' is a clear title for a paper in Nature. 'Pathogenesis of human cytomegalovirus infections: the role of bone marrow stem cells and vascular endothelial cells' is an informative title for a doctoral dissertation. On the other hand, 'Vasoactive intestinal peptide as a modulator in the neuroimmune axis; the influence of stress' is not very good, in part because it does not state that the work was performed on human monocytes.

Try to avoid paltry phrases such as 'The effect of …' or 'Results from …'. If possible, also avoid abbreviations and colloquialisms and do not overburden the title with strings of modifiers, as in 'tobacco mosaic virus transformed long-passaged cell lines' or '... growth from peripheral blood human megakary-ocyte progenitor cells'.

Abstract

The abstract should be comprehensible without reference to the body text; most readers read only the title and sometimes the abstract of a paper. Only those especially interested will continue to the figures, tables, introduction or maybe more. You should start with the background, purpose or principal problem. Continue with the principal means of attack, then the most reliable and certain results, and finish with a popularly understood conclusion. Be sure to study the abstracts of the journal to which you will submit your paper, and follow their style. If you have the choice, I believe that it is better not to include data or P-values in an abstract. It should be obvious that you have not made statements which are not backed up by significant data. If you repeatedly use unusual abbreviations, they should be defined upon first use; for example, 'The plasma concentration of interleukin 1 (IL-1) increased after ...'.

After the abstract, or alternatively on the title page of the manuscript, in compliance with the Instructions to authors, include the keywords or index words not in the title; preferably select words from the Medline (PubMed) Medical Subject Headings (MeSH) database.

Introduction

The introduction should state the scientific problem and why it should be addressed. Try to avoid statements of the obvious and platitudes in the first sentences, as regrettably often used in journal papers. Keep your audience in mind. With a few sentences, introduce your theme and set it in perspective.

A short historical review may be suitable, preferably with references to overview articles. Here, as elsewhere in the paper, meticulously, clearly and honestly state the findings of referenced works. It is all too easy to make several statements followed by a reference that appears to support all of the statements, although it only underpins the last of them. Likewise, here as elsewhere, you should know the field sufficiently well to credit those who so deserve, as most often it was they who made the discoveries.

Pretend that you are a storyteller, spinning out a thread that is easily followed and leads to the what and the how of what you intend to do. Usually, you should say something about the relevance of the project and about why the problem has not been addressed, to your knowledge. Some writers end an introduction with a summary of the principal finding, but I maintain that is so overly repetitious in the standard paper as to be unnecessary.

A suitable length for an introduction is one or two double-spaced, 12 point type manuscript pages.

Materials and methods

Here again, it is wise to study the style of the journal to which you will submit your paper. This section is your 'handbook'; it will be read only by those especially interested in your topic. Use several subheadings to guide your reader to the points of interest to be pursued further.

The human test subjects or the experimental animals should be described in the first paragraph. Test subject data (gender, age, weight, etc.) may be presented in a table or in a flow diagram that shows those enrolled, dropouts, etc. Inclusion and exclusion criteria, written informed consent, etc., should be described (see Chapters 1, 4, 8 and 10). Moreover, any approval granted by an ethics committee must be stated. The experimental animals should likewise be described in terms of species, gender, weight or age, food and drink and supplier's name and address, and sometimes also by whether they are inbred or outbred, specific pathogen free (SPF), transgenic and, if so, by what means. Here also, ethic approval must be explicitly described, such as by stating that 'All animal experiments were approved by the Experimental Animal Board/Committee under the Ministry of ... and conducted in conformity with the European ... Convention for the Protection of Vertebrate Animals used for Experimental and other Scientific Purposes', or a shorter version of the statement. Randomization procedures must be stated for both man and mouse. If the variables measured are subject to diurnal variations, observation times should be stated.

The next subsection can have a subheading such as Chemicals, Reagents or Biological materials. It comprises a listing, preferably alphabetical, of preparations used, with the trade names, producers and their addresses in parentheses, such as cyclophosphamide (Sendoxan™ [alternatively, Sendoxan®]; Orion, Espoo, Finland). Whenever the quality of a biological material varies, you should state the batch, lot or code number. If the preparation used was a gift, name the donor in parentheses (kind gift from Dr N.N., the XX Institute, Cambridge, UK) and also thank him or her in the Acknowledgements.

For previously published descriptions of procedures, name the procedure with a reference to its original description: 'In short/brief, leucocytes were isolated with a density gradient method … (ref.)'. However, modifications, as well as new methods, should be described in detail sufficient to enable other scientists in the field to replicate the experiments. Remember to state all essential physiochemical parameters, such as temperature (18–20°C is preferable to 'room temperature'), pH, osmolarity, concentrations (preferably in SI system units, such as mmol/l), gas pressure, solution volume, storage conditions, centrifuging data [G or g is preferable to rotations per minute (rpm)].

It may be practical to include a section on Experimental design in which you describe the course of the trial or experiment. Further, you may allow yourself a section entitled Methodological considerations, in which you put forth results of methodological research germane to your methods and which you prefer not to disturb the story of the Results section.

Usually, the last subsection is Statistics or Statistical methods, in which you describe data presentation and processing. You may state that the data are presented in medians (or in arithmetic means) with their accompanying 95% confidence intervals. Further, the methods used in hypothesis testing (including the name and supplier of any computer application used), significance level (P-levels and any modifications of them in multiple testing, as with the Bonferroni correction, in which the P-level is equal to the quotient of 5% and the number of hypothesis tests. This means that if you compare test and control data at five different times, and the hypothesis opens for more than just one time of interest, the P-level should be set to 1% to be in line with the ordinary 5%). You also should state whether the testing is one-sided or two-sided, whether parametric or nonparametric methods were used, and whether you have conducted a power analysis. In descriptions of clinical, epidemiological and qualitative research, this section often comprises a more extensive account than customary in reports of laboratory activities. Such extensive accounts include distinctions between explanatory (independent) variables and effect variables, choice of effect measure, and choice of univariate, bivariate or multivariate analyses (see Chapters 4 and 11).

This section should be written in the passive voice, as what has been done is more important than who did it.

Results

This section says what you found, in logical order. Tell your story or stories, with subheadings if allowable, in brevity and in the past tense. Try to mention the most important results first, and place pivotal words early in sentences. Negative findings and trivial findings may be mentioned towards the end. Journals often recommend that the results section of a paper reporting clinical trials should begin with an overview of the test subjects. In my opinion, such overviews are so unexciting that they should be included under Materials and Methods, as mentioned above, but follow the journal's style guide if it says that they belong here.

This section is not the place to discuss findings, but occasionally you must briefly present findings to clarify to the reader why the experiment was conducted, what the results mean and how they lead to the next experiment.

Occasionally you may start the Results section by referring to an experiment that documents the premises of the main enquiry.

The Results section can be telegraphic. Concise language often is best. Remember, some journals have considerable page charges! Hence, I prefer the concise 'Blood pressure remained constant (Fig. 1)' to the more verbose 'It can be seen from Figure 1 that blood pressure remained constant …'.

In the text, do not repeat data already included in cross-referenced figures and tables. However, you should refer to the data presentation to document the description of the course of the experiment and the differences observed, as between test and control groups. Biological significance is just as important as statistical significance. So if it is doubled or halved, declines or increases linearly or exponentially with time, such a description is in order. In my opinion, it is better to state P-values together with figures or tables than in the text of the Results section.

A figure or table that summarizes the raw data of all replicated experiments is more convincing than the presentation of a single experiment with the ancillary explanation that 'three replicate experiments gave similar results'. Whenever the interexperimental variation is large, you may be obliged to present the data this way. In any case, state that you have selected representative, not typical, experiments, as after all, you do present the best experiments. Alternatively, you may present a summary of all experiments, provided that you normalize data, such as by setting the median or the mean for the control group at 100%. (But then you must state the absolute value of the 100% figure, such as in a footnote: 'The 100% values ranged from 32 to 56 mmol/l'.)

Murphy's supposed law of nature states that anything that can go wrong will go wrong. But you should have good reason for excluding atypical results, and the exclusion criteria should be set before project work starts. So, you may need to describe the common characteristics of the atypical results at the end of the Results section.

Discussion

The Discussion section interprets results against the backdrop of the consensus of a scientific field. It also sets forth the relevance of the findings. As elsewhere, it should be phrased in the past tense, but the general conclusions should be in the present tense. A good rule of thumb is to start the discussion with a recapitulation of the principal findings in summary. In this way, you can interpret the results in light of the main problem that was presented in the Introduction of the paper. What do the findings mean?

How do your results fit in with the common body of scientific knowledge? Which previous findings, yours and those of others, support your findings, and which deviate from your results? As mentioned above, it is essential that you are familiar with the literature of the field and know who prioritizes orthodox opinion. Here you should cite the first, most important publications and not overview articles. (This may present a dilemma if the initial publication was qualitatively weak, but the next from the same group of authors more fundamental, as you must economize on the number of references in your paper.)

Occasionally the meaning of your findings may not be obvious. There are several solutions to this. Start with the most plausible interpretation. Hans Selye allegedly said: 'Our facts must be correct. Our theories need not be, if they help us to discover important new facts'.

It may be relevant to point to the necessary and sufficient conditions for the findings you have presented and to explain why others have not found what you have found. Perhaps the experimental set-up, the animal model, etc., differed.

Often you must in an unassuming way say what is new in and important about your paper to get it published. (Your findings should be both new and important, so that you do not get the response once given by a referee who quoted Dr Samuel Johnson's famed remark: 'Your manuscript is both good and original; but the part that is good is not original, and the part that is original is not good.') Honesty pays!

Occasionally you will need to discuss groups of observations concerning dissimilar phenomena or parts of problems. You might then start with the most important. If you cannot decide what is most important, use the order of the presentation of results. If the journal layout so permits, you can put subheadings on these discussion capsules.

The final section may contain a few summarizing or concluding sentences, perhaps with a rough sketch of further research. If allowed by the journal style, these could be entitled conclusions.

Common faults of discussions:

- Repetitions from the Introduction (intended to introduce the background of the project).
- Excessively long account of the Results section.
- Overwritten discussion (in which the author seeks to justify a place in history or show a command of the literature), which may be speculative. The discussion should be of the actual findings and not expand beyond them. Avoid excessive generalization.
- Faulty gradation of reality. Indeed, you must point to what is new and important and to how the results may be generalized and used, but do not exaggerate.
- Discussion of trivialities. However, negative results might advantageously be discussed.
- Too few explanations. Disagreements with the findings of others should not be swept under the rug. Resist the temptation to criticize colleagues personally; preferably, discuss their incongruous findings with a tone of understatement.

As can be seen, writing the Discussion section is like sailing between Scylla and Charybdis. Use common sense; here, as elsewhere in science, it is the best guide. A suitable length for the discussion section is three double-spaced, 12 point type manuscript pages.

Tables and figures in general

You have your raw data, and you have chosen the audience for your paper. First, ask yourself which data *must* be presented and which data may be replaced by qualitative mention in text. Might a summary of these data [means, or medians, with their 95% confidence interval, or standard error of the mean (SEM), and the number n of data points] be sufficient? If the data are vital or copious, you may choose to present them in tables or graphs. Some types of data are certainly not summarized in numbers alone, such as the results of gel electrophoresis or DNA microarrays.

A usual rule for choosing between tables and figures is: use tables whenever exact values or comparison with the data of others is essential; use figures when the relationship between an independent variable (x) and a dependent variable (y) is the principal point, as in an event over time. In other words, figures ease understanding of main points, and tables provide more comprehensive documentation.

You know that many readers will only read the title and abstract and scan your illustrations. This will influence your choice of form of presentation, curves or bar charts, photographs of gels or recordings of action potentials, etc. Remember that tables and figures with titles, captions and footnotes principally should be self-explanatory.

Ask yourself about the average level of knowledge of your audience, and express titles, terminology and abbreviations accordingly. Moreover, these graphic elements must comply with the journal 'Instructions to authors', and in sizing, you must take account of whether the journal is printed in one- or two-column format. The summarizing of results and the preparation of illustrations are usually the first steps in compiling a manuscript, where you may discover points for which you must perform more experiments or otherwise collect more data.

Let your imagination run free; sketch several versions of tables and figures. The goal is a maximum of information in a minimum space, without overloading, ambiguity or deceitful layout. Occasionally tables or figures may be divided in two, to promote understanding. One approach for plots is a composite figure of two or more frames above each other, with the same x-axis in all frames. Remember that figures in a paper often are ill-suited to lectures (PowerPoint presentations) or posters and that tables in a paper are always unsuitable for other uses, so as you prepare figures and tables for a paper, make other versions for lectures or posters if need be.

Tables and figures should be self-explanatory, so often you must strike a compromise between brevity and comprehensiveness in captions and footnotes. The details of methodology should be given under Materials and Methods, but occasionally you must include sufficient details on methods to promote understanding of data presented. Abbreviations, which you should avoid as much as possible, should be written out in full upon first use or defined in a dedicated section in the paper. Moreover, you must define locus (centroid), uncertainty measure (and occasionally measure of spread), number of replicate analyses and replicate experiments. For example, 'Means with their 95% confidence intervals are shown; $n = 7$–8 replicate determinations in each case. A total of three independent experiments with similar results was performed.' If you present normalized data, state the level of raw data corresponding to 100% as well as the range of raw data. When such information is given in the first figure caption or footnote to the first table, you often can save space in the rest of the data presentation by writing 'See legend to Figure 1 for further information'.

When you have finished the drafts of the tables and figures, check to see whether they are presented in a uniform manner. Are the headings of the columns showing the same sort of data the same in different tables? Likewise, have axes of the same variable been labelled the same in different figures? Go carefully through all the illustrations, with an extra eagle eye on the tables, looking for superfluous information. Are there any words, parts of pictures, non-essential data that can be deleted or moved to figure or table captions or footnotes? Again, ask a capable, critical, non-specialist colleague to go through your illustration material.

Finally, check your data presentation material for internal consistency as well as for consistency between it and statements in the body text. Readers will have a poor impression of your work if in the Results you make a contention that is not obvious in the tables and figures or if the text and the tables or figures are contradictory.

Finally, the tables and figures must be numbered in their order of appearance in the paper. In a margin you may mark the desired location of data presentations ('Figure 1 approx. here').

Tables

A typical table looks like this, but in all cases, its format should comply with that stated in the 'Instructions to authors'.

ROW HEADING	COLUMN HEADING		
	COLUMN 1	COLUMN 2	COLUMN 3
Row 1	Data field		
Row 2			
Row 3			

Footnote a.

Footnote b.

Footnote c.

The caption for the table should convey its main message, say what it shows. Avoid vague, dull phrases, such as 'The effect of …'. Use the footnotes to convey more information on table content and interpretation.

The same sort of data should be presented in columns, not in rows. The columns should be arranged so that comparison of data sets is easy, for example, in adjacent columns in the table. It is convenient to start with the control data or the normal data to the left and up in the table. The average reader will absorb the information in the table from left to right, and from top to bottom, so that order is the most educational. (See Box 13.3.)

Box 13.3 Good tables
- Note: Clarity and clear column headings.
- Almost never the same table style in a paper as created with presentation graphics software, such as Microsoft's PowerPoint. Tabulations of slides should be simple without too many footnotes.
- Same sorts of data under each other (in columns, not in rows)
- Avoid conveying a false impression of precision such as by listing more than, for example, *two significant digits.*

- Do not include sequences of numbers that can easily be calculated from other numbers in the table.
- Proofread the table against the original draft.
- Caption and footnotes: *succinct. Describe what is shown* in the caption. Often wise to describe purpose of the table. State the uncertainty measure or spread measure, such as standard error of the mean (SEM), 95% confidence interval, standard deviation (SD). State the *P*-values and the range of the control values that you have set to 100% in normalization.

Data that easily can be calculated from the raw data presented or monotonic data (such as multiple occurrences of 0, 100 + or −) may be moved to footnotes or to the Results section (see Box 13.4). Be meticulous and uniform with the number of significant figures (see Box 13.4) and with the use of units.

How should you choose the number of figures so as to avoid creating a false impression of high precision? The number of significant digits is equal to the number known with some degree of confidence, so my choice is to be guided by the uncertainty measure, SEM or 95% confidence interval. I choose two significant digits for the confidence interval (or $4 \times$ SEM). Therefore, in text or tables, you do not write 1.1224 (1.0015–1.2468) $\cdot 10^9$ cells/l, for this interval range has four significant digits (245.3 million/litre). Presentation to the correct number of decimals then is 1.12 (1.00–1.25) $\cdot 10^9$ cells/l (median, 95% confidence interval, $n = 20$), alternatively 1.12 (1.00, 1.25) nl^{-1}. But a correct example (undesignated) of a high-precision measurement might be 1.1224 (1.1215–1.1268), as the width of the confidence interval (0.0053) now has only two significant digits, while the mean – correctly – has five significant digits. Some authors go farther than me in contending that whenever data are listed for comparison, as in a table, they should have no more than two significant figures (Ehrenberg 1977). This means, for instance, that statements of percentage $\geqslant 10\%$ never have decimals.

SI system units should be used as consistently as possible, and powers of 10 should, whenever practical, be replaced with SI abbreviations (micro, nano, kilo, mega, etc.) (see Box 13.5).

Box 13.4 A poor table

Table Z. Cellularity and differential counts of peritoneal exudate cells, accumulated 1 hour after injection of chemoattractant FMLP

DOSE OF CHEMOATTRACTANT (μG)	TREATMENT	TIME (H)	TOTAL CELLULARITY (× 10⁶)	DIFFERENTIAL COUNTS (%)			
				MACROPHAGES	LYMPHOCYTES	NEUTROPHILS	EOSINOPHILS
Saline	0	1	5.8 ± 3.2	76.6 ± 8.6	16.5 ± 8.6	6.1 ± 4.7	0.7 ± 1.3
FMLP	25	1	14.0 ± 5.6	63.4 ± 8.6	26.8 ± 8.8	9.1 ± 5.2	0.7 ± 0.8
FMLP	50	1	21.1 ± 7	63.9 ± 11.7	23.0 ± 9.5	12.1 ± 7	1 ± 0.8
FMLP	100	1	25.5 ± 5.4	64.5 ± 10.1	16.8 ± 7.7	17.6 ± 8.8	1.2 ± 1.3

Redundancies in the left part of the table should be removed; for example, column 1 (reading column) should be deleted, column 2 should be headed 'Dose of FMLP' (preferably given in μmol or nmol, since MW of FMLP, which is a microbial peptide, is known), column 3 should be deleted. The variability (SEM are given) of replicate analyses (the number of which should be given in the legend or in the field) indicates that the data should be presented without decimals (as has in fact been done inconsistently, for part of three entries), according to the 'significant digits' convention. The data furthermore demonstrate the inappropriateness of parametric methods (i.e. calculation of SEM) applied to data that are not normally distributed: Taken at face value, cell counts can apparently have negative values here!

Box 13.5 SI system

Système International d'Unités (SI units): some examples

Names and symbols for basic SI units

PHYSICAL QUANTITY	NAME OF SI UNIT	SYMBOL FOR SI UNIT[a]
Length	metre	m
Mass	kilogram	kg
Time	second	s
Electric current	ampere	A
Thermodynamic temperature	Kelvin	K
Luminous intensity	candela	cd
Amount of substance	mole	mol

[a] Symbols for units do not take a plural form and should not be followed by a full stop, e.g. 5 cm, but not 5 cms or 5 cm. (except at the end of a sentence)

Special names and symbols for some derived SI units

PHYSICAL QUANTITY	NAME OF SI UNIT	SYMBOL FOR SI UNIT	DEFINITION OF SI UNIT
Energy	joule	J	$kg\,m^2\,s^{-2}$
Force	newton	N	$kg\,m\,s^{-2} = J\,m^{-1}$
Power	watt	W	$kg\,m^2\,s^{-3} = J\,s^{-1}$
Pressure	pascal	Pa	$kg\,m^{-1}\,s^{-2} = N\,m^{-2}$
Electrical charge	coulomb	C	$A\,s$
Electrical potential difference	volt	V	$kg\,m^2\,s^{-3}\,A^{-1} = J\,A^{-1}\,s^{-1}$
Electrical resistance	ohm	Ω	$kg\,m^2\,s^{-3}\,A^{-2} = V\,A^{-1}$
Frequency	hertz	Hz	s^{-1}
Concentration	mole per litre	$mol\,l^{-1}$ $(mol\,dm^{-3})$	
(Radio)activity[a]	bequerel	Bq	s^{-1}
Radiation dose	gray	Gy	$J\,kg^{-1}$ ($1\,Gy = 100\,rad$)

[a] $1\,\mu Ci = 37\,kBq$

Prefixes for SI units: the following prefixes may be used to indicate decimal fractions or multiples of the basic or derived SI units

FRACTION	PREFIX	SYMBOL	MULTIPLE	PREFIX	SYMBOL
10^{-1}	deci	d	10	deca	da
10^{-2}	centi	c	10^2	hecto	h
10^{-3}	milli	m	10^3	kilo	k
10^{-6}	micro	μ	10^6	mega	M
10^{-9}	nano	n	10^9	giga	G
10^{-12}	pico	P	10^{12}	tera	T
10^{-15}	femto	f			
10^{-18}	atto	a			

Note that expressions such as mg/kg/d and mmol/ml/s are not correctly written; the designations should be mg/kg · d (or $mg\,kg^{-1}d^{-1}$) and mmol/ml · s (or $mmol\,ml^{-1}s^{-1}$)

Good style, as well as most journal style guides, dictates aligning decimal points under each other in a column; the same applies to the plus/minus sign $\pm$ in front of SD or SEM figures. Inside borders of tables should not be used. Occasionally, white space may be used to separate data presented, but otherwise the space between columns of data should be as small as practical, without disrupting overall appearance or legibility.

Figures

Russian writer Ivan S. Turgenev (1818–1883) is reported to have said that a glance at a picture gives what a book needs a hundred pages to depict. Refrain from trying that. Some authors do try, with their presentation graphics applications, but remember that an illustration should be quickly understood. Nonetheless, I agree with Tufte (1983) that 'Graphical excellence is that which gives to the viewer the greatest number of ideas in the shortest time with the least ink in the smallest space.'

A checklist for good figures is presented in Box 13.6. First, you should choose the data to be presented in each figure, and then choose the most suitable type of figure. Curves (Box 13.7) are good for dynamic comparisons and for illustrating courses, such as weight (y) as a function of concentration (x) or time (x). A bar chart (Box 13.8) is well suited for presenting discontinuous variables or ratios.

Box 13.6 Good figures

- *Function*: Simple; illustrate a single point.
- *Accuracy*: No disagreement with the text, neither of data nor of terminology.
- *Composition*: Use the space available.
- *Contrast*: Highlight difference between test and control, if it is a point.
- *Legibility*: Avoid overloading a figure; just a few curves or bars in each. Draw figures large and consider reduction.
- *Choice* of figures or tables: use a figure to illustrate the relationship between two variables, such as the changes in concentration (y) with time (x).
- *Uniformity* both in creating figures and in their text. Mark coordinate axes with units and magnitudes; they may be broken at discontinuities (even from the origin) or for changed scales. Use familiar abbreviations. Add arrows pointing to the essential features in complex photographs. Figure captions in the figure: preferably in slides; often not allowed in set type.
- *Figure caption*: Succinct. Describe what is shown, preferably in the first sentence of the caption, which may be in italics, like a subtitle. See also the last item in the bullet list of Box 13.3.

Box 13.7 Good and less good curves

Note: (1) the various ways of labelling the axes; (2) the misleading impressions that can be created by manipulation with the scales (panels b and c vs d); (3) the acceptability of axis displacements; (4) the marking of discontinuities; (5) the explanation of the symbols in the figure space (panels a and c); (6) the displacement of symbols (panel d) or variability bars (panel c) to avoid overlap; (7) the utility of combining panels (e.g. panels a and b); (8) the use of the SI system and avoidance of powers of 10 (panel e); (9) the better readability of words written in lower case letters than in equal-height upper case letters (on y-axis of panel c); and (10) the dispensability of the horizontal finials of the variability bars (shown in panel a).

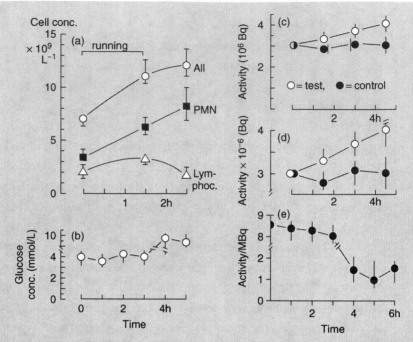

Figure F Five panels illustrating various ways of constructing line graphs, some of them not recommended

Box 13.8 Bar charts

Upper panel: A conventional column chart. You must be sure that the left- and right-hand scales unequivocally refer to the left- and right-hand columns, respectively. Median values are shown, with their 95% confidence intervals (estimated with a non-parametric method and therefore asymmetric around the medians) and number of replicate values. Alternatively, the *y*-axis designations may run vertically upwards, along the axes.

Middle panel: Variant design, with space between the columns (not having the same width as the columns). SE(M)s are shown as measures of uncertainty of mean values.

Lower panel: A third variant, with slightly overlapping columns and only 1 SE(M) drawn for each column.

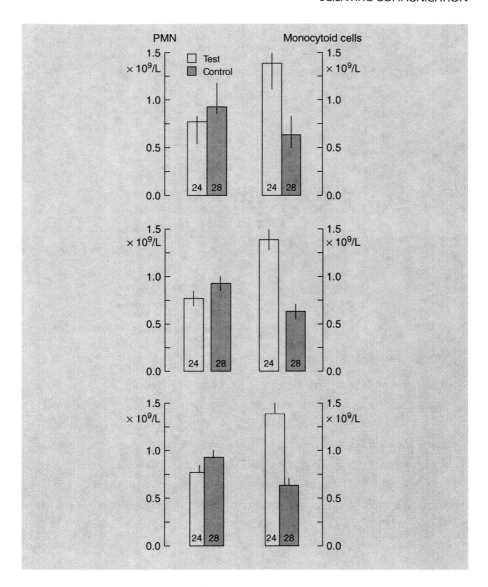

Statistical comparisons are often more obvious in comparisons of magnitude than in the figures of a table. Scatterplots (Box 13.9), with or without regression lines, are often the best way to illustrate relationships between *x* and *y* variables. In any case, if the data are amenable, make a scatterplot for your own use.

Box 13.9 Scatterplot with a regression line

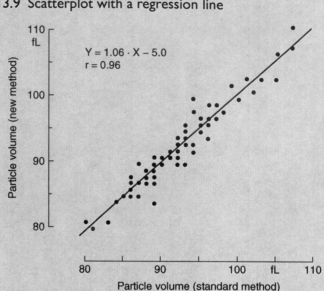

Figure H Example of a scatterplot.

The regression line (here very close to the line of identity, $x = y$), its formula and the correlation coefficient (r) are indicated. Y = particle volume recorded with new method; X = particle volume recorded with standard method. Sometimes the line should be omitted and the formula and r-value given in the text or the figure legend. The P-value for testing $r = 0$ should also be shown; alternatively, information on the 95% confidence intervals of the coefficients may also be presented (slope: 1.00–1.13; intercept: –11 to +1).

Occasionally it may reveal interesting trends in the data, such as deviating subgroupings, that you otherwise might have overseen.

The advantage of photographs used to be that they could confirm the authenticity of a phenomenon, but now digital photographs may be manipulated to give false impressions. An editorial in a recent issue of Nature (Anon 2006) calls this 'beautification' and goes on to explain that 'the data are legitimately acquired but then processed to yield an idealized image. … a form of misrepresentation. … the Nature family of journals has developed a concise guide to approve image handling.' We may find wit in the historical remark of a professor, *video, sed non credo* ('I see it, but I do not believe it'), but nowadays nonetheless admit its wry truth. Line drawings are usually clearer and convey points better. Pictures supported by sketches are often a good combination. Photographs of gels, gel prints, etc., must occasionally be retouched or otherwise processed; if so, the processing used should be stated in Materials and Methods or in Supplementary information. Be sure adequately to scale and explain added graphic items, such as designations, letters and arrows. Microscope photographs should have a scale in the original submitted, so that the indication of size is preserved, regardless of whether the photograph is enlarged or reduced by the journal. Rules available for depiction, together with histological preparations can be used to show picture scale. The method of colouring must also be stated in the figure caption.

Do not overload figures. Have no more than three curves in a graph, or four if they are far from each other and do not cross each other several times. The curves may be identified by different symbols (circles, triangles, square, open or solid) or by different line styles (solid, dashed, etc.), but not both. Do not show dissimilar parameters (such as body temperature and sedimentation rate) on the same y-axis: think rather of using a left axis and a right axis, or draw several panels under each other in the same graph. Consider simplifying axis designations (and supplementary explanations of them in the caption) and whether you can delete, for instance, every other number from the axes, and let the graduations stand. Designations appearing near curves or in symbol explanations in the figure space can promote understanding and are excellent in slides, but are not always allowed in print. The same holds for a short title in the figure space above the figure.

Resist extravagance. Can two figures be combined into one? Can unessential parts of a photograph be cropped out? Might the axis scales be more economical? Have you ensured that the axis and symbol identifications are so brief and well placed that the curve(s) fill the column width? The axes need not extend beyond the maxima of the data plotted. Make the numerals, letters and symbols so large that they are easily deciphered even when you reduce the figure to the smallest size that the journal may use. Changes of scale should be indicated by discontinuity marks on curves and axes; however, it is best if the curves are plotted to avoid breaks that easily may connote disorder (example in Box 13.7). It should be easy to see if an axis does not start at zero (or 1.0 for a logarithmic scale). Axes may be displaced to avoid data points falling on them. The x-axis may be deleted in bar charts. Let the spaces between bars be less than their widths. Here, as elsewhere, see how figures are drawn in a journal renowned in your discipline.

Acknowledgements

In the short Acknowledgements section you should extend thanks for financial support, such as by stating that: 'Support to H.B.B. from the Norwegian Cancer Society and the Research Council of Norway is gratefully acknowledged. We also thank N.N. and M.M. for excellent technical assistance'. Occasionally it is appropriate to underscore the contributions of assistants.

Quality control

Have you followed the structure of Box 13.2? If not, you should say why not. Most important, check all data, figures, tables, text and references cited. Reference control is simple and straightforward using Reference Manager®, ProCite®, PubMed® and similar database-based applications. Likewise, spell checkers and grammar checkers are useful aids, but they cannot replace conscientious review. The error in 'the colleges raised there voices' will not be spotted by a spell checker. Have a goal for your review. Can long sentences be divided into two or three sentences? Can you weed out superfluous words, expressions, sentences (Box 13.1)? Are there any sentences with too many prepositions, nouns that can be replaced by verbs, passive voice that can be replaced by active? Are all verbs in the correct tense? Have you really tried to improve your writing style, such as by avoiding redundant use of your favourite verb? A thesaurus is a great help here. 'Record' might be replaced by monitor, measure, analyse, assess, examine, determine, study, investigate, scrutinize, evaluate, explore or probe. Are all the references named in the text in the reference list at the end, and vice versa? Check once more with the 'Instructions to authors'; be your personal reviewer by reading, and *require* that your co-authors also thoroughly read the entire manuscript. Now you can start thinking about submitting the manuscript.

Submission cover letter

A typical submission letter might read:

> Dear Editor,
>
> I have enclosed a manuscript in quadruplicate, 'Membrane …', by H.B., R.H. and K.L. I hope you may find the paper suitable for publication in Exp. Hematol. The investigation concerns …, which we feel falls within the scope of your journal. All data presented in the article are original data of the authors. The data have not been published previously. We now submit the article to your journal only. The manuscript has been reviewed and approved by my co-authors. The authors hereby declare that they have no competing financial interests.
>
> Yours sincerely

Many journals require that you submit the manuscript electronically in a PDF file. Adobe Acrobat programs for generating PDF files are available on the Internet.

What should you do if your manuscript is provisionally accepted? Manuscripts are seldom accepted without changes. You change those aspects of the critique with which you agree as well as those that you feel do not diminish the value of the manuscript. Thereafter, you follow the editor's instructions. This may involve answering each referee on a separate sheet, point by point, with references to places in the text where changes have been made and the extent of each of them. If a referee has done a conscientious job or has provided valuable hints on improvements, you should acknowledge his or her contribution. Be polite, even if the referee, in the light of anonymity or due to unfamiliarity, has not grasped your points. Then you can carefully explain why the suggested changes have not been made.

Co-authors

Co-authorship is a sensitive subject. The situation is asymmetrical. The scholarship holder or research fellow thinks that he or she has done the work, perhaps late into the night, and has neglected his or her family. The adviser, who has built up the laboratory, worked in procedures and quality control, acquired funds for research, and moreover probably hatched the idea for the project and followed the data collection, is aware of his or her own indispensable role. Clearly, both should be co-authors, the research fellow first, as he or she has drafted the manuscript. The situation worsens when the head of the institute, or a colleague who has merely performed a particular analysis or delivered a particular reagent, insists on being listed as a co-author, as often happens. And this happens despite rules to the contrary that the International Committee of Medical Journal Editors (ICMJE) has put forth. (In 1978, a group of editors of general medical journals met informally in Vancouver, British Columbia, Canada, to establish guidelines for the format of manuscripts submitted to their journals. The group became known as the Vancouver Group. It subsequently expanded and became the ICMJE, now with a website at www.icmje.org, where the rules are revised when appropriate.)

Authorship should solely be based on:

- significant contributions to the idea and design, *or* data collection, *or* analysis and interpretation of data
- compiling of the manuscript *or* critical review of the paper's intellectual content
- approval of the version of the paper to be published.

Each co-author must have contributed sufficiently to assume responsibility for appropriate parts of the content of the paper. More specific requirements are usually stated on the journal's website.

13.3 POSTERS

Most meetings and symposia have poster sessions. A researcher may feel thwarted at not being able to present a paper and must be satisfied with presenting at a poster session. But the poster session is no subculture, and justifiably not. If you manage to interest a session audience in your subject and your results, the informal discussion of your posters can be more thorough and more rewarding than the nervous session that you may endure after giving a paper. Moreover, you can gain friends and colleagues for life.

Regrettably, poster sessions are often poorly organized, so that you may have to cut down on a lunch break or first meet late in the evening, when you are weary. Your work may then be undervalued. A poster session may be expedient for the researcher who must present something to have travel expenses covered. However, poster sessions can be superb, when the audience has the opportunity to study the posters, undisturbed by simultaneous sessions and with the posters of a particular sector grouped together.

Take time in planning. Sketch alternative layouts on letter or A4 sheets. You will have received accurate details on dimensions and on whether the posters should be in portrait or landscape orientation. Think of a newspaper page. Set the title, problem or purpose description and the final conclusion in large letters. The most common mistake is to overload the poster with information.

Be daring: choose a shorter, more catchy and challenging title than you would have used on a paper. Figures should go through the same simplification process that you would use in taking those of a paper to slides, and tables should be far simpler than those of a paper. Your message must be discernible by a weary watcher, without strain. Therefore, you can delete measures of spread and confidence intervals and just state significant differences, such as with asterisks. You may also curtail the method description, or provide more detailed information in smaller type than used elsewhere on the poster, and you can use handouts to provide supplementary details for those interested. Sometimes a small version of the poster can serve as a handout. If anyone wants to know the standard deviation of your measurements, they can ask, and you get into a discussion.

The Physiological Society in London has published guidelines for producing posters, some of which are thought-provoking or challenging. The space available is limited, so your poster should have just one main point. All information should be readable at a distance of 2 metres, which means that the amount of information is limited. Have no more than six illustrations that together take up no more than 50% of the poster. Limit the prose on the poster to a minimum. The message of the poster should be obvious without supplementary verbal explanations. Acronyms, abbreviations and slang should be avoided. The order of reading the material presented should be obvious.

In addition, I have found that a three-column poster in the customary portrait orientation is a good pattern. The tall letters of the title should be at least 4 cm high, subheadings at least 1.5 cm high, and text in the first and last sections – Introduction (problem description) and Conclusion – at least 1 cm high. Colours are commonplace, but ponder them well, as there should be an educational reason

Box 13.10 Typical poster sketch

Organic Life in Outer Space
By NN, MM, OO, Institute of Astrobiological Research, University of Oslo

Problem: *Do life forms similar to those we know from the earth exist on outer space planets, and can modern super-spectroscopic techniques identify such organisms?*

Fig. 1. Signal spectra received by apparatus X from planet Y indicate presence of organic molecules
....
....
....
....

Fig. 2. Decomposition of spectroscopic data shows that the source includes a new type of nuclear acid........................
....
....
....
....

Handout: Take a copy!
xxxxxxxxxxxxxxxxxxxxxx

Table 1. Decoding of nuclear acid sequences obtained from outer space

Footnotes
....
....
....

Box 1. Mathematical modelling suggests the presence of higher forms of life on planets in outer space
....
....
....
....
....
....
....
....
....
....
....
....
....
....

Table 2. New amino acids and their preponderance in extraterrestrial organisms

Fig. 3. Reconstruction of insectoid from planet Y.
....
....
....

Fig. 4. Tentative anatomy of ET from planet Y.
....
....
....
....

Conclusions
Life exists on planet Y, its highest developed form being the famous ET, but with a smaller brain and bigger hands than generally appreciated.

for their use. For instance, comprehension is eased if all the curves or bars representing test group data have the same colour in all illustrations, while control group data curves and bars have another. Colours may also be used to highlight key points: the problem you started with, its solution (conclusion) at the end. Let the figure and table captions convey the results that tell your story.

Increasingly, computer applications, such as PowerPoint and the Goliat® poster printer are used to produce posters. Regrettably, one then risks overloading a poster with beautiful, colourful items and too much text. As Goethe advised: '*In der Beschränkung zeigt sich erst der Meister*' ('Constraint is the hallmark of the master') (Goethe 1802). (The Adobe InDesign application can be used for poster production; a guideline is available on the Internet.)

Even though your poster must be transported, inconveniently rolled up, take it as hand baggage.

A good mnemonic rule is POSTER: P for 'prepared and planned', O for 'one main theme', S for 'simple pictures', T for 'tables minimal', E for 'explains itself' and R for 'readable at 2 metres' (Brown 1996).

A sample poster sketch is shown in Box 13.10.

REFERENCES

Anon (2006) Not picture-perfect. Nature 439: 891–892.

Brown BS (1996) Communicate your science! … Producing punchy posters! Trends in Cell Biology 6: 37–39.

Ehrenberg K (1977) Rudiments of numeracy. Journal of the Royal Statistical Society A 140: 277–297.

Goethe JW (1802) Das Sonnet, written for the opening of the new theatre at Lauchstaedt on 26 June 1802, second part, line 13.

Tufte ER (1983) The Visual Display of Quantitative Information. Graphic Press, Cheshire, CT.

FURTHER READING

Briscoe MH (1990) A Researcher's Guide to Scientific and Medical Illustrations. Springer, New York.

Ebel HF et al. (1987) The Art of Scientific Writing. From Student Reports to Professional Publications in Chemistry and Related Fields. VCH, Weinheim.

Gowers E (1971) The Complete Plain Words. Penguin Books, Middlesex.

Gustavii B (2000) How to Write and Illustrate a Scientific Paper. Studentlitteratur, Lund.

Simmonds D, Reynolds L (1994) Data Presentation and Visual Literacy in Medicine and Science. Butterworth-Heinemann, Oxford.

Strunk WS, White EB (1979) The Elements of Style. Macmillan, New York.

Style Manual Committee/Council of Biology Editors (1994) Scientific Style and Format. The CBE Manual for Authors, Editors, and Publishers (6th edn). Cambridge University Press, New York.

SUCCESSFUL LECTURING

Heidi Kiil Blomhoff

14.1 INTRODUCTION

Today, we are surrounded by computers, mass media and audiovisual aids, yet paradoxically, the simple spoken word remains a cornerstone of communication. There are daily reminders aplenty, as the success or failure of politicians, lawyers and other public people depends on what they say and how they say it. Verbal skills are no less important for scientists in presenting our research results and hypotheses for more or less interested audiences. Indeed, the root of the word 'lecture' is the Latin *lectura*, meaning 'to read'. Unquestionably, verbal communication skills remain vital.

Certainly, we know from experience that even the most highly motivated listener often has difficulty grasping everything presented in a lecture. We also know that attentiveness varies during the course of a lecture, as illustrated in the attentiveness curve for a typical 45 minute lecture, as shown in Figure 14.1.

At first glance, the high-to-low swing of the attentiveness curve may discourage a lecturer, but it is actually a valuable aid to preparing a lecture. The highs of the curve show that a lecturer may convey a message most easily in the first few minutes and towards the end of a lecture. The course of the curve implies that there is always room for improvement in lecturing.

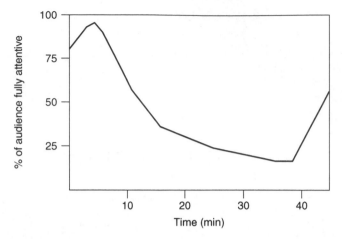

Figure 14.1 Attentiveness during a 45 minute lecture

The goal of this short chapter on the lecture method is to put forth hints about and advice on being a better lecturer, so that you may actually enjoy lecturing, and not least so that an audience will enjoy listening to you. Success in lecturing requires preparation, preparation and more preparation. You must prepare the lecture, and you must prepare and practise your delivery of it.

14.2 PREPARING THE LECTURE

In advance of preparation, think of the two essential aspects of it. First, you must start well in advance, and secondly, you must gauge your audience (Box. 14.1).

> **Box 14.1 Preparing the lecture**
> - Take time
> - Gauge your audience according to:
>   - type of lecture
>   - audience level of knowledge

Take time to prepare

The good lecture is a result of a maturing process in which both the material to be presented and the way it is presented improve with the time taken to prepare.

Devote a few weeks to preparation; not continuously – none of us has time for that – but begin well in advance, so the material can mature in your subconscious. A lecture prepared the evening before is seldom successful. The earlier you prepare the lecture, the more assured you will be about the message it is to convey, the more time you will have to practise, and the more confident you will be of your delivery. Indeed, even the most skilled of lecturers might heed a remark attributed to Winston Churchill: 'If you want me to speak for two minutes, it will take me three weeks of preparation. If you want me to speak for thirty minutes, it will take me a week to prepare. If you want me to speak for an hour, I'm ready now.'

Gauge your audience

Many lecturers live with the pretence that audiences are impressed by a wealth of details supported by intricate hypotheses. On the contrary, listeners are happiest when the material is so simple and clearly delivered that they can follow all parts of a lecture. Accordingly, as lecturers, it is vital that we gauge our audience. We must consider both the type of lecture to be given and the level of knowledge of the audience.

First of all, as a lecturer, you must be aware of the type of lecture to be given, whether it is to be a university course lecture, a talk at a review seminar or a presentation of your own work at an international meeting, as the type of lecture dictates the way you assemble it. A university course lecture or a seminar talk entails few of your own data, while a presentation at a meeting usually comprises only your own results set in the perspective of a research sector.

In assessing the level of knowledge of your audience, you should determine whether they are experts in your field, biomedical researchers in many fields, clinicians or students. Audience expertise is decisive in determining the level of detail, the amount of background information given and the amount of data presented. The broader the audience, the more background information must be included and the more results must be explained on the way. A lecture for experts in your own field may include far more results and put forth many more hypotheses than would a lecture for listeners with little background in your research field. Remember: in lecturing, it is almost impossible to underestimate your audience; as a rule, we overestimate.

14.3 LECTURE CONTENT AND FORM

The rest of this chapter focuses on a scientific lecture presentation of research results. We lecturers in biomedical and clinical research are fortunate in that we usually can build our lectures around illustrations. In a lecture, illustrations may be used in the introduction, in the presentation of background material, in the presentation of results, and not least in the final summary of results. Even though illustrations ease the structure of the scientific lecture, there are general ground rules to follow.

Amount of material

Assessing the right amount of material is one of the greater challenges. Lectures usually are built around illustrations in the form of slides, computer presentations and transparencies, of which no more than 25–30 should be included in a 45 minute lecture. The exact number varies, of course, with the complexity of the material presented. A clue to the right amount of material can be gleaned by reading the lecture aloud to yourself. You should speak at a rate of no more than 80–100 words a minute, so the time you take is a guide to determining the right amount of material. That said, it is wise to include leeway by having an amount of material that takes 10% less time to present than you are allocated, as unless you are not too nervous, live delivery always takes longer than reading aloud to yourself (see Box 14.2).

> **Box 14.2 Match amount of material to time allocated**
> - Maximum 25–30 illustrations per 45 minute lecture
> - 80–100 words per minute
> - aim for 10% less time than allocated

Lectures versus journal papers

Scientific lectures differ from scientific journal papers. Compared to a journal paper, a lecture should place greater emphasis on introductory and background information, less emphasis on describing methods (unless the lecture is on the development of a methodology) and less emphasis on discussion of data collected. A usual approach for a 10 minute lecture is summarized in Box 14.3.

Box 14.3 Lectures and journal papers

Compared to a paper, a lecture has:

- more introductory and background information
- shorter description of methods
- shorter discussion

10 minute lecture:

- 3 minute introduction or background information
- 5–6 minutes on results
- 1–2 minutes on summary and conclusions

Lecture structure

Start with the basics and go from the simple to the more complex (Box 14.4). The obvious goal is for as many listeners as possible to follow the lecture from its start, and that is easiest if it starts with the basics. Overestimating the audience from the start may irritate and discourage, so much so that attention is lost for the rest of the lecture (remember the attentiveness curve of Figure 14.1). See the following remarks on the Introduction.

Box 14.4 Lecture structure

- Start with the basics
- Build from the simple to the more complex
- Logical sequences of short points
- Build the lecture around illustrations
- Reach conclusions on the way
- Repeat essential points

The logical structure promotes understanding; do not jump back and forth between topics, but try to have a sequence of short points. Do this most easily by building the lecture around computer presentations, slides or transparencies. With the attentiveness curve of Figure 14.1 in mind, it is wise to include summaries and conclusions on the way. This strategy brings back the listeners who have lost the thread of the lecture and makes it easier for all listeners to follow the lecture and remember your points. It is also wise to repeat essential points several times in the course of a lecture.

Introduction

The introduction arguably is the most important part of the lecture, so work hard on it. A good introduction rouses interest and, as mentioned above, most easily conveys your message. Most of the audience will follow well in the first few minutes, so if you first awaken attention, much has been won.

A chairperson may introduce you and your topic, but if not, you may do that yourself at the start of your lecture. Here it is best to be succinct, before you launch into the relevant background details of your topic. The background details should enable the listeners to understand the subsequently presented data and should link directly to the approach to the problem presented. Take time in delivering the background details. The shorter the lecture, the greater the proportion that should be devoted to background details. The approach to the problem should be presented as precisely and briefly as possible, as point by point. If you can point to matter-of-fact importance or meaning, do so (see Box 14.5).

Box 14.5 Introduction
- Vital! Rouse audience interest
- Start by introducing yourself and your topic
- Take time for the background details
- Present the approach to the problem precisely, briefly, point by point

Illustrations

As mentioned above, it is beneficial to build lectures around illustrations. You may use computer presentations, slides or transparencies, but one rule applies to all: Keep It Short and Simple (KISS) (Box 14.6).

Box 14.6 Illustrations (computer presentations, slides or transparencies)

KISS – Keep It Short and Simple

- Few points per picture
- Figures are preferable to text
- Figures are preferable to tables
- Colours and symbols are fine, but avoid excesses. Think of legibility
- Use large fonts and clear typefaces, such as Arial or Comic Sans, with upper case and lower case letters; do not write all capitals
- Slides and computer presentations in landscape orientation; transparencies in portrait

First of all, few points per picture encourage the audience to try to follow what you have to show. Next, as a rule, it is better to present a message in a simple figure than in a bulleted list (which is the alternative). Most people more easily understand a graphic figure with explanations than a lecturer who goes through text, as can be seen by comparing the impact of Figure 14.2 with the text of Box 14.7. Further, a figure has more impact than a table (compare Figure 14.3 with Box 14.8).

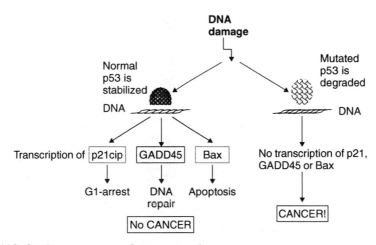

Figure 14.2 Graphic presentation of genome guardian

Box 14.7 Figure or text?

p53 – genome guardian

- p53 – a transcription factor induced upon DNA damage, through protein stabilization.
- p53 induces transcription of p21^{cip} (CDK inhibitor). Cells arrest in G1.
- p53 induces transcription of gadd45. Damaged DNA repaired.
- p53 induces transcription of bax (apoptosis-inducing protein). If the DNA damage is not repaired, the cell dies of apoptosis. Cancer development hindered.

Upon mutation of p53:

- p53 not stabilized upon DNA damage.
- p53 does not transcribe p21cip, gadd45 or bax.
- The cell replicates the damaged DNA, and the damaged cell does not die of apoptosis. Cancer may develop.

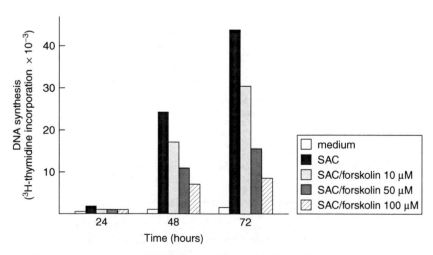

Figure 14.3 Graphic presentation of effect of forskolin on DNA synthesis

Box 14.8 Tabular presentation of effect of forskolin on DNA synthesis

Table: Effect of forskolin on DNA synthesis in B-cells.

STIMULATION OF CELLS	DNA SYNTHESIS (^{3}-THYMIDINE INCORPORATION)		
	24	48	72 (HOURS)
Medium	150	440	1 560
SAC	1 400	24 050	46 800
SAC/forskolin 10 μM	1 050	17 360	30 654
SAC/forskolin 50 μM	850	12 406	16 388
SAC/forskolin 100 μM	650	8 040	10 446

Font size and typeface also are vital. Choose large fonts, do not use all capitals (ALL CAPITALS), and choose a clear typeface, such as Arial or Comic Sans, that is readily legible at distance. Orientation should be natural: slides and computer presentations in landscape, transparencies in portrait.

Concluding a lecture

As illustrated in the attentiveness curve of Figure 14.1, often some listeners can be regained towards the end of a lecture. At this point, many listeners may suddenly realize that they have missed greater parts of the lecture (sound familiar?). So here you have a chance to fix that failing by summarizing the key results as well as their relevance to the initial approach to the problem. Thereby, the wheel has come full circle, and even those who have been attentive only at the start and the end of the lecture will have received your message (see Box 14.9).

The summary and presentation of conclusions must be short and precise, and preferably point by point; simple cartoons work well here. Your message is underscored if you conclude the lecture by sketching the future prospects for your research, perhaps with their transfer value to other research areas (generalization). Do not forget to list your co-authors!

> **Box 14.9 Concluding a lecture**
> - Vital! Conclusions are remembered
> - Tie the conclusion to the introduction: the wheel goes full circle
> - Summary and conclusions
> - Short sentences (point by point)
> - Simple figures
> - Further prospects?
> - List of co-authors

14.4 MANUSCRIPT

Now you have finished structuring the lecture. Before you start working on delivery, you should decide what sort of manuscript you will use. My advice to beginners is to have a manuscript to hand, but not to use it. It is reassuring to know that the manuscript is readily accessible (Box 14.10).

> **Box 14.10 A manuscript is reassuring!**
> - In compiling a manuscript:
>   – start with the 'whole text'
>   – highlight or underline the principal points
>   – write a short version
>   – write a catchword manuscript
> - For a good catchword manuscript:
>   – write only on one side of a sheet of paper
>   – use large fonts, preferably marked in various colours

One of the best approaches to writing a lecture manuscript is to start with the 'whole text', that is, everything that you might say, in verbal form. Then highlight or underline the principal points, before going on to a short version. Finally, compile a catchword manuscript (example in Box 14.11). You are free to choose the type of back-up manuscript, but I recommend that you use a catchword manuscript if you need an aid in lecturing. Write the catchword manuscript only on one side of a sheet. One handy trick is to use smaller, stiffer filing cards. Another useful trick is to write with large fonts and use colours to highlight different sections.

Box 14.11 Manuscript: whole text and catchword manuscript

MANUSCRIPT: WHOLE TEXT

First, *thanks* for inviting me to present my results at this meeting. In *our laboratory* we have worked for years with *regulating proliferation and cell death (apoptosis) of lymphoid cells,* and we have focused upon studying *how elevated levels of cAMP influence these processes in normal versus malignant lymphocytes.*

cAMP is an important physiological regulator of normal lymphocytes. Induction of cAMP, such as by prostaglandins, is known to inhibit proliferation of lymphocytes, and it has been shown that stimulation of normal lymphocytes causes secondary increase of cAMP level. cAMP acts as a *secondary messenger* as activator of *cAMP-dependent protein kinase (PKA),* and PKA is known to inhibit cellular activation of lymphocytes by activating tyrosine kinase *csk* that in turn inhibits the central tyrosine kinase *lck.*

Normal lymphocytes isolated from peripheral blood are in a *rest phase, called G0,* and for the cells to divide, they must be *activated so they enter the G1 phase and subsequently through the S phase, G2 and M.* The principal control point of this process is the *restriction point (R) in G1,* in which cells elect to continue in the cell cycle or remain in G1. For cells to pass the R point, they must receive signals from *mitogens, such as in the form of growth factors.* The R point itself is known to involve *phosphorylation of a tumour suppression protein called pRB.* Phosphorylation of pRB causes *release of a transcription factor (E2F),* which is necessary for a series of S phase-specific genes to be transcribed.

Phosphorylation of pRB is a *closely regulated process* that involves activation of the *cell cycle machinery.* pRB is sequentially phosphorylated by various cyclin-dependent kinases *(CDKs),* and these CDKs are in turn

---

activated by specific cyclins, and inhibited by specific inhibitors called *CDK inhibitors (CKIs)*.

We have previously shown that cAMP via PKA causes cells to *arrest in the G1 phase* of cell cycle, and the aims of the work I will present today are the following:

- How can increased levels of cAMP cause G1 arrest in normal lymphocytes?
- How does cAMP influence the cell cycle machinery that regulates the R point in G1?

The following components of the cell cycle machinery are investigated:

- pRB phosphorylation
- levels of various cyclins in G1
- levels of various CKIs in G1
- activity of various CDKs in G1
- free versus bound E2F

CATCHWORD MANUSCRIPT

- Thanks for the invitation
- Our lab: Proliferation and apoptosis in lymphocytes
- Special interest: Role of cAMP
- cAMP: physiological regulator in lymphocytes, prostaglandins – inhibit proliferation
- cAMP: Secondary messenger, PKA
- Csk activated, lck inhibited
- Normal lymphocytes in G0
- Activation into G1, S, G2/M
- R point in G1 – mitogens/growth factors
- R point in G1 – phosphorylation of pRB
- pRB phosphorylated – free E2F – S-phase genes
- Phosphorylation of pRB – closely regulated

- Cell cycle machinery:
  - pRB phosphorylated by CDKs
  - CDK activated by cyclins
  - CDKs inhibited by CKIs
- We have previously shown: cAMP causes G1 arrest

CURRENT AIMS:

- How can increased levels of cAMP cause G1 arrest in normal lymphocytes?
- How does cAMP regulate the cell cycle machinery at the R point in G1?
- Phosphorylation of pRB in G1
  - Level of G1 cyclins
  - Level of CKIs in G1
  - Activity of CDKs in G1
  - Free versus bound E2F

14.5 DELIVERING A LECTURE

Preparation

As mentioned above, it is wise to start preparing well in advance (Box 14.12). When you have finished preparing a lecture, it is time to practise its delivery. The more time you devote to practise, the more confident you will be as a lecturer. Again, thorough preparation is essential. Just think of how much actors and actresses rehearse their roles!

Box 14.12 Delivering a lecture
Prepare well:

- Mentally: learn to master nervousness
- Rehearse: take time, get feedback
- Be familiar with the lecture room and audiovisual aids

Mental preparation – nervousness

Most people are nervous before they perform, be they actors, musicians or lecturers; it is a natural, human reaction. However, what is vital is that you do not allow nervousness to overpower, but rather learn to master it. Nervousness is marked by accelerated heart rate, butterflies in the stomach, shaky legs and a dry mouth. These symptoms are due to an increased adrenaline level and, to a moderate degree, are a healthy reaction for a lecturer. The adrenaline level gives you the kick you need to do just a bit more. But you need to learn to master your natural nervousness, by using a few simple tricks (Box 14.13).

Box 14.13 Nervousness
- Everyone is nervous, more or less
- A higher adrenaline level gives you an extra, useful kick
- Learn to master nervousness:
  - be well prepared
  - be positive: boast to yourself!
  - keep a manuscript at hand (but use it only in an emergency)

The most important and easiest way to combat nervousness is to know that you are well prepared. If you have devoted enough time in preparation, your lecture will have lain maturing in your subconscious, and you will have sufficient time to rehearse your delivery. Simple steps such as these increase confidence and combat nervousness. Boast to yourself. Say to yourself that you are the one in the lecture hall that knows your topic best and that you have been asked to lecture because of your professional expertise. Think positive thoughts, such as it being a joy to tell others about and discuss your interesting results.

A manuscript at hand also diminishes nervousness. If the 'worst' happens and you miss the thread of your presentation, just leisurely pick up the manuscript. Then you can continue lecturing, aided by the manuscript, or just glance at it and then continue lecturing as if nothing happened. That certainty should make you confident. As long as you have a manuscript to hand, you cannot go wrong!

Rehearse

The value of rehearsing the delivery of a lecture cannot be underestimated. Through rehearsing you gain confidence and combat nervousness. Rehearse aloud, so you are accustomed to hearing your own voice. Use students or colleagues as an audience and accept all the feedback they offer. Time yourself, so you know that you have rightly gauged the time required to cover your material. One good trick is to mark your manuscript at the halfway point and check against it, so that you learn to adjust delivery speed on the way.

Familiarize yourself with the lecture room

Many lecturers, even the more experienced ones, overlook the importance of being familiar with the lecture locale. Undoubtedly, you have seen lecturers delay seminars or lecture sessions because they are unfamiliar with the audiovisual aids, such as microphones, remote controls or laser pointers. Embarrassing pauses result, and you often see that the lecturer is unnerved. So, always take time to check such matters in advance (Box 14.14). You then appear more experienced and less nervous, and you do not cause unnecessary delays to a programme. Not least, this applies to computer presentations, for which the technologies still create problems for lecturers and meeting organizers alike. A small detail: turn off a laser pointer when you are not using it to point at a board or screen. A laser beam can injure eyes if you let it dance over the audience as you lecture.

> **Box 14.14 Familiarize yourself with the lecture room**
> - Order the AV aids you need
> - Check that they work
> - Learn how to operate the lighting and the sound system
> - Laser pointer: a fine aid, but learn to use it correctly

Non-verbal communication

Before you go forth to lecture, be aware that the overall impression you give has a significant effect on how well the audience receives your lecture. It has been said that an audience gains its first impression of a lecturer within 90 seconds, and the greater part of that impression regrettably depends not on what you have said, but rather on non-verbal communication. Hence the so-called 55–38–7% rule (Box 14.15): 55% of the impression you give depends on your stage presence, your eye contact, your movements, your appearance, your clothes, etc.; 38% of the impression comes from your voice, that is, your pitch, intonation and regularity; a mere 7% comes from what you say! Perhaps this is discouraging, but it is all the more reason to choose your style carefully.

> **Box 14.15 Non-verbal communication: the 55–38–7% rule**
> - 55%: your stage presence, eye contact, movement, appearance, clothes, etc.
> - 38%: voice (pitch, intonation, regularity)
> - 7%: what you say, your message

The delivery itself

You have prepared a lecture in all possible ways, you know the lecture room and its audiovisual aids, so you now are ready to deliver the lecture. Again, there are small tricks that can improve a lecture, both for you and for your audience (Box 14.16). Often a relaxed posture (within reasonable limits) helps. If you seem at ease, your audience may take the cue, and the atmosphere will be friendly. Take a few deep breaths before you start your delivery in a relaxed, informal tone. This helps you to relax and combats any nervousness, and you gain control of your voice.

> **Box 14.16 Delivering a lecture – I**
> - Assume a relaxed posture
> - Take a deep breath before you begin, but do not sigh
> - Wait for the audience to quiet down before you start
> - Start with a relaxed, informal tone
> - Gain eye contact with the listeners, including those in the back rows
> - Let your glance wander over the audience, but do not focus on a single 'victim'

It is best to wait for the audience to quieten down before you start. Distracted listeners give you, the lecturer, an uninspiring start and, as mentioned above, it is best that all are attentive from the beginning.

Consciously keep eye contact with the audience. If the lighting does not blind, let your gaze wander over the audience, so that you have looked at everyone in the course of a lecture. Be sure not to choose a 'victim', a single person to whom you seem to lecture.

Refer to a manuscript if you wish, but do not read your lecture from it. That is easiest if you have prepared a catchword manuscript. Rapid speech is hard to understand, so talk at no more than 100 words per minute. As you undoubtedly know from experience, it is easier to follow a lecture when the lecturer varies intonation, takes brief breaks, underscores key points, and overall tries to draw the audience along throughout the lecture. An audience will be more motivated if the lecturer smiles and seems enthusiastic. Remember that the lecturer is the salesperson for the message of the lecture! Again, there is a fine line. An audience will be favourably impressed by a lecturer who is certain, but unfavourably impressed if the certainty becomes arrogance (Box 14.17).

Box 14.17 Delivering a lecture – II

- If you have a manuscript, do not read it directly
- Maximum of 80–100 words per minute
- Speak loudly and clearly; vary intonation
- Take brief breaks and underscore essential points
- Do not talk with your back to the audience
- Point at the board or screen only when necessary
- Smile: show enthusiasm!
- Be certain but not arrogant

Allowing sufficient time for questions is also worthwhile. Answering questions well is itself an art, and that art may be problematical at international meetings where it may be difficult to understand the questioners. Politeness is the solution; there is nothing wrong with asking that a question be repeated.

Another check is for you, the lecturer, to repeat the question in your own words, as you have understood it. Thereafter answer, keeping in mind that the question and your answer go to the entire audience, not just to the listener who asked. Avoid arrogance in answers. An answer such as 'I believe I covered that earlier in the lecture' makes the questioner feel stupid and exposed to contempt, but in fact you may be responsible for the person not having grasped what you said. Instead, you might say: 'Sorry that I didn't make that sufficiently clear, but the …'. In this way, you lower the threshold for others to pose questions.

On the whole, this applies when you are lecturing as it otherwise does in communication between people. Try to be as positive, polite and humble as you can. But here the comparison with ordinary conversation ends, because in the environment of a lecture, you are trying to 'sell' a message. You must be completely honest and informative in your accounts of results and hypotheses, but it is also essential that you try to appear certain in your presentation. It is always easier to convince an audience of your message if it is confidently and well delivered. These skills must be learned, and the only way to learn is to prepare well when asked to lecture, and then practise, practise, practise …. Good luck!

FURTHER READING

Bakshian A (1998) American Speaker. Your Guide to Successful Speaking. Georgetown Publishing House, Washington DC.

Benestad HB (2000) Communicative skills – a matter of negligence? Experimental Hematology 28: 1–2.

Kenny P (1998) A Handbook of Public Speaking for Scientists and Engineers. Institute of Physics Publishing, Philadelphia.

Pihl A, Bruland ØS (2000) Oral presentation in science and medicine – an art in decline? Anticancer Research 20: 2795–2800.

GUIDE TO GRANT APPLICATIONS

Bjørn Reino Olsen, Petter Laake and Ole Petter Ottersen

15.1 INTRODUCTION

The need for financial support for research projects becomes part of the harsh reality early in the career of a biomedical scientist. In many countries, graduate students are often expected to apply for financial support, and at the postdoctoral level, the writing of competitive fellowship applications to national and international agencies and foundations is now commonplace in countries with a societal commitment to biomedical research.

Beyond the level of postdoctoral training, continued success in research for junior as well as senior investigators at universities, institutes and hospitals requires a constant, intense effort to acquire research funding. Writing successful grant applications is as essential to scientific success as having good ideas, testing them by experimentation and reporting discoveries in published papers. Among these activities, for many scientists, grant writing is like a dark cloud over a bright landscape: necessary, but stressful and unpleasant.

One reason for the stress associated with writing proposals for funding to support work that is planned but not yet done, is that innovative research cannot be really planned; only the general conditions for it to happen can be optimized. In some ways we deceive ourselves when we believe that the process of scientific discovery (at least in basic biomedical research) can be planned to the extent required by many funding agencies that call for applicants to define and predict a timeline for discoveries (called *milestones* or *deliverables*). Therefore, writing applications that conform to the required formats frequently involves making promises that investigators know cannot be kept. In the area of innovative basic biomedical research, this dishonesty, necessary to obtain funding, is at odds with the open and honest mindset of the most talented researchers, generating stress.

A second reason why grant writing can be stressful is that the review of a grant by scientific peer-groups or administrators involves an evaluation of personal ideas and intuitive concepts about the unknown and how it should be explored. In a proposal, the investigator puts forth personal speculations and hypotheses to be tested by future experiments and must expose his or her scientific 'soul', so to speak, to convince reviewers that the ideas put forth are sufficiently important and innovative to warrant investment. In many cases, a rejection of a proposal is tantamount to a rejection of these personal ideas. That can shatter an investigator's confidence as a scientist much more forcefully than does a rejection notice for a paper submitted to a professional journal. Rejection of a submitted manuscript amounts to a disagreement between its author(s) and reviewer(s) or editor(s) about experimental evidence and how it is interpreted. Usually, such differences of opinion can be overcome by improving the evidence and rewriting the paper or by submitting the manuscript to another journal. A rejected grant application usually may be revised and resubmitted; although this increases the likelihood of being funded, it does not guarantee that funding will be awarded (see below).

Finally, most research grants include *indirect costs* (overheads) that pay for the institutional costs of maintaining the infrastructure (space, electricity, water, gas, etc.) that makes research possible. Hence, there is pressure on investigators to obtain grants for reasons that are more financial than scientific. Coupled with a continuous stream of *requests for applications* (RFAs or CALLs) from funding agencies in specific disease areas or on fashionable technologies, it has become too easy for departmental and institutional administrators to put pressure on scientists as to what they should do, urging them to follow the 'money-trail' instead of their 'idea-trail'. Increasingly, many institutions link the research space allocated to investigators to the amount of indirect costs that they generate through grants. The result is institutional pressure on scientists to shift their focus from doing innovative science to obtaining large grants.

That said, it is possible to structure the process of grant writing so it becomes part of the creative aspect of scientific discovery. For many types of large grant, this may be difficult, but for smaller, investigator-initiated grant proposals, we believe that a few selected strategies help to make grant writing an intellectually rewarding exercise.

15.2 GETTING STARTED

Students and postdoctoral fellows once struggled to write proposals for funding, with little or no help, unless their advisor(s) or the principal investigator (PI) were sufficiently experienced to offer guidance at all stages of the process. In many countries, the process now is easier. University-based workshops and courses on grant writing are being given, and although such formal training may never fully supplant personal mentoring of a trainee by a dedicated senior investigator, it certainly can help to demystify the process of grant writing. Recognizing the importance of early introduction to the grant writing process, major research centres, such as the Harvard Medical School in Boston, USA, and the Karolinska Institute in Stockholm, Sweden, have long organized one-semester courses for graduate students. These courses include the writing and critiquing of proposals, and at the Karolinska Institute, the courses apparently end with the participants submitting a proposal to a Swedish funding source (Kreeger 2003).

Additional help is available from the sources of funding; major granting agencies offer specific instructions both in print and online. For example, Britain's Wellcome Trust (http://www.wellcome.ac.uk) offers a short article for those who apply for a grant for the first time: 'Grantmanship: signposts for competitive grant applications', comprising a short list of questions that applicants should ask themselves before submitting the application (Box 15.1). The National Institutes of Health (NIH) in the USA offers a variety of informational services through the Office of Extramural Research (OER, http://grants1.nih.gov/grants/oer.htm) and organize seminars by NIH experts on grant proposal writing at major research centres in the USA. The National Science Foundation (NSF) in the USA organizes NSF Regional Grants Conferences (see http://www.nsf.gov/events/ for more

Box 15.1 Questions to ask before you submit
- Is the work novel, exciting and necessary?
- Are you repeating experiments already done by others?
- Have you justified the requested funding for various aspects of the study?
- Have you included preliminary data to support the feasibility of the proposed studies?
- Is the size of the study sufficient to detect what you are looking for?
- Have you read the instructions carefully and filled out the forms correctly?
- Have you checked for spelling and grammatical mistakes?

information). The National Institute of Allergy and Infectious Diseases at the NIH has compiled an instructive, annotated example of a well-written R01 grant application and a comprehensive monograph, 'All about grants tutorials' (available at http://www.niaid.nih.gov/ncn/grants/default.htm).

Moreover, several books and journal articles cover the general principles of grant writing and discuss features that may be specific to grants within different fields of biological and biomedical sciences (Box 15.2).

Box 15.2 Useful publications on grant writing
- McCabe LL, McCabe ERB (2000) How to Succeed in Academics. Academic Press, New York, Chapters 4–6.
- Inouye SK, Fiellin DA (2005) An evidence-based guide to writing grant proposals for clinical research. Annals of Internal Medicine 142: 274–282.
- Gill TM et al. (2004) Getting funded. Career development awards for aspiring clinical investigators. Journal of General Internal Medicine 19: 472–278.
- Goldblatt D (1998) How to get a grant funded. BMJ 317: 1647–1648.

- Reif-Lehrer L (2000) Applying for grant funds: there's help around the corner. Trends in Cell Biology 10: 500–504.
- Kessel D (2006) Writing successful grant applications for preclinical studies. Chest 130: 296–298.
- Koren G (2005) How to increase your funding changes: common pitfalls in medical grant applications. Canadian Journal of Clinical Pharmacology 12: e182–e185.

15.3 THE POSTDOCTORAL FELLOW AND JUNIOR SCIENTIST

With all these electronic and print resources available, there is no shortage of answers to questions about how to write a grant application. However, at an early stage of a research career, a winning proposal is not only one that reviewers rate highly for its exciting and significant research plan, but also one that scores high in the training experience and scientific standing of the advisor or mentor and in the quality of the institutional training environment. Postdoctoral fellowship applicants who propose research projects that do not depart significantly from their previous research and do not take place in a new environment (another institution, state or country) are viewed unfavourably. For example, postdoctoral applicants who propose continuation of previous PhD research projects with no change in research direction usually are ranked lower, both for NIH fellowships and for fellowships from international organizations such as the *Human Frontier Science Program*.

Consequently, postdoctoral fellowship applicants should devote time and energy in deciding the direction that they would like their scientific career to take after graduate school. This is a critical decision, not only because of what it means for getting fellowship applications funded, but also because postdoctoral research usually develops into a lifelong research commitment. As with all crucial decisions, it is difficult to define a set of advisory rules that fulfils the needs of all cases. However, as success requires commitment, undoubtedly a decision based on what one finds fascinating is unlikely to take the research in the 'wrong' direction. Once the decision starts to take shape, the choice of funding agency for a fellowship application should be considered carefully. Gaining support from a prestigious, highly competitive funding source can help to open doors to laboratories of outstanding mentors and institutional environments that help to stimulate significant research accomplishments.

Having a recognized mentor in a strong institution with a significant commitment to research is crucial to gaining awards to support the transition to independence (such as the new Pathway to Independence Award from NIH in the USA), because the experience and commitment of the mentor, the mentoring plan and the institutional commitment together weigh heavily in the evaluation of the grant proposal. For the junior investigator, deciding where to submit the very first research grant application is also important. Funding organizations have distinct missions and they are likely to reject proposals that are inconsistent with their remits. In the USA, NIH supports both medical research and basic science projects, but NSF rejects proposals dealing with causes of disease, diagnostic and therapeutic medical procedures, drugs and animal disease models. In many other countries, national research support for biomedical and more fundamental biological research is likewise divided between different funding entities. Good examples are the Medical Research Council (MRC) and the Biotechnology and Biomedical Research Council (BBSRC) in the UK. To ensure that a grant proposal matches the mission of a potential sponsor, the applicant needs to seek out the themes that various funding agencies are willing to consider. Submitting a good proposal to the wrong agency in the hope that it may work arguably is a waste of time, as it will probably just be rejected. Fortunately, there are extensive web-based resources that provide the details of worldwide sources of funding as well as of the types of grants available (Box 15.3). Whenever the information provided by these or other sites is insufficient to reach a decision, a call to the agency grants office may be helpful. Some funding agencies, such as the European Union (EU), provide lists of projects that have obtained funding in previous calls. Such lists may help to identify important success factors. For large organizations, such as the NIH in the USA, where the overall research support programme is divided among numerous institutes, the various institute programmes often overlap. A telephone call to programme directors at the institutes in question may then be helpful before deciding which institute is thematically the most appropriate target for the proposal.

Box 15.3 Useful resources for proposal writers, adapted from (Reif-Lehrer 2000)

Biotechnology and Biological Sciences Research Council (BBSRC)	http://www.bbsrc.ac.uk/	UK's leading funding agency for academic research in the non-medical life sciences
Computer Retrieval of Information on Scientific Projects (CRISP) Database	http://crisp.cit.nih.gov	Contains descriptions of all projects funded by NIH
Community of Science (COS)	http://www.cos.com (main COS Home Page); http://fundingopps2.cos.com/	Worldwide funding opportunities database; available only to researchers at COS member institutions. Weekly e-mail funding alerts keep researchers abreast of new opportunities in their disciplines
COS	http://www.cos.com/services/	Database of researcher profiles from over 190 leading universities. Available only to researchers at COS member institutions
DFG	http://www.dfg.de/en/research_funding/index.html	Provides a description of the various types of funding available through the Deutsche Forschungsgemeinschaft
EMBL Heidelberg	http://www.embl-heidelberg.de/training/index.html	Contains description of advanced training programmes at EMBL and information about international fellowship programmes

(Continued)

Box 15.3 (*Continued*)

EMBO	http://www.embo.org/fellowships/index.html	Contains descriptions of long-term and short-term EMBO fellowships, applications and guidelines
European Science Foundation	http://www.esf.org/	Has funding programmes in many scientific fields including medical, life and environmental sciences
European Union	http://europa.eu.int/comm/research/	Provides an entry to the research programmes, multinational networks and business-orientated research strategies organized and funded through the EU
Funders Online	http://www.fundersonline.org/, http://www.fundersonline.org/grantseekers/	Allows searches of Europe's OnLine philanthropic community (foundations and corporate funders)
Grantsnet	http://www.grantsnet.org (e-mail: grantsnet@aaas.org)	Website for young biomedical scientists (undergraduate to just beyond postdoctoral training). Is part of Science's 'Next Wave'. Grantsnet has a large database of fellowships, links to websites of funding organizations and information about (and tips from) previous recipients of funding
Grantsnet 'Global Links'		For scientists working in the UK, connects to funding databases around the world. 'Next Wave' has sites for the UK, Canada and Germany

Human Frontier Science Program	http://www.hfsp.org	Contains description of programmes for postdoctoral fellows and young investigators
Inserm	http://www.inserm.fr/eu/home.html	Provides an entry site for the programmes supported by the Institut National de la santé et de la recherché médicale (Inserm)
International Grants Finder	http://www.nature.com/	Database, maintained by Nature, for locating grants available in scientific fields worldwide. Updated annually using information extracted from Macmillan Reference Ltd, UK's Grant's Register
Medical Research Council (MRC)	http://www.mrc.ac.uk/	Provides an entry site for the wide range of medical research opportunities that are funded by the MRC in the UK
NIH website	http://www.nih.gov/	Provides an entry to all the information that is available through the individual NIH Institutes
Science's 'Next Wave'	http://sciencecareers.sciencemag.org/career_development	Provides career information and reviews of useful resources for young scientists
Wellcome Trust	http://www.wellcome.ac.uk/funding	Maintains a database of UK organizations supporting biomedical research

Note that most research institutions offer advice on how to identify appropriate funding sources. Major universities have significant expertise in-house, while smaller institutions often draw on the expertise of professional consultants with insight into the operation of specific funding agencies. In the EU Framework Programmes (FPs) (Section 15.8), in particular, it is expected that the application addresses a number of issues that extend beyond the realm of science and into the economic and political spheres. For example, in the recent FPs, applications have been required to describe the 'added value' of the proposed research, including the social and economic impact that the results will have in the European community. The seventh FP, starting in 2007, has a requirement to show the 'expected value'. Expected value will be demonstrated via publications in high-impact journals and international high-level conferences, and/or via patents. A failure to score high on such issues may easily bring down an application that scores well on its scientific merits (Section 15.8).

Targeting a proposal to a specific institute at the NIH may significantly affect its likelihood of success. This is because the various NIH institutes have different budgets and make different decisions about how they allocate resources between individual investigator grants and large multiproject Program and Centre grants. Consequently, the funding rates for junior investigators vary from institute to institute. Whenever congressional funding for NIH is ample, this difference is of less importance. But when funds are meagre, the dissimilarity in funding rates between institutes can translate to the difference between funding and no funding of a particular grant. Once the funding source is identified and grant writing has started, a recurring question that inevitably comes to mind is what can be done to maximize the chance of being funded. With funding rates of 10–20% of submitted applications for grants to federal agencies such as NIH and NSF in the USA, this question looms large in the minds of junior investigators trying to secure their first research grants. Another way to put the question is to ask what kinds of mistake most commonly contribute to bringing reviewers' evaluation scores down to the non-fundable level? The National Institute of Allergy and Infectious Diseases (NIAID) at NIH has put together a list that highlights the most common negative phrases used by reviewers to describe non-funded applications (Box 15.4 summarizes some of these; for the full list, visit the NIAID website).

> **Box 15.4 Common reasons cited by reviewers for an application's failure to gain an award**
>
> - Problem not important; study not likely to produce useful information
> - Study based on shaky hypothesis/data; alternative hypotheses not considered
> - Unsuitable methods; controls not included
> - Problem more complex than the investigator appears to realize
> - Too little detail in research plan; no recognition of potential pitfalls
> - Overambitious research plan
> - Direction or sense of priority not clearly defined; lack of focus
> - Lack of new ideas; fishing expedition; method in search of problem
> - Rationale for experiments not provided
> - No consideration of statistical power; lack of statistical power; insufficient consideration of statistical needs

15.4 WHAT GOES INTO A SUCCESSFUL GRANT APPLICATION?

The first comment in Box 15.4, 'Problem not important', indicates that an applicant has failed to convince the reader of the proposal that the problem *is* important in the context of contemporary understanding. Obviously, it is unconvincing for an applicant to state that a problem is significant. A critical analysis of current evidence and an identification of gaps in contemporary understanding must be presented, so that the problem to be studied logically emerges as paramount and that addressing and solving it along the experimental lines described in the application will help to move the field forward. As discussed in Chapter 7, a problem not only should be interesting; it must be important, so that the answer to it is significant (see Box 7.2 in Chapter 7). The problem should permit a clear description of overall goals and specific aims, contain testable hypotheses, and allow the selection of efficient and relevant methods for exploring these hypotheses.

The effort expended in writing a succinct, compelling story about the context and rationale of the research planned becomes part of the intellectual exercise that is requisite to initiating worthwhile research. The investigator should view it as the challenging initial phase of the research process and not just as a step necessary to obtain funding, so as to be more able to convey her or his excitement to the reader and reviewer. The grant writing period is an exciting time for brainstorming. However, it also should focus on and include leeway for speculation and intellectual 'long jumps' as well as take several steps back. It is a time for taking in the big view of the intellectual landscape, but also for zooming in close on crucial details. The applicant who manages to view the process in this light will be stimulated by and actually may enjoy the writing of a grant application.

Applications are assessed by reviewers who often have limited time available, because they must read several grant applications in a brief period. Therefore, an application should be well organized and written so the reader can grasp the flow of ideas and understand the experiments and potential outcomes. Simple diagrams summarizing the workflow of the project may prove helpful. The writing should be edited and formatted to enable the reader to follow easily the main points and thereby evaluate rapidly the significance and novelty of the proposed work without being mired in excessive detail. Simple, preliminary experiments should be identified and described along with preliminary results, to help to instil confidence in the soundness of the proposed ideas and eliminate unnecessary speculation. Minor experimental details should not be mentioned, but all essential details should be included. The experimental strategies and the rationale for selecting some methods and not others must be explained.

A wise strategy to preclude the common 'mistakes' listed in Box 15.4 is to ask friends and colleagues to read and critique the draft of the grant application. It may also be beneficial to seek the advice of a professional consultant (see Section 15.3). However, this is effective only if sufficient time is allowed for revising and rewriting before the submission deadline. It may also be helpful to put the document aside for a few days and then read it again through the eyes of a potential reviewer. By addressing matters that the reviewer(s) may question (based on comments in Box 15.4), the applicant may preclude questioning of them when the application is finally reviewed. For example, it is better to acknowledge any weaknesses in the preferred experimental plan and offer alternative back-up strategies than for the reviewer(s) to point them out later. Reviewers often list discussions of potential weaknesses and alternative plans among the strengths of an application. In addition, acknowledging that the results of the planned experiments may prove the hypotheses wrong is likely to be viewed favourably, provided that the applicant can convince the reviewer that it is important to find out, one way or the other.

The title and abstract are often left to the last minute and accordingly receive the least attention. This is a mistake. A concise title and a clear abstract together enable the reader to understand readily the studies proposed and serve to attract attention. This is particularly important whenever reviewers meet as a group to discuss and priority rate a large number of proposals (as in the 'Study Sections' at NIH), because group members who do not serve as in-depth reviewers of a particular application may nonetheless be asked to rank its priority. When the reviewers assigned to a particular application summarize their evaluations, the rest of the group may quickly scan the abstract and diagrams and figures in the experimental plan to assess the application. To ensure that the abstract reflects all aspects of the proposal, it is useful to think of it as a means of providing answers to six germane questions: (1) What is proposed? (2) Why is it important? (3) What has been done already by the applicant and others? (4) How will the study be done? (5) What will be learned? (6) What are the qualifications that provide the applicant with the competitive edge relative to other research groups that seek to address the same scientific problems?

15.5 THE INVESTIGATOR-INITIATED RESEARCH GRANT

All successful grant applications have research plans that reviewers find exciting, significant and well documented, be they applications for graduate or postdoctoral fellowships, mentored awards for transition to independence or independent research awards. However, the extent to which evaluation of the science contributes to the overall rating varies according to the nature of the application involved. For research awards, evaluation of the scientific plans is of primary concern, although other aspects are also considered in assessing the overall merit of the grant proposal. For NIH Research Project Grant (R01) proposals in the USA, the review criteria depend on whether the application was submitted in response to a special initiative, such as an RFA or a program announcement (PA) in a specific area of biomedical research, or whether it was investigator initiated. The standard NIH review criteria for an investigator-initiated R01 grant are summarized in Box 15.5. Reviewers are requested to keep these criteria in mind

> **Box 15.5 Review criteria for investigator-initiated NIH R01 grants**
> - *Significance*: Does the study address an important problem?
> - *Approach*: Are the conceptual or clinical framework, design, methods and analyses adequately developed, well integrated, well reasoned and appropriate to the aims of the project?
> - *Innovation*: Is the project original and innovative?
> - *Investigators*: Are the investigators appropriately trained and well suited to carry out the work?
> - *Environment*: Does the scientific environment contribute to the probability of success?

in reviewing applications, but there is no direct numerical relationship between the criteria and the overall rating. An application need not score high in all criteria to be given a high priority score; thus the final rating is an assessment of the overall merit of the grant. However, the most reliable route to a high (fundable) score is to take all the criteria into account in writing an application. In the EU system there is a threshold score, and a typical value might be 80% for the quality or 60% for impact. The proposal will be considered further only if the scores on such criteria are higher than the preset thresholds (see Section 15.8). Obviously, an applicant should identify and address any threshold scores before submitting a proposal.

The R01 type grant is the original, historically oldest grant mechanism used by the NIH in the USA. Today, the R01 grant is one of several types of grant available for the support of health-related research and training activities, yet it still accounts for a substantial portion of the federal funding for biomedical research. As for all research project grants, R01 grants are awarded to sponsoring institutions on behalf of a PI to facilitate research projects conceived by the PI. The awards are usually for periods of three to five years. In accepting an award, the sponsoring institution assumes the financial responsibility for the grant and for providing the laboratory and other facilities required for the research project to be carried out.

The procedures for submitting grant applications, the relevant deadlines, and the ways in which applications are reviewed and funding decisions are made differ among countries and among funding agencies. Covering all the relevant rules and policies is beyond the scope of this book. As an illustrative example, we will describe the scenario of an R01 application in the USA that results in funding of an investigator-initiated research grant. We believe that the R01 grant mechanism is germane for two reasons. First, the sequence of steps, from submission to funding, is typical of the sequences for similar grants in most countries. Hence, with some modifications, the process for an R01 grant is applicable to most situations. Secondly, the peer-review process used for evaluation of R01 grants in the USA has long served American biomedical research well. The process is not perfect, yet it is useful and is imitated elsewhere as countries in other parts of the world strengthen their mechanisms for funding research.

R01 applications are submitted to the Center for Scientific Review, the NIH entity responsible for assigning applications to an institute with a mission that best fits the proposed research topic, and to a scientific review group (SRG or 'Study Section'). Each application is given an ID number, and the applicant is notified that the application has been assigned to a specific Study Section for scientific merit evaluation and an institute or centre for funding consideration. For questions about Study Section assignments or other questions that may arise before the Study Section review takes place, the applicant is provided with the name and contact information of a Scientific Review Administrator. For questions that may arise after the Review, the applicant is directed to programme staff of the assigned institute or centre.

Getting the application assigned to the right institute and Study Section is important and applicants have some control over this. They can request assignments at the time of submission of applications, and they can request a reassignment should the Center for Scientific Review assign the application to an Institute and Study Section that they feel are not in their best interests. Study Section rosters are available on the web, and an applicant who feels that the members of the Study Section to which the application has been assigned do not have the required expertise to review the application adequately, or have competitive conflicts of interest, can request reassignment to another SRG.

Study Sections, comprising scientists with relevant expertise, are responsible for evaluating applications for scientific and technical merit, and meet (usually at NIH) to discuss the assigned applications. Note that this initial review is a peer review, with scientists evaluating the proposals of their colleagues. Health policy and funding issues are not considered at this stage; instead, the goal is to identify proposals that scientifically are the most promising and exciting. To accomplish this as rigorously as possible with a steadily increasing number of grant submissions, after brief discussions, grants applications that a Study Section deems to be non-fundable because of major scientific weaknesses are put aside (triaged) without further consideration. The remaining grants are reviewed in depth. Two reviewers (primary and secondary) provide written reports on each grant, and this is followed by a discussion by the entire group. Copies of the written reviews and a summary of the group discussion, plus the priority score, are sent to the applicants (as Summary Statements) and to the assigned Institutes.

Within the institutes, a second level review takes place at a meeting of the institute-specific National Advisory Council, composed of scientists and laypersons with special interests in health-related issues relevant to the mission of the institute. Council members review the appropriateness of the Study Section reviews and weigh in on issues that may affect funding of applications, such as mission relevance, programme goals and availability of funds. This system of separating assessment of scientific merit from public health-related political consideration has proven to be effective in maintaining high scientific standards for NIH-supported research.

Should an application not be funded on the first try (the most usual outcome), applicants can revise the application and try again. Should the first amendment also be unsuccessful, a second amendment can be submitted. If that fails, the proposal can be resubmitted as a new application. Over 50% of NIH applicants will eventually be funded after revisions and resubmissions.

15.6 MULTIPROJECT GRANTS

The days when investigators working alone in their laboratories made startling discoveries are for the most part over. The complexities of modern biomedical research require teamwork, with disparate team members contributing varieties of technical or theoretical skills to a research project. In the case of research projects supported by R01-like awards, the PI may direct a team of graduate students and postdoctoral fellows, collaborate with several co-investigators and have several consultants who provide expert advice and specialized reagents.

There are other grant mechanisms for larger team efforts, including grants to support Program Projects, Specialized Centres or Centres of Excellence, and multiproject Networks. Program Projects, supported by some NIH institutes, comprise several (more than three) PI-directed projects focused on different, but complementary, objectives within a greater common goal, usually combined with specialized Cores that provide technical, administrative and molecular/cellular/clinical resources for the programme. An advantage of a Program Project is that the built-in mechanism for funding of Cores provides collaborating research groups with resources to enhance their collaboration and embark on research projects that are more complex than those that can be carried out by individual R01-funded laboratories.

Centre grants are used in many countries as mechanisms for stimulating collaboration among a large number of scientific teams and clinicians, to enhance research activities that can result not only in improved understanding, but also in detection, treatment and prevention of human diseases. The Centre concept may work particularly well whenever the collaborating teams of scientists and clinicians have previously worked together before being awarded a Centre grant. Investigators who work well together without a grant that so requires have responded to scientific need. Then a Centre grant provides a funding base for a natural extension and expansion of collaborative studies that already have been started. Whether Centre grants in specific areas of biomedical research truly stimulate excellence based on entirely new collaborations is less clear. This is because PIs frequently tailor applications to compete for the available funding, but will continue to do more or less what they were doing before a grant is awarded. A financial factor also works against starting new collaborative projects. To be competitive for funding of large Centre grants, applications must define ambitious goals and objectives. That they are overly ambitious, given the available funding, usually becomes obvious only after a grant has been awarded and the funds have been divided among all participating groups, only to be found insufficient to accomplish the planned work. The risk of ending up in such a situation can be curtailed if the number of groups included in a given proposal is tailored to the anticipated size of the grant. As a rule of thumb, for each participating group, the budget should accommodate a minimum of one full-time researcher plus reasonable running costs.

For larger networks, there may be similar problems associated with imposed collaborations. It is unfortunate if collaboration fails to promote innovative science. For this reason, network grant mechanisms probably work best when they aim at clearly defined goals, such as the development of new technologies or clinical research objectives. In these cases, it is essential to assemble teams covering many different technical approaches and often necessary to generate and study large depositories of patient data and tissue samples from large populations.

Accordingly, NIH has an Office of Portfolio Analysis and Strategic Initiatives (OPASI, http://opasi.nih.gov/) that serves as an 'incubator' for trans-NIH initiatives and support of priority projects and thereby meets the needs for stimulation of collaborative research along such lines, and helps to identify and support future research needs. Through its Division of Strategic Coordination, OPASI manages the process by which new initiatives in the NIH Roadmap are implemented. Implementation of Roadmap themes (see Box 15.6) is to a large extent based on network- and centre-based activities.

Box 15.6 NIH Roadmap initiatives (http://nihroadmap.nih.gov/)

1. Theme: New Pathways to Discovery
   - Building blocks, pathways, and networks implementation group
   - Molecular libraries and imaging implementation group
   - Structural biology implementation group
   - Bioinformatics and computational biology implementation group
   - Nanomedicine implementation group

2. Theme: Research Teams of the Future
   - High-risk research implementation group
   - Interdisciplinary research implementation group
   - Public private partnerships implementation group

3. Theme: Re-engineering the Clinical Research Enterprise
   - Clinical research implementation group

NIH also offers the NIH Director's Pioneer Award (NDPA), which is intended to capture those highly innovative investigator-initiated ideas that have potential for appreciable impact, but are judged to be too risky and too far off the beaten path to be funded through the R01-type mechanism. Applications for an NDPA are brief in that they consist primarily of an essay of three to five pages and three reference letters (see http://nihroadmap.nih.gov/pioneer/). The NDPA is intended to be a supplement to, not a substitute for an R01-type grant. It is designed to support creative individuals rather than a project, based on the assumption that if highly creative scientists are given some unrestricted money, significant findings may result.

15.7 INTERNATIONAL RESEARCH COLLABORATIONS

Science knows no borders, and international cooperation between scientists is not only common, but also responsible for some of the most significant advances in biomedical research. Unfortunately, obtaining funding for international collaborations is not always straightforward. In part this is because agencies that fund research in most countries primarily have a remit of promoting national research programmes. Another reason is that rules that regulate research involving humans and animals are not always exactly the same in different countries, and this may generate problems when material collected in one country is used for research by a collaborator in another country. However, these are problems that can be solved; it may simply require more effort than in cases where collaborators are within the same institution or in the same country.

A PI does not need to be a citizen or permanent resident of the USA to apply for an NIH research project grant (R01). Hence, investigators in other countries can obtain independent funding from NIH for projects that may complement or match a project of an American collaborator. Another possibility is for a PI in the USA to include a foreign collaborator and her or his project as a subcontract on an NIH grant. Thus, in principle, NIH funding for research knows no borders. However, in the case of a foreign grant or foreign component of a US grant, one has to convince the reviewers that the foreign research is unique and does not duplicate research that can be done as easily or better by investigators in the USA.

NIH also has a centre specifically established (in 1968) to address global health problems by supporting innovative and collaborative research and training programmes. This is the John E. Fogarty International Center for Advanced Study in the Health Sciences (FIC) (http://www/fic/nih/gov/), currently supporting research and training programmes in over 100 countries and involving some 5000 scientists in the USA and other countries. For example, the FIC Global Health Research Initiative Program for New Foreign Investigators provides partial salary and research support in behavioural and social science or basic science for NIH-trained investigators who are returning to their home countries; and the FIC International Research Collaboration Award supports collaborations between NIH-supported American scientists and researchers in developing countries. Like NIH, NSF also has an office, Office of International Science and Engineering (OISE), for facilitating research overseas. OISE supports international training programmes for scientists of all levels and funds postdoctoral fellowships for training in international research.

In the UK, BBSRC has established an *International Relations Unit*, which promotes contacts with international scientists, provides advice on funding opportunities for collaborative projects, supports international visits, helps BBSRC-funded institutes and universities to identify international sources of scientific expertise, and contributes to international science policy. Several types of grant are available for international travel and collaborations for investigators who are either supported by BBSRC or affiliated with BBSRC-supported institutions. These include awards to travel abroad [International Scientific Interchange Scheme (ISIS)], International Fellowships for 'high-profile' researchers to visit the UK for periods of up to one year, grants to support international workshops (particularly those that establish further links with the USA, Canada, EU member states, Japan, China and India), and Partnering Awards for BBSRC-supported scientists to establish collaborations with researchers in Japan and China. The BBSRC website (see Box 15.3) also contains links to other international funding agencies and grant mechanisms, such as the EU FP, Human Frontier Science Program, European Science Foundation, Co-operation in Science and Technology (COST), British Council, Royal Society and Alexander von Humboldt Foundation.

For African scientists, South Africa's National Research Foundation (NRF) website contains useful information about opportunities for international collaboration and the availability of grants and fellowships administered by the NRF's International Science Liaison office (http://www.nrf.ac.za/funding/international.stm). For investigators in Australia, the Australian Research Council (ARC) (http://www.arc.gov.au) provides support for international collaboration and network building by funding Research Awards, International Fellowships and Internationally Coordinated Initiatives. For international collaborations in clinical trials, funding is available in cases where the investigators involved are each eligible for funding by the MRC (UK), Veterans Administration (USA) or Canadian Institutes for Health Research.

Researchers in the Nordic countries should note that in 2005 a new funding agency, NordForsk (www.norden.org/forskning/sk/nordforsk.asp), was established to promote seamless research across the countries, particularly in scientific disciplines in which they already have leading positions. So far, NordForsk has rather limited resources at its disposal. Thus, the funding from this agency should be regarded as seed money or add-on funding meant to facilitate collaboration between Nordic research groups. NordForsk already supports three Nordic Centres of Excellence through its Molecular Medicine programme.

Finally, several charities devote significant funds to international biomedical research. For example, the Wellcome Trust (UK) supports international biomedical research in developing and restructuring countries, provides funding for tropical and clinical research in developing countries, and supports international networks and partnerships. This support is available through fellowships, project and programme grants, and other targeted mechanisms. In the USA, the Bill & Melinda Gates Foundation (http://www.gatesfoundation.org) is emerging as one of the world's most powerful charities, with funding of global health programmes as a major focus. The foundation supports research and clinical initiatives in several priority disease areas, including child health, HIV/AIDS, malaria, tuberculosis and vaccine-preventable diseases. It also will consider proposals for support of research that is likely to achieve fundamental breakthroughs in three areas: (1) science and technologies to make advances against diseases in developing countries; (2) technologies that can serve as platforms for accurate and affordable diagnostic tools; and (3) application of advances in genetics and molecular biology to global health problems.

15.8 THE EUROPEAN UNION'S SEVENTH FRAMEWORK PROGRAMME

The EU FP is intended to promote international collaboration across Europe. Legally it is intended to boost the competitiveness of European industry, and to tackle problems related to the pursuit of European policies.

Framework Programme participation with Community funding is available to organizations in EU countries, countries that are candidates for EU membership, EFTA-EEA countries and, depending on the particular FP, other countries. Some FP funding is available for organizations elsewhere, provided that their participation in a project is justified as essential for achieving its objectives.

Starting in 2007 with the seventh Framework Programme (FP7), the European Research Council (ERC) will evolve into a funding body similar to the NSF in the USA. Consequently, a greater share of the research and development (R&D) budgets of European countries is now channelled through the organizations of the European Research Area (ERA). Hence, young researchers are encouraged to relate to the EU and ERA funding systems. It is the experience of most grantees that the benefits of being a member of a strong European network justify the time it takes to run the network and to report the results.

The EU's FP7 was launched in 2007 and exceeds the previous FPs in terms of time-horizon and total budget. In FP7, most research projects relevant to health researchers are located in the thematic area Health (budget €6 billion) with other opportunities in the Food/Biotechnology (€2 billion), Environment (€2 billion), Information and Communication Technologies (€9 billion), and Nanosciences (€3.5 billion). Another programme is the People (i.e. training of young and established researchers, plus international fellowships) programme, often known as the Marie Curie programme. In FP7 the budget allocation is about €4.5 billion.

It is the main objective of the Health research area to help to improve the health of Europe's citizens by developing the knowledge (and its use) needed to overcome major health scourges, with the potential to reduce the high healthcare costs related to the treatment and care of patients. Among other things, translational R&D and validation of new therapies are emphasized (see Box 15.7).

Box 15.7 Components and characteristics of the EU FP7 research
       projects

The projects focus on objective-driven research and development, i.e. clearly defined scientific and technological objectives, aiming at significant advances in the established state of the art.

Research teams often are multidisciplinary in their composition.

Projects are put together with an eye on how the knowledge advanced by them will be used, and therefore include aspects such as activities relating to the protection and dissemination of knowledge, activities to promote the exploitation of the results, and, when relevant, 'take-up' actions.

Projects include training of researchers and other key staff, research managers, and potential users of the knowledge produced within the project.

There is a focus on project management activities.

Many new projects (large or small) are expected to be evaluated in two stages: short proposal first, followed by a longer submission for those shortlisted in stage 1.

EU projects within the FP7 will be evaluated according to various criteria, depending on type of proposal. The evaluation criteria for collaborative projects are listed in Box 15.8. Prospective applicants should check the websites accessible via the FP7 portal at http://cordis.europa.eu/fp7, and especially the website http://cordis.europa.eu/fp7/dc/index.cfm?fuseaction=UserSite.CooperationDetailsCallPage&call_id=10 that contains important information on the application and evaluation process.

Box 15.8 Evaluation criteria applicable to collaborative project
       proposals for FP7

1. S/T Quality – 'Scientific and/or technological excellence'.
   • Soundness of concept, and quality of the objective.
   • Progress beyond the state-of-the-art.
   • Quality and effectiveness of the S/T methodology and associated work plan.

2. Implementation – 'Quality and efficiency of the implementation and the management'.
   - Appropriateness of the management structure and procedures.
   - Quality and relevant experience of the individual participants.
   - Quality of the consortium as a whole (including complementarity, balance).
   - Appropriateness of the allocation and justification of the resources to committees (budget, staff, equipment).

3. Impact – 'Potential impact through the development, dissemination and use of project results'
   - Contribution, at the European and/or international level, to the expected impacts listed in the work programme under the relevant topic/activity.
   - Apptopriateness of measures for the dissemination and/or exploitation of project results, and management of intellectual property.

15.9 SUMMARY AND PERSPECTIVE

Grants and other support mechanisms are available to students, postdoctoral fellows and investigators in most countries for training and research in biomedical sciences. Knowing as much as possible about where to apply and how to write a good application are crucial aspects of the process by which a grant can be successfully obtained. However, in a climate of increasing competition for grant funding, two simple rules must be kept in mind: First, to receive a grant, you must apply. Secondly, if on the first try you don't succeed, try again!

REFERENCES

Kreeger K (2003) A winning proposal. Nature 426: 102–103.
Reif-Lehrer R (2000) Applying for grant funds: there's help around the corner. Trends in Cell Biology 10: 500–504.

INDEX

Lightning Source UK Ltd.
Milton Keynes UK
UKOW030405181212

203794UK00017B/463/P